The Science of Imaging
An Introduction

About the Author

Graham Saxby spent 27 years in the RAF, first as a photographer carrying out almost every possible kind of assignment; then, after being commissioned as a technical education officer, as Officer Commanding Photographic Science Flight at the RAF School of Photography. On leaving the Service in 1974 he joined the staff of what is now the University of Wolverhampton, teaching educational technology and, later, modern optics. His researches in display holography earned him an international reputation in the field. Now formally retired, he operates as a freelance editor of technical books and as a consultant in optical and photographic matters. He is the author of a number of books on holography and photography, and is a regular contributor to photographic journals.

The Science of Imaging
An Introduction

Graham Saxby

Institute of Physics Publishing
Bristol and Philadelphia

© IOP Publishing Ltd 2002

All rights reserved. No part of this publication may be reproduced, stored in a retrieval system or transmitted in any form or by any means, electronic, mechanical, photocopying, recording or otherwise, without the prior permission of the publisher. Multiple copying is permitted in accordance with the terms of licences issued by the Copyright Licensing Agency under the terms of its agreement with the Committee of Vice-Chancellors and Principals.

The author has attempted to trace the copyright holders of all the figures reproduced in this publication and apologizes to them if permission to publish in this form has not been obtained.

British Library Cataloguing-in-Publication Data
A catalogue record for this book is available from the British Library.

ISBN 0 7503 0734 X

Library of Congress Cataloging-in-Publication Data are available

Commissioning Editor: Nicki Dennis
Production Editor: Simon Laurenson
Production Control: Sarah Plenty
Cover Design: Frédérique Swist
Marketing Executive: Laura Serratrice

Published by Institute of Physics Publishing, wholly owned by The Institute of Physics, London

Institute of Physics, Dirac House, Temple Back, Bristol BS1 6BE, UK

US Office: Institute of Physics Publishing, The Public Ledger Building, Suite 1035, 150 South Independence Mall West, Philadelphia, PA 19106, USA

Typeset by Academic + Technical Typesetting, Bristol
Printed in the UK by J W Arrowsmith Ltd, Bristol

To the memory of Michael Langford
A gifted and dedicated teacher, and a much-valued friend

Contents

Preface	xv
Acknowledgments	xvi

Chapter 1 The nature of light — 1

Models for the behaviour of light	1
Box: Maxwell and electromagnetism	2
Electromagnetic radiation	2
The electromagnetic spectrum	3
Polarisation	4
Interference	5
Diffraction	8
Box: The grating condition	9
The Airy diffraction pattern	9
Reflection and refraction	10
Box: Snell's Law	11
Total internal reflection	11
Prisms	11
The pinhole camera	13
Development of a lens	13
Digging deeper	14

Chapter 2 Photometry, lighting and light filters — 15

Photometric units	15
Luminous intensity	15
Luminous flux	16
Box: Planck's equation and retinal sensitivity	16
Illuminance	16
Inverse Square Law	17
Luminance	18
Reflectance	18
Luminous energy	18
Luminous efficacy	18
Spectral power distribution	19
Colour temperature	20
The mirek scale	20
Types of light source	21
Photographic light filters	22
Polarising filters	25
Box: Applications of polarising filters	25
Circular polarisation	26
Digging deeper	26

Chapter 3 Visual perception — 27

The eye and evolution	27
Optics of the eye	27
Short and long sight	28

The retina	29
Rods and cones	29
Sensitivity range	30
Visual pathways	30
Box: The Weber–Fechner Law	31
Box: Neural processing of the visual signal	32
Visual fields and binocular vision	32
Colour perception	33
Seeing a range of colours	34
Constancy	35
Visual illusions	35
Perception and imaging	37
Digging deeper	38

Chapter 4 Camera lenses — 39

A model for the geometry of camera lenses	39
The simple lens	39
The lens laws	39
Real and virtual images	40
Depth of field	40
Box: Hyperfocal distance	42
Depth of focus	42
Gaussian optics	42
Telephoto lenses	43
Retrofocus lenses	44
Varifocal and zoom lenses	44
Angle of field	45
Lens aberrations	46
Aspheric surfaces	49
Fall-off	49
Box: Lens coating	50
Specialised lenses	50
Perspective	51
Box: The Scheimpflug rule	53
Digging deeper	54

Chapter 5 Resolution in optical systems — 56

Testing for resolving power	56
Diffraction limitation	57
The Rayleigh criterion	57
The inadequacy of resolving power measurements	58
The modern approach to image quality	58
Box: Analysis of a square wave	59
Modulation	60
The optical transfer function	61
The MTF of an 'ideal' lens	62
Box: OTF and spread functions related	62
Cascading of transfer functions	62
Granularity and pixel size	63
Conclusion	64
Digging deeper	64

Chapter 6   Images in colour   65

   Early attempts   65
   Lippmann photography   65
   Box: Lippmann's desaturated colours   66
   The Young–Helmholtz theory of colour perception   67
   Additive colour synthesis   68
   Quantifying colour: the CIE chromaticity diagram   68
   Box: Measurement systems for colour   69
   Other scales of colour measurement   69
   Subtractive colour synthesis   70
   Colour separation negatives   70
   Colour prints from separation negatives   71
   Tripack colour transparencies   72
   Prints from transparencies   73
   Infrared emulsions   73
   Polaroid colour   75
   Colour negative–positive systems   76
   Box: Colour masking   76
   Cross-processing   77
   Digging deeper   77

Chapter 7   Still cameras   79

   Early cameras   79
   Shutters   79
   Types of camera   81
   Viewfinders   84
   Rangefinders and focus finders   86
   Automatic focus control   87
   Automatic exposure control   89
   Flash synchronisation   90
   Camera shake   91
   Image motion compensation   92
   Digging deeper   93

Chapter 8   Motion and high-speed photography   94

   Persistence of vision   94
   Early experiments   94
   The modern cine camera   95
   Slow motion and time lapse   96
   High-speed cine   96
   Mirror and drum photography   97
   Smear and streak photography   98
   Lighting for high-speed photography   99
   Stroboscopy   100
   Digging deeper   101

Chapter 9   The photographic process   102

   The uniqueness of silver halides   102
   The latent image   103

The Science of Imaging

Speed and inherent contrast	105
Colour sensitivity of emulsions	106
Development	106
Constituents of a developer	107
Box: Oxidation and reduction	107
Box: Developing agents and molecular structure	108
Fixing, washing and drying	110
Printing	110
Colour emulsions	110
Processing of colour emulsions	111
Digging deeper	112

Chapter 10 How photographic films behave — 113

Box: Hurter and Driffield's task	113
Practical units of measurement	114
How the H & D curve is produced	114
Box: The ISO speed index	114
Average gradient ($\bar{G}$)	115
What the H & D curve can tell us	115
Box: A do-it-yourself H & D curve	115
Effect of varying the development time	116
Effect of varying the developer composition	117
Tonal reproduction of a film	118
Box: Comparison of two films	118
Exposure latitude	119
Variation of speed and contrast with wavelength	120
Effects of after-treatment	120
Reciprocity Law failure	121
H & D curves for colour negatives	121
H & D curves for transparency materials	121
Push processing	122
Speed criteria for colour materials	122

Chapter 11 How photographic print materials behave — 123

A cautionary note: can sensitometry harm your image?	123
The H & D curve for print materials	123
The density range of a paper	124
Contrast control of prints	125
Variable contrast papers	125
Cascading H & D curves	126
Colour print papers	129
Digging deeper	130

Chapter 12 Image modification — 132

Contrast control	132
Shading, dodging and burning in	132
Unsharp masking	133
Adjacency effects as unsharp masking	134

Box: Adjacency artefacts	135
Adjacency effects in colour emulsions	136
Colour masking	136
Undercolour removal	137
Unbalanced lighting	137
Correction filters	138
Skewed curves	138
Digital manipulation of images	139
Rectification	140
Other image effects	140
Artefacts in digital images	142
Digging deeper	142

Chapter 13 Digital recording of images — 144

The digital principle	144
Digital recording of luminance	144
Box: The Nyquist criterion	145
Box: Bits, bytes and binary arithmetic	146
Principles of electronic information storage	147
Getting the image out of the camera	150
Colour in a digital camera	151
Compression	152
The future for digital cameras	152
Scanners and scanning methods	153
Frame grabbers	154
Digging deeper	154

Chapter 14 Halftone, electrostatic and digital printing — 155

Continuous tones with printer's ink	155
The halftone principle	155
Printing in colour	157
Electrostatic copying	157
Printers	159
Digging deeper	160

Chapter 15 Television — 161

Beginnings	161
The electron beam tube	162
The television camera	164
Transmission and reception of a TV signal	166
Microwave relay transmission	169
Satellite transmission	170
The signal	170
The TV receiver: types of display	171
Visual display units and computer monitors	171
Digital television	172
Aspect ratio and high-definition television	172
Digging deeper	173

The Science of Imaging

Chapter 16 Video recording and replay systems	**174**
Magnetic tape recording	174
Box: Ferromagnetism, hysteresis and a.c. bias	174
Tape recording principles	175
Box: Noise reduction systems	177
Videotape recording techniques	177
The VHS format	177
Digital video recording	178
Digital videotape	178
Camcorders	179
Digital videodiscs	179
Box: Before the CD	179
Digging deeper	181
Chapter 17 Three-dimensional imaging	**183**
How we see depth	183
The limitations of stereo pairs of images	184
Early stereoscopic images	185
Stereoscopic camera formats	185
Aerial reconnaissance and survey photography	186
Hyperstereoscopy in aerial photography	187
Hypostereoscopy	188
Viewing methods for stereo pairs	189
Viewing without optical aids	190
Coincident image stereograms	190
Autostereoscopic systems	191
Stereoscopic cinema and television	192
Further developments in stereo projection	193
Integral imaging	193
Digging deeper	194
Chapter 18 Holography	**196**
Coherence	196
Off-axis holograms	197
Box: How a hologram works	198
Processing a hologram	201
The real image	201
Transfer holograms	203
Contact copies	204
Focused-image holograms	204
Rainbow holograms	205
Pulse laser holograms	205
Embossed holograms	206
Holographic stereograms	206
Holograms in natural colours	207
Holographic interferometry	208
Holographic optical elements	209
Box: Zone plates	210
Computer-generated holograms	211
Digging deeper	212

Chapter 19   The high and the low **214**

    Panoramic images   214
    Box: Why 'fisheye'?   214
    Pinhole photography   217
    Aberrations of a pinhole   218
    Box: Making a pinhole   219
    High-level aerial and satellite photography   219
    Photomacrography and photomicrography   220
    Box: Resolution criterion   222
    Microimaging   226
    Underwater photography   226
    Digging deeper   228

Chapter 20   Medical and scientific imaging **230**

    Ultraviolet and fluorescence photography   230
    Endoscopy   231
    Radiography   232
    Tomography and scanning systems   235
    Analysis of scanning outputs   237
    Ultrasonic imaging   238
    Thermal imaging   239
    Schlieren photography   241
    Digging deeper   243

Appendix 1   Logarithms: what they are, what they do **246**

Appendix 2   The meaning of pH **248**

Appendix 3   The Fourier model for image formation **249**

Index **256**

Preface

When my publisher suggested that I should write a book on the science of imaging, my first thought was that this must have been done already a dozen times or more. However, I roughed out a plan, and then, in order to see what the competition was, did a literature search. To my surprise it turned up only six books. Four were out of print. The other two were nothing like what I had in mind: they were advanced textbooks; and both omitted certain areas I felt were essential in a comprehensive book. There was nothing for beginners.

Over the years I have had to read a great many textbooks, and I have usually been cautious when coming across one with the word 'Introduction' in its title. All too often the first chapter turns out to be a recap on the principles of tensor calculus, or a brief review of quantum electrodynamics, after which it gets down to the serious stuff. To rub salt in, the blurb usually witters about its being 'suitable for first-year students'.

This book is different: it really is an introduction. I wasn't able to follow Stephen Hawking's example and write a whole book with only one equation in it (not that that makes *A Brief History of Time* any easier reading); but I have tried to keep the equations to the irreducible minimum, and mostly they are quarantined in boxes which you can ignore (if equations make you nervous) without losing the thread. Chapter 4 does have a few, though: these are unavoidable if you are to understand how a lens works.

It has always been my firm belief that a textbook should be the next best thing to a live tutor at one's elbow. In this respect the Open University's texts are impressive. I have not hesitated to follow their example both in their personal approach to the reader (so often absent from traditional textbooks), and in their practice of using wide margins to accommodate notes, small diagrams and parenthetic remarks.
I hate footnotes as much as printers do; and comments that are placed within the text break its continuity. So my own asides have gone into the margins, and if you are not interested in the fact that Charles Wheatstone invented the concertina, or that Joseph Fourier was appointed Governor of Egypt by Napoleon, you can ignore the marginal note and read on.

Lists of references can be useful if the discussion of a particular topic has whetted your appetite for more information; but in practice such lists rarely give an indication of the quality or scope (or even reliability) of the sources. I have tried to be more helpful, by including at the end of each chapter a section entitled 'Digging Deeper'. Rather than giving an exhaustive and perhaps less than discriminate bibliography, I have chosen the books that seem to me the best ones to lead on from what I have written in each chapter. These days there is a tendency among publishers to allow a book to go out of print as soon as it is selling fewer than a hundred or so copies a year. This has been the fate of many texts I would personally have regarded as indispensable. Fortunately, it is always possible to borrow a copy through the public library system, or through a university or other academic library. There are also book dealers, often with their own websites, who specialise in hunting down out-of-print books. With that in mind I have not hesitated to recommend one or two seminal books written many years ago that still deserve a place in any enthusiast's personal reference library.

G.S.

Acknowledgments

It would be dishonest to pretend I had written everything in this book off the top of my head. It is true that much of it represents my own experience. But whereas in the fifteenth century it was possible to possess all the knowledge there was in the world (this was said of Leonardo da Vinci, and, much later, of Johann Wolfgang Goethe), by this twenty-first century it has become impossible for anyone to claim a comprehensive knowledge even of comparatively narrow subject matter. And imaging is a very broad subject.

You can get a good idea of some of the sources I have needed to consult while working on the book, from the 'Digging Deeper' sections at the end of this chapter. I owe a debt of gratitude to all these authors. The practical image-makers to whom I also owe a large lunch are headed by Sidney Ray. Apart from the inspiration his books have provided, he has contributed a number of crucial illustrations for my text. David Burder has been equally helpful, contributing a vast amount of stereoscopic material, and information to match. Jon Tarrant and Ron Graham went out of their way to provide me with some tricky material. Adrian Davies sent me some excellent examples of specialised imagery. And David Pizzanelli of Light Impressions was instrumental in the provision of the embossed hologram included in the book, courtesy of John Brown and Light Impressions.

I have to thank Rolls Royce for permission to use Ric Parker's interferograms, and Fountain Press for allowing me to use the micrographs of colour réseaux and other diagrams from the late Jack Coote's splendid swansong *The Illustrated History of Colour Photography*. A big bouquet, too, for Bob Gibson, who not only found me some arcane details of transmission antennas, but also sorted out a number of problems with my computer, which had swallowed some of my early chapters and added insult to injury by refusing to print the symbol $\bar{G}$. I must also mention my three cats, without the attentions of whom this book might have been finished in half the time.

Chapter 1 The nature of light

Models for the behaviour of light

For thousands of years people have wondered what light really is, and have tried to construct models predicting its behaviour. In the seventeenth century Sir Isaac Newton put forward the concept of 'energy', and used it as a fundamental property of objects in motion. He also ascribed it to such things as unwinding springs, burning gases, sound and – important in our context – light. Not everyone agreed with him at the time. But today, when we operate switches, drill teeth and even weld metal with light beams, there is no longer any doubt.

Newton himself believed that light consisted of particles like tiny bullets, travelling at enormous speed. This model does predict much of the more obvious behaviour of light such as reflection and the formation of shadows, and refraction, too, if one makes some dodgy assumptions. Indeed, the entire system of what we now call geometrical optics is based on the idea of the rectilinear propagation of light.

Newton's contemporary Christiaan Huyghens suggested that the behaviour of light, particularly with regard to refraction, could be accounted for better if light consisted of waves like sound waves. Newton strongly opposed this theory, and this disagreement led to a permanent antagonism between the two men.

It has been suggested that Newton's famous remark about having stood on the shoulders of giants was partly a jibe at Huyghens, as well as at Robert Hooke, with whom Newton had also fallen out. Both men were short in stature.

When the polarisation of light was discovered, it became necessary to modify Huyghens's model of longitudinal waves (which cannot be polarised) to transverse waves (which can) (Figure 1.1).

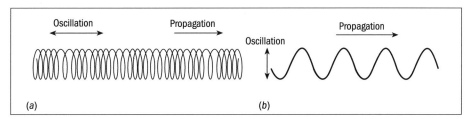

Figure 1.1 Longitudinal and transverse waves: (a) Slinky spring; (b) rubber rope.

The wave model provided a reasonably good account of diffraction and interference, as the principle assumed that each point on a wavefront was a source of 'wavelets', the envelope of which would form the new wavefront (Figure 1.2).

By the middle of the nineteenth century the speed of light had been established to within a few per cent, but there still seemed to be no logical reason for this particular speed, until James Clerk Maxwell postulated a connection between electricity and magnetism, and light. By combining two fundamental constants of electricity and magnetism he obtained an expression that gave a value to the speed of electromagnetic propagation (see Box). A few years later Heinrich Hertz's experiments with radio waves showed that there was a whole electromagnetic spectrum, of which light was a part.

When the photoelectric effect was discovered and quantified by Albert Einstein, it became clear that a continuous electromagnetic wave in Maxwell's terms, whose energy would depend only on its intensity, could not possess the appropriate energy.

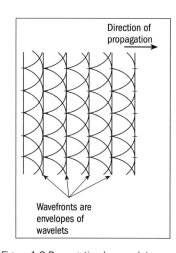

Figure 1.2 Propagation by wavelets (Huyghens principle).

The Science of Imaging

James Clerk Maxwell was a brilliant theoretical physicist who established the kinetic theory of gases, postulated the structure of Saturn's rings and made notable contributions to thermodynamics and to the theory of colour perception (producing the first-ever colour photograph). In the formulation of his electromagnetic equations, perhaps his greatest contribution to modern science, he was the first person to apply the newly minted vector calculus. This paved the way for Heinrich Rudolf Hertz (1857–1894) to discover and identify radio waves, and to show that they behaved in a manner similar to light. Both men died tragically young.

Maxwell and electromagnetism

Maxwell found that the speed of electromagnetic propagation c was related to the permittivity ε_0 and permeability μ_0 of free space in the simple relationship

$$c = 1/\sqrt{\varepsilon_0 \mu_0}$$

Maxwell's work led to four famous equations that describe the behaviour of electromagnetic radiation. To appreciate them mathematically you need a fair understanding of vector calculus, but when translated into simple prose they go something like this:

- The distribution of electric charges creates electric fields.
- Magnetic fields that change in time can also produce electric fields.
- Magnetic fields are continuous, with no beginning and no end.
- Both electric currents and electric fields that change in time can produce magnetic fields.

(I am indebted to Milo Sholt of the Open University for this insight.)

It was for his work on the photoelectric effect and the principle of photon emission and absorption that Einstein (1879–1955) was awarded the Nobel Prize for Physics in 1921, and not, as many people think, for his much more famous relativity theories. Max Planck (1858–1947) discovered that energy consists of fundamental indivisible units, which he called *quanta*. For this discovery he was awarded the Nobel Prize for Physics in 1918.

In order to obtain the right answers Einstein suggested that light was not a continuous wave, but was emitted in the form of tiny pulses of light energy (which he called *photons*), and that the energy carried by a photon was directly proportional to its frequency. This concept fitted Max Planck's quantum theory, and the photon took its place among fundamental particles. A photon was considered as having zero rest mass, and could therefore travel 'at the speed of light' without violating the laws of special relativity.

Electromagnetic radiation

Electromagnetic radiation is so much part of our daily lives that we scarcely ever think about it. But now we need to look at its nature a little more closely. Why 'electromagnetic'?

Michael Faraday (1791–1867) was undoubtedly the greatest experimental physicist of the nineteenth century. But as well as discovering the relationship between magnetism and electricity and designing the first electrical generator and motor, he made a large number of practical discoveries in organic chemistry (he was the first person to isolate benzene). He was a bookbinder by training, and had no mathematical ability: all his successes were the result of intuitions.

The first person to appreciate the connection between electricity and magnetism was Michael Faraday. One of his experiments is very simple to repeat:
If you pass an electric current through a piece of copper wire it will set up a magnetic field, and if you take a small compass needle and hold with its spindle parallel to the wire it will align itself with the direction of the field. If you reverse the direction of the current through the wire the needle will reverse its direction too (Figure 1.3).

If the wire is carrying an a.c. mains current, this changes its direction back and forwards again 50 times a second (60 in the USA). This is said to be a frequency of 50 hertzes (Hz). If the compass needle were small and light enough it would follow these reversals too. It is electromagnetic energy that moves the needle. The strength of the field becomes less as you move away from the wire, but the effect is still present, and would still be there even if you were to go millions of miles away. Now, in basic electrical theory we tend to think of a magnetic field as being set up instantaneously everywhere, but Maxwell showed that it is not. If your compass needle happened to be 300 000 kilometres from the wire, it would sense the existence of the field one second

The nature of light

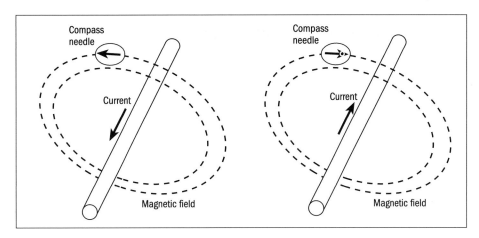

Figure 1.3 Reversing the current reverses the direction of the magnetic field.

later. If our a.c. were 1 Hz instead of 50 Hz such a compass needle would always be exactly one cycle behind. A needle at half this distance would be half a cycle behind, and so on. If you could freeze the system at a particular instant it would look like Figure 1.4.

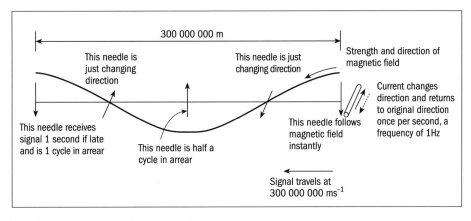

Figure 1.4 Instantaneous state of the magnetic field.

As the distance 300 000 km contains one complete cycle, we say that the *wavelength* is 300 000 km (3×10^8 m). The number of cycles per second is the *frequency*; we can say, therefore, that

$$\text{velocity} = \text{wavelength} \times \text{frequency}$$

The electromagnetic spectrum

The electromagnetic field set up by a 50 Hz a.c. is not very strong. If you place a compass needle under an electricity pylon it will barely quiver. But as frequency increases, so does the energy. To obtain high frequencies you need to use an electronic device called an oscillator, and if you adjust this to give a frequency of, say, 198 000 Hz (198 kHz) the effects can be detected many miles away.

This is almost the lowest frequency that can be used for radio broadcasting. The radio spectrum continues up to a frequency of around 1 000 000 000 Hz (10^9 Hz or 1 GHz),

In fact, if you superimpose a microphone signal on it you will be broadcasting (illegally) on the same frequency as Radio 4 (long wave).

The Science of Imaging

which represents the (surface-based) TV band. Here the radiation takes on some of the properties shown by light: it can be reflected, and it travels in straight lines; large objects such as tower blocks and hills can obstruct and reflect the beam. Above this region, corresponding to wavelengths of only tens of millimetres, are microwaves, which, of course, possess enough energy to cook potatoes. They are used in radar and in satellite communications, as well as in ground based links, and when used for such purposes require parabolic reflectors to concentrate the beam on the receiving antenna. The next step up is the beginning of the infrared (IR) region, at around 10 mm wavelength.

Calling it a keyboard is not a mere whimsy. Each octave on a (musical) keyboard represents a doubling of sound frequency. In Figure 1.5 each interval represents a multiplication of electromagnetic frequency by 10. Both scales are *logarithmic*. There is more about logarithms in Appendix 1; the concept underlies much of the nature of visual perception (see Chapter 3).

Electronic devices cannot generate shorter wavelengths than these. But there are IR lasers. Lasers produce single frequencies, just like oscillators, at wavelengths that range from IR wavelengths right through the visible spectrum into the ultraviolet (UV), and there is even an X-ray laser operating at less than 10 nanometres wavelength (1 nm = 10^{-9} m). At present we cannot generate shorter wavelengths such as gamma radiation at single frequencies, though the electron laser is already becoming more than just a dream. Figure 1.5 summarises the electromagnetic spectrum. You will notice that on this gigantic keyboard the visible spectrum spans barely a single octave.

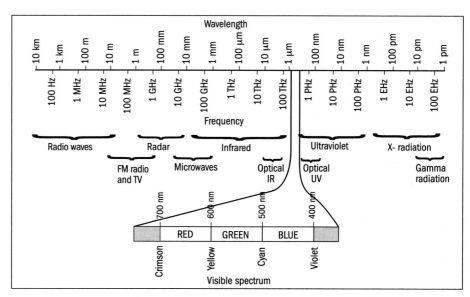

Figure 1.5 The electromagnetic spectrum.

We don't use single frequencies much in making images. Only radar and holography demand such highly disciplined radiation. For most imaging purposes we need to use the whole visible spectrum as generated by fairly ordinary light sources, including daylight.

Polarisation

An important characteristic of light that is predicted by the transverse wave model is *polarisation*. If you were to take up the end of the rubber rope of Figure 1.1b and shake it in random directions, waves would still travel along it, but the plane of vibration would fluctuate wildly. This is an *unpolarised* wave. If you shake the rope in one direction only, the vibration will be in a single plane, and this is a *linearly polarised* wave. Ordinary water waves are linearly polarised: although the energy spreads out from the source in all directions, at any one point the water is only going up and down. To produce polarised light from an ordinary source you need a polarising filter, which is described in Chapter 2. Laser light is polarised, and polarised light also occurs naturally in skylight and in light reflected from a shiny non-metallic surface.

If you take your rubber rope and give it a rolling motion by moving your hand in a circle, you will still produce a wave, but of a different kind. It is certainly not random. Is it another form of polarisation? Yes, it is: it is called *circular polarisation*. The direction of polarisation rotates through 360° for every wavelength. This type of polarisation doesn't seem to occur in nature, but it does have a place in certain physics experiments, and even in photography (Figure 1.6).

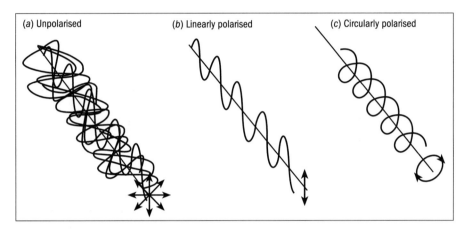

Figure 1.6 Polarisation of light: (a) unpolarised; (b) linearly polarised; (c) circularly polarised.

Interference

Two waves of the same wavelength passing through the same space in the same direction are said to *interfere*. (This has nothing to do with the sort of interference that sometimes mars TV and shortwave radio.) Their amplitudes (i.e. the heights of their crests above the median) will add algebraically. If the crests coincide, the resultant wave has an amplitude that is the arithmetic sum of the two amplitudes. If the crest of one wave coincides with the trough of the other, the resultant will be the arithmetic difference between the two amplitudes. If these are equal they will cancel, and in space there will be no disturbance at all (Figure 1.7).

These two conditions are described respectively as 'in phase' and 'in antiphase'. In between ('out of phase'), the algebraic sum of the two amplitudes may be more or less than the average of the two amplitudes, depending on whether the phase relationship is nearer to in-phase or antiphase. The position of the crest of the resultant will be somewhere in between the other two. The phenomenon is called *interference*. It is qualified by the terms 'constructive' or 'destructive' depending on whether the resultant is greater or less in amplitude than its components.

A striking demonstration of interference occurs when a beam of monochromatic (single wavelength) light passes through two closely spaced slits on to a screen. This is

The Science of Imaging

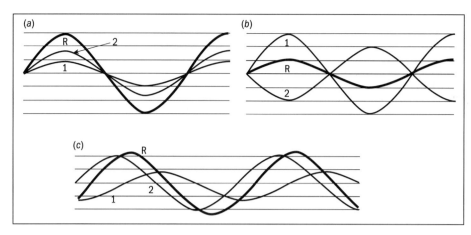

Figure 1.7 Interference between two transverse waves (1 and 2), and resultant (R): (a) in phase; (b) in antiphase; (c) out of phase.

Thomas Young (1773–1829) not only confirmed the wave nature of light, but explained polarisation and additive colour synthesis (see Chapter 6). He was also largely responsible for bringing the Rosetta Stone to England and deciphering its Egyptian hieroglyphics.

known as the *Young's slits* experiment after its originator, Thomas Young. Where the two beams overlap, the screen shows a series of evenly spaced spots or *fringes*. These mark the angles at which the difference in distance from the two slits to that point on the screen is either a whole number of wavelengths (0, 1, 2, 3 etc.) for a bright fringe, or an odd number of half-wavelengths ($\frac{1}{2}$, $1\frac{1}{2}$, $2\frac{1}{2}$, $3\frac{1}{2}$ etc.) for a dark one. This is illustrated in Figure 1.8.

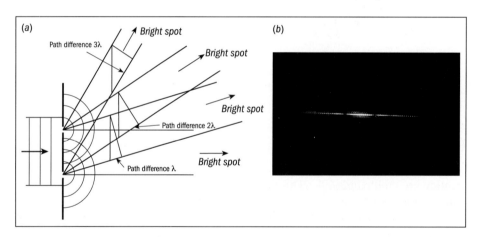

Figure 1.8 (a) Formation of bright spots where optical path differences are integral multiples of wavelength; (b) photograph of fringes.

The best light source for this demonstration is a small laser such as a diode laser pointer, as the beam is very bright (the laser being still 160 years in the future, Young had to make do with sodium light). You can replicate the demonstration if you make a double slit by scratching two fine lines, as close together as you can, with a needle on a piece of fogged black photographic film, or, better, make a high-contrast photographic negative of two parallel lines drawn 2 mm apart, from a distance of around 1 m. When you shine your laser pointer through these slits at a white wall in a darkened room you will see Young's fringes clearly. You can replicate the experiment on a large scale using water waves in a bath. Make the two wave sources by dipping your fingers in the water several times a second, with your fingers 10–20 cm apart.

You will be able to see the interference pattern radiating outwards from the two sources in alternate lines of still and disturbed water.

You may have guessed (especially if you did the bath experiment) that there is a relationship between the separation of the Young's slits and the separation of the spots. It is an inverse relation: halve the slit separation and you double the spot separation. There is also a direct relation between the spot separation and the wavelength of the light for any given separation. So if you illuminate the slits with blue light the spots have a smaller separation than if you illuminate them with red light. In fact, if you illuminate the slits with white light (which contains all wavelengths from 400 to 700 nm) you get a drawn-out spectrum instead of the spots (Figure 1.9a, colour plate).

Interference occurs in many natural situations. The iridescent colours of beetles, tropical butterflies' wings, oil films and soap bubbles are all the result of interference. The type involved here is known as *thin-film interference*, and it happens when the space between two adjacent transparent surfaces is of the order of half a wavelength of light. The light is reflected more or less equally from the first and second surfaces of the scale, chitin layer or liquid film, and light with a wavelength equal to twice their spacing will have both these reflected beams in phase and will be reflected strongly, whereas light of other wavelengths will be out of phase and will be weakened or suppressed (Figure 1.10).

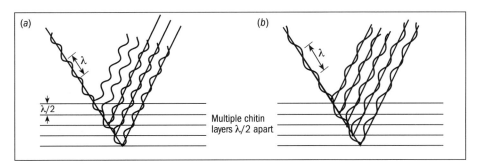

Figure 1.10 Bragg reflection: (a) light of wavelength twice the spacing is reflected in phase and reinforced; (b) other wavelengths are reflected out of phase and are suppressed.

One effect of interference that can be particularly irritating to photographers is called *Newton's rings* (or, more correctly, *Newton's fringes*, as they are not always circular). Newton was the first to describe the effect, though he was unable to suggest a satisfactory explanation. If you place a surface of low curvature, such as a very thin lens, in contact with a polished glass plate on a white sheet, and illuminate it from above with a small light source, you will see a set of concentric coloured fringes. With monochromatic light these will be better defined (Figure 1.11). The innermost of these represents the position where the spacing between the adjacent surfaces is one half-wavelength; the second represents a spacing of $1\frac{1}{2}$ wavelengths; and so on. With white light the fringes quickly become muddled together, but if you use monochromatic light you can see many more fringes. Lens makers use these fringes as a guide to accuracy when testing a lens against a pattern, but to the photographer they are simply a nuisance, occurring as irregular patches in slides mounted between glasses, or on enlargements made with the negative held in a glass stage. The colours you see in soap bubbles and in oil films on water are also Newton's fringes, this time between internal rather

These irregular coloured fringes are often called *Fizeau's fringes*, after Armand Hippolyte Louis Fizeau (1819–1896), who worked with Louis Foucault and first applied the Doppler principle to the measurement of the movement of stars. He also made the first reasonably accurate determination of the speed of light in 1849, and contributed to the development of photography in astronomy.

The Science of Imaging

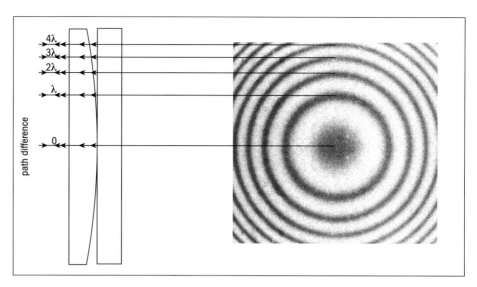

Figure 1.11 Formation of Newton's rings (monochromatic light). The central spot is dark because there is a 180° phase reversal at the lower air–glass surface, so that destructive interference occurs when the path difference for the two reflected rays is 0, λ, 2λ, 3λ etc.

than external surfaces. Interference also plays a large part in the anti-reflection coating of lenses (Chapter 4), and it is fundamental to the making of holograms (Chapter 17).

Diffraction

Diffraction is a wave phenomenon closely related to interference. It describes the way a beam of light behaves when it passes through an aperture, or past an obstruction. There are several models that describe diffraction with varying degrees of success. The best is the Fourier model, which deserves, and has received more than once, a whole book to itself. There is a short account of it in Appendix 3, which may whet your appetite. However, the Huyghens wave principle is simpler, and is adequate for our present purposes.

If you consider a wavefront passing through a very small aperture only one or two wavelengths wide, then beyond it the waves will spread out in all directions, as the aperture acts like a point source, in effect a single wavelet. However, if the aperture is a slit that is somewhat wider, say a few tens of wavelengths, a screen placed on the exit side will show what looks much like an interference pattern, with spacing similar to that given by a double slit. How does this happen? Well, to explain it you need to do a little constructive thinking. Imagine the single slit aperture to be made up of a large number of mini-slits, all touching. Now take the mini-slit at the extreme left, and match it to the one next to the midline of the aperture. Think of these two mini-slits as a pair of Young's slits. They will form a double-slit interference pattern. Now consider the adjacent pair of mini-slits. They will also form an identical pattern, and so on until you reach the right hand edge of the main slit. So you will have a pattern like a Young's slits pattern, but with a fringe spacing corresponding to that given by a pair of Young's slits spaced at half the width of the single aperture; and so it proves (Figure 1.12).

The nature of light

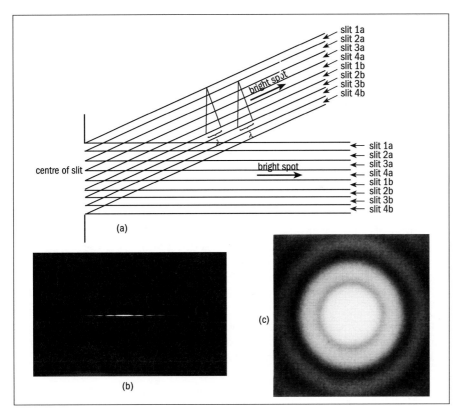

Figure 1.12 (a) The diffraction pattern formed by a single slit aperture can be explained by considering the aperture to be a number of pairs of Young's slits separated by half the aperture width; (b) diffraction pattern for a single slit; (c) Airy diffraction pattern for a circular aperture.

The grating condition

The relationship between the grating spacing (the *pitch* or *spatial period*) and the spot spacing follows directly from Figure 1.8. For the bright fringes the angles at which the beams emerge are given by

$$\sin\theta = n\lambda/d$$

where θ is the angle to the normal, λ the wavelength, d the period of the grating and $n = 0, 1, 2, 3 \ldots$. If the screen is a distance x from the grating, the first-order fringe will be formed at $x \sin\theta = \lambda x/d$, the second-order fringe at $x \sin\theta = 2\lambda x/d$ and so on. For a single slit of width d, the fringes will form at $2\lambda x/d$, $4\lambda x/d$ etc.

Appendix 3 shows how this is done using the Fourier model.

George Biddell Airy (1801–92) was Astronomer Royal for 46 years. He was also Director of Cambridge Observatory. He introduced many innovations into astronomy. He also established the border between Canada and the USA, assisted with the laying of the first transatlantic cable and helped with the design of the chimes of Big Ben.

The Airy diffraction pattern

It is somewhat more complicated to work out the diffraction pattern for a two-dimensional figure. One such figure is very important in imaging: a circular aperture. Its diffraction pattern was first described by Sir George Airy in connection with the images of stars, and it is named after him. It is a central disc (the *Airy disc*) containing more than 80 per cent of the light, surrounded by rings. Its diameter D to the first zero

is given by
$$D = 2.44\lambda x/d$$
where D is the diameter of the Airy disc, d the aperture diameter and x the screen distance.

In a telescope or camera system, d represents the focal length and d/x is simply the *f*-number of the lens or objective. (For an explanation of '*f*-number' see Chapter 4.)

Reflection and refraction

In oblique reflection each part of the wavefront arrives at the reflecting surface at a different time, so that the envelope of the wavelets sets off in a new direction that is symmetrical with respect to the surface. In practice we don't use the surface as our reference, because it may be (indeed, usually is) curved. Instead, we use the perpendicular at the reference point, called the *normal*. We can use a simple geometrical model here. Instead of the tedious business of drawing wavefronts we can simply draw a line showing the direction in which they are travelling, a *ray*. Reflection is then a straightforward matter. The angle the entering (incident) ray makes with the normal is called the *angle of incidence*, and the angle the emerging (reflected) ray makes with it is called the *angle of reflection* (Figure 1.13).

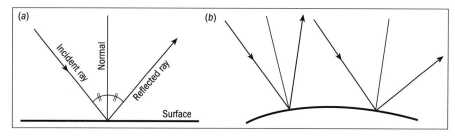

Figure 1.13 Reflection (a) at a flat surface; (b) at a curved surface.

The two laws of reflection are:

- The incident and reflected rays are in the same plane as the normal.
- The angle of incidence is equal to the angle of reflection.

These laws also apply to curved surfaces, and here it is much easier to follow the path of a light beam by using rays rather than wavefronts.

The term *refraction* refers to the change in direction of a light beam entering a substance having optical qualities different from the one it has left. The speed of light discussed earlier was actually the speed of light in empty space. In air this is very slightly less, and in liquids and transparent solids it is very much less. When a beam of light enters a glass block obliquely, the wavefront is slowed down, and successive crests become closer together. There is thus a change in direction of the wavefront towards the normal. On emergence from a parallel surface the process is reversed, and the beam resumes its original direction (Figure 1.14). The change in direction is called *refraction*. Again, we can simplify the geometry by using the ray model.

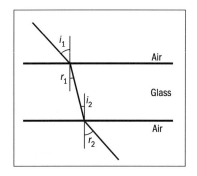

Figure 1.14 Refraction in a parallel sided glass block: $i_1 = r_2$ and $i_2 = r_1$.

As in reflection, the incident ray, the refracted ray and the normal are all in the same plane. For small angles the angle of incidence bears a constant relationship to the angle of refraction (i.e. the angle the refracted ray makes with the normal). It is the same as the ratio of the speeds of light in the two media, and is called the *refractive index*. This constant is of immense importance in lens design. For angles of incidence larger than a few degrees we have to use a more precise relationship, and this is shown in the Box.

Snell's Law

Snell's Law follows directly from the observation that light changes its speed when it passes from one optical medium to another of different optical density (this term refers to its light-slowing powers, not its physical density). It states that where the refractive indices of the first and second media are n_1 and n_2 respectively, and i and r are the angles of incidence and of refraction respectively, then

$$n_1 \sin i = n_2 \sin r \quad \text{or} \quad \sin i / \sin r = n_2/n_1$$

If the first medium is air, n_1 is approximately 1. (The refractive index of empty space is exactly 1.) The refractive index of an optical medium with respect to air is often represented by the Greek letter μ (Figure 1.15).

Snell's Law leads directly to the formula for the critical angle. At this angle $\sin r = 1$, so that

$$n_2/n_1 = \sin i$$

and for glass to air,

$$\sin i = 1/\mu_{glass}$$

Willebrord Snell (1580–1626) was a trained lawyer, and became Professor of Mathematics at Leyden University, where in addition to his work in optics he developed the method of survey by triangulation. He also made an accurate determination of the radius of the Earth.

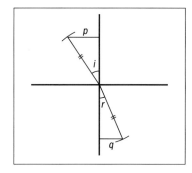

Figure 1.15 Construction for Snell's Law: $p/q = \mu = \sin i / \sin r$.

Total internal reflection

If you pass a beam of light obliquely through a glass block into air, the angle of refraction will be greater than the angle of incidence, and will increase faster than the angle of incidence. So at a certain angle, the *critical angle*, the emergent beam will have an angle of refraction of 90°, and will travel along the surface (Figure 1.16). If you increase the angle of incidence beyond the critical angle the beam cannot escape. Instead it is reflected. This effect is known as *total internal reflection*, usually abbreviated to 'TIR'. It is more efficient than a conventional mirror, which can manage at best no more than about 98 per cent reflectance. As the critical angle for glass is less than 45°, TIR is used in prismatic reflectors (*roof* or *Porro prisms*) for binoculars and for retroreflectors (corner cubes).

Prisms

When a beam of light passes through a parallel-sided glass block the emergent beam is parallel to the incident beam, but is displaced from it a little. However, if the second

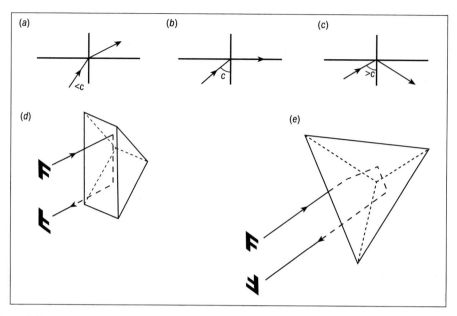

Figure 1.16 Total internal reflection: (a) glass to air; (b) at the critical angle; (c) total internal reflection (TIR); (d) Porro prism for image inversion; (e) corner cube reflector.

surface is not parallel to the first, as in a prism, the beam undergoes a second deviation (Figure 1.17). The combined angle of deviation depends on the angle between the prism's faces (the *refracting angle*) and the angle of incidence at the first surface. The deviation is at a minimum when the incident and emergent beams are symmetrical to the prism. This is another property that is useful in optical design.

A prism has another important effect on white light. The refractive indices of transparent media are greater for shorter wavelengths than for longer ones, and a prism, like a diffraction grating, spreads white light into a spectrum (Figure 1.9b, colour plate). Notice that the spectrum produced by a prism is not linear in its spread, and runs in the opposite direction to that produced by a diffraction grating. Newton was the first person to make a proper examination of the spectrum produced by a prism. Perhaps because of the mystical importance of the number seven, he assigned seven hues to the spectrum, namely red, orange, yellow, green, blue, indigo and violet. This may seem somewhat confusing today, as the dye indigo (used to colour denim jeans) is a desaturated colour that's not in the spectrum at all. Modern colorimetry (see Chapter 2) suggests that the spectrum of white light can be better described as

In Newton's day 'indigo' meant a colour more like royal blue, and what he termed 'blue' was more greenish than the way we visualise it today.

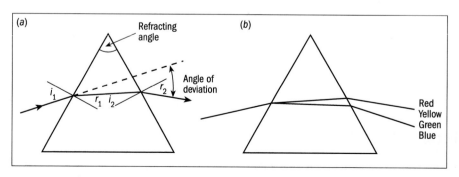

Figure 1.17 (a) Path of light through prism; (b) dispersion of light by prism (see Figure 1.9c, colour plate).

consisting of five basic hues: red, yellow, green, cyan and blue-violet (usually called simply 'blue').

The pinhole camera

It was recorded as early as the second century AD, in the Middle East, that if a small hole was made in the outside wall of a dark chamber, an inverted image of the outside scene would appear on the opposite wall (Figure 1.18).

Today we call this device a *pinhole camera*. The image is in full colour, but is not very bright. Nor is it very sharp, owing to the finite size of the pinhole, and if you reduce the diameter of this too far, diffraction takes over and the image becomes even less sharp. There is thus an optimum pinhole size. There is more about pinhole cameras in Chapter 19.

Development of a lens

In the form of magnifiers and burning glasses, lenses have been known for a very long time. In the fourteenth century some unknown person was smart enough to couple a burning glass with a camera obscura aperture and obtain a bright sharp image at its focus. In order to see how a lens focuses, we may look on it initially as a series of prisms (Figure 1.19).

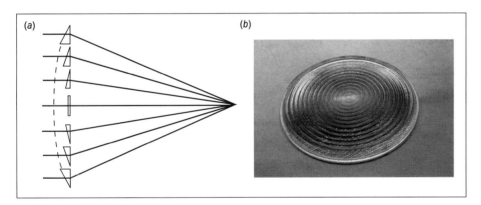

Figure 1.19 (a) A lens as a series of concentric ring prisms; (b) a Fresnel lens as used in a spotlight.

You need to think of these prisms as cross sections of a ring of glass, of course, because a lens is circular. In fact, lighthouse lenses are traditionally made in this way, to save weight. They are called *Fresnel lenses*. This type of stepped lens is also used in spotlights, and as field brighteners in overhead projectors and camera viewfinders. As a rule lenses have smooth rather than stepped profiles, but you can think of a lens as being a very large number of concentric ring prisms, each giving just the right amount of deviation to bring light to a focus at the same point.

When you use a lens to burn a hole in a piece of paper, you are focusing an image of the sun on the paper, and the distance between the lens and the paper when the image is tiniest is called the *focal length*. (The German word for focal length is *Brennweite* ('burning distance').) If you move the lens a little farther from the paper you can bring into focus nearer objects such as trees. Chapter 4 discusses focusing relationships.

Figure 1.18 The principle of the pinhole camera was known in the Middle East before AD 200.

This is a direct consequence of the rectilinear (straight-line) propagation of light. This principle became the basis of an artist's aid called the *camera obscura*.

In the Archaeological Museum in Heraklion, Crete, there is a beautiful rock crystal lens dating from Minoan times, i.e. before 1500 BC (Figure 1.20).

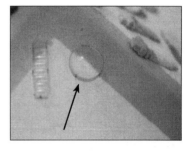

Figure 1.20 Minoan rock crystal lens, ca. 1500 BC (Museum of Archaelogy, Iraklion, Crete) (actual diameter is about 25 mm).

Augustin Fresnel (1788–1827) was a civil engineer working for the French government, and when Napoleon returned from Elba, Fresnel was put under house arrest. With nothing useful to do, he worked out the transverse wave theory of light and the mathematics of diffraction. He invoked polarisation to explain the double refraction of certain crystals. After he was freed and went back to work he won a government award for his lighthouse lens designs.

A good shape for a lens surface is approximately part of a sphere, and as this is also the easiest curved surface to shape, most lenses have spherical surfaces. However, a spherical shape is not exactly correct, and later in Chapter 4 we shall see some of the consequences of this discrepancy, and how lens designers minimise them.

A curved mirror behaves similarly to a lens in most respects, and it has certain advantages that make it particularly suitable for generating some types of optical image. Again, a spherical shape is very nearly (but not quite) ideal.

Digging deeper

When it comes to the physics of light, you can dig as deep as you like; but the deeper you dig the tougher it gets. There are many standard textbooks up to degree level. *Light*, by Michael Sobel (University of Chicago Press, 1989) is an excellent non-mathematical survey of the whole field of light and optics. *Insight into Optics*, by Oliver Heavens and Robert Ditchburn (Wiley, 1991), is a good introduction to all branches of optics, with fairly rigorous but not too difficult mathematical backgrounds. *Modern Optics*, by Robert Guenther (Wiley, 1990), is a full degree-level course. The standard work on optics is *Principles of Optics*, by Max Born and Emil Wolf. The last edition these two outstanding teachers actually wrote was the 3rd edition in 1965, but it has been regularly updated since then. It is now published by Cambridge University Press, the most recent edition being dated November 1999. It is not an easy read.

Chapter 2 Photometry, lighting and light filters

Photometric units

The formal name for the measurement of light is *photometry*. It is part of the wider system of measurement of electromagnetic radiation, which is called *radiometry*; but as it is concerned with visible radiation its units have to be associated with the visual process. In all aspects of imaging technology we need terms to define such matters as light sources, quantity of illumination, reflection, transmission and so on, in terms that can readily be related to one another and to units of mechanical and electrical energy. In the past there were three different systems, FPS, CGS and MKS, resulting in a proliferation of named units, to the confusion of students and the exasperation of their tutors. Unfortunately, some academic texts still employ these outdated systems. In this book I use only SI units, with the occasional centimetre or inch where it is obviously called for.

FPS was foot, pound, second, the old Imperial system. CGS was centimetre, gram, second, the old metric system. MKS was metre, kilogram, second, the forerunner of SI (Système International d'Unités).

Luminous intensity

The SI unit of *luminous intensity* is the candela (cd). Although still described as a fundamental SI unit, its original definition was unsatisfactory, and it is now defined in terms of the lumen (as 1 lumen per steradian). The candela is a unit of power, and thus has the same physical dimensions as mechanical or electrical power, namely energy emitted (or work done) per second. The SI unit of power is the watt, and you might expect the candela to be interconvertible with it; but this

There are 2π radians in a circle and 4π steradians in a sphere (see next page).

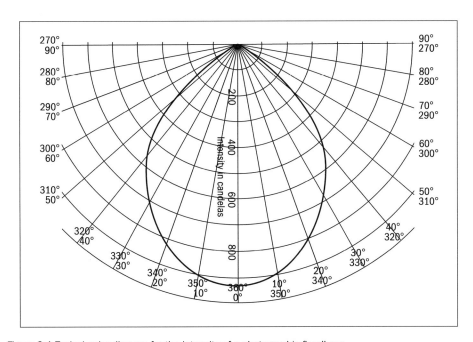

Figure 2.1 Typical polar diagram for the intensity of a photographic floodlamp.

The Science of Imaging

is in fact possible only for monochromatic (single wavelength) light. So the candela was originally defined in terms of white light, using a standard light source.

If you want to draw up a complete specification for a light source such as a photographic spotlight you have to do so by means of a polar diagram (Figure 2.1). But you can't gauge the overall power of a floodlamp, a flash head or even a car headlamp simply by studying a polar diagram. You have to integrate all the values over a complete sphere to do this. You can, however, specify *mean beam intensity* over a specified angular area, say a 10° × 20° rectangle. No source (except possibly an aerial photoflash) is uniform over a complete sphere, but we use this hypothetical uniform source to relate the candela to a more generally useful unit, the lumen.

Flux is an old-fashioned word dating from Newton's time, and meaning 'flow'. I have no idea how or why it has survived.

Whereas linear angles are measured in radians (rad), solid angles are measured in steradians (sr).
1 steradian is represented by a portion of a sphere of 1 metre radius and 1 square metre area. To find a solid angle you take the projected area and divide it by the square of the radius (Figure 2.2).

Luminous flux

Except for special purposes we are not interested in measuring light output in just a single direction. We are more interested in the aggregate output of a light source designed to cover a particular solid angle. The lumen is now effectively a fundamental SI unit, the others being the metre, kilogram, second, ampere, kelvin and mole. We shall meet them all in due course. It is linked directly to the equivalent radiometric unit, the watt. It is defined to be such that 1 watt of radiation at a frequency of 540×10^{12} Hz produces a flux of 683 lumens. In practice we use the corresponding wavelength of 555 nm.

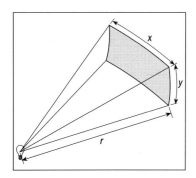

Figure 2.2 Solid angle: here the solid angle is xy/r^2.

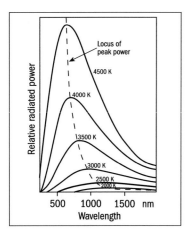

Figure 2.3 Spectral power distribution for a black body radiator.

Planck's equation and retinal sensitivity

The radiant energy emitted by a hot totally non-reflective surface (usually dubbed a *black body*) has the form of a continuous spectrum. When energy is plotted against wavelength for different temperatures the result is a family of curves (Figure 2.3), broadly similar in shape, their maximum value both increasing and moving towards shorter wavelengths as the temperature increases. Max Planck found an equation that generated the family of curves, as well as its theoretical justification. It is

$$M_c = c_1 \lambda^{-5}/[\exp(c_2/\lambda T) - 1] \, \text{W m}^{-3}$$

(watts per square metre per wavelength in metres) where
$c_1 = 3.74183 \times 10^{-16}$ W m^2 and $c_2 = 1.4388 \times 10^{-2}$ m K. (T is in kelvins.)

From this and the sensitivity versus wavelength curve for the human retina (Chapter 3) we can relate lumens to watts for any visible wavelength, the maximum value being 683 lm W^{-1} at 555 nm.

Illuminance

A hypothetical uniform source of 1 candela emits 1 lumen per square metre (lm m^{-2}). When this light falls perpendicularly on a surface, the amount of illumination, called the *illuminance*, is 1 lux (lx). If the light beam is oblique, with

Photometry, lighting and light filters

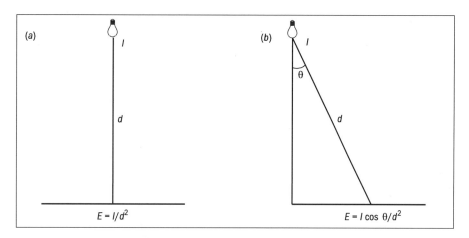

Figure 2.4 Lambert's Law: (a) perpendicular illumination; (b) oblique illumination.

an angle of incidence θ (theta), the illuminance is multiplied by $\cos\theta$ (Lambert's Law, Figure 2.4).

Inverse Square Law

The illuminance on a surface illuminated by a small source is proportional to the intensity of the source and inversely proportional to the square of its distance from the source. This is because when you double the distance you quadruple the area covered, and when you triple it you multiply the area ninefold (Figure 2.5). So the full expression for calculating the illuminance E at a distance d from a source of intensity I at an angle of incidence θ is given by

$$E = I\cos\theta/d^2$$

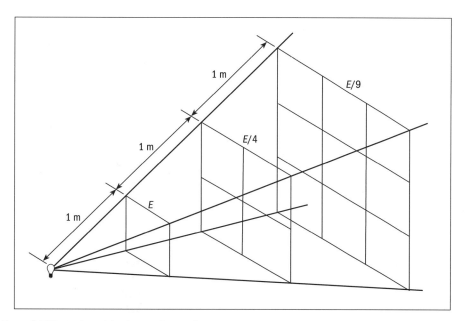

Figure 2.5 Illustration of the Inverse Square Law.

Luminance

If you look at a 200 candela filament lamp it seems very bright. If you look at a 200 candela fluorescent light it seems much less bright. The same amount of light is being emitted, but from a much larger surface area. The quantitative term for brightness is *luminance*, and you obtain it by dividing the luminous intensity by the projected area of the luminous surface. As with luminous intensity, you have to specify the direction. The unit of luminance is the candela per square metre ($cd\,m^{-2}$).

Reflectance

You can specify the luminance of an illuminated surface in the same way. You can also calculate this, if you know the illuminance, and the *reflectance* of the surface in a particular direction. There are two kinds of reflection:

- *Specular* 'Speculum' is Latin for 'mirror', and specular reflection is direct reflection from a shiny surface, obeying the laws of reflection. The incident light does not penetrate the surface, so the reflected beam is the same colour as the incident beam (Figure 2.6a).

- *Diffuse* In diffuse reflection the light penetrates the surface and is modified spectrally by any pigment present in the material. On emergence it is partially scattered. You can represent the pattern of scattering by a polar diagram (Figure 2.6b). If the material is perfectly diffusing, the reflected light obeys Lambert's Law, and the reflectance is said to be *Lambertian*.

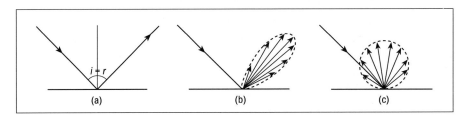

Figure 2.6 Reflection at a surface: (a) specular; (b) partly diffuse; (c) totally diffuse. Lengths of arrows show relative intensities.

Luminous energy

The luminous energy emitted by a photoflash is measured in lumen seconds (lm s). The intensity varies throughout the duration of the flash, and we obtain the total luminous energy by measuring the area under the time–intensity curve (Figure 2.7). The measurement is usually made between the rising and falling one-third peaks, and the time interval between them is the effective flash duration. (It usually contains around 90 per cent of the total luminous energy.)

Luminous efficacy

For a given light source the luminous energy emitted is a fixed proportion of the electrical energy fed into it. If we were in a position to make a direct comparison, i.e. watts out ÷ watts in, the answer would be simply the luminous (or rather

Photometry, lighting and light filters

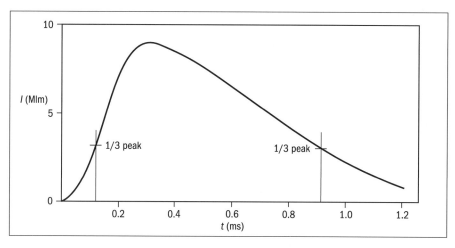

Figure 2.7 Typical intensity-time graph for an electronic flash.

radiant) efficiency. You can do this for monochromatic sources such as lasers, and state the output efficiency as a simple percentage. You can convert watts to lumens, too, for single wavelengths, based on the sensitivity curve for photopic (bright light) vision (Chapter 3). At 555 nm (yellow-green), 1 watt corresponds to 683 lumens. For other wavelengths the figure falls as the wavelength moves farther away from this central point. For light sources with a broad spectrum there is no single conversion figure. Instead we quote the *luminous efficacy* in lumens per watt ($lm\ W^{-1}$). The luminous efficacy of a tungsten filament lamp is around $20\ lm\ W^{-1}$ and of a xenon discharge tube about $40\ lm\ W^{-1}$.

Photographic flash heads are rated in joules (J). A joule is a fairly large amount of energy (a 10 joule pulse laser can blow paint off a wall!), so it seems odd that a studio flash should be rated at 200 joules or more. In fact, it is the electrical *input* that is being quoted. And this has to be multiplied by the luminous efficacy to find the output in lumen seconds.

Spectral power distribution

The simplest way to depict this is by percentage figures for red, green and blue content, or as a simple column chart (Figure 2.8a). A more accurate description is

The curves of Figure 2.8c are produced by a spectrophotometer, a device that scans the intensity via a monochromator (i.e. wavelength by wavelength) and plots the values as a graph.

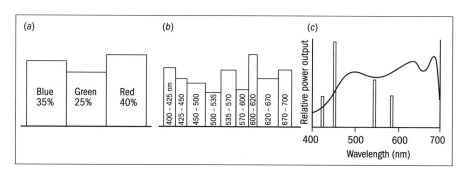

Figure 2.8 Depiction of spectral power distribution: (a) primary hue content; (b) histogram; (c) power output vs wavelength for fluorescent lamp. The bars are the superimposed line spectrum.

The Science of Imaging

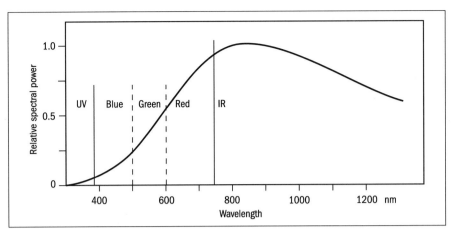

Figure 2.9 Spectral power distribution of a filament lamp (3400 K).

by a histogram (Figure 2.8b). The third and most accurate method is to show a full plot of power versus wavelength (Figure 2.8c).

An incandescent light source has a spectral power distribution close to that of a black body at approximately the same temperature, following the Planckian energy distribution system (Figure 2.3).

Colour temperature

As the Planckian distribution provides a precise description of the spectrum of an incandescent light source, we can specify the spectral power distribution of any source of this type by stating the equivalent temperature in kelvins (see marginal note on next page). This figure is called the *colour temperature*. A halogen lamp as used in overhead projectors has a colour temperature of 3400 K. The true temperature of the filament is a little lower, which is just as well, as 3400 K is uncomfortably close to the melting point of tungsten. The sun, too, is an incandescent body: standard (or 'mean noon') sunlight is established as 5400 K, a figure obtained from a series of measurements made in Chicago in the 1920s.

This figure is somewhat low, and modern colour films for use in daylight are balanced for 5500 K.

The mirek scale

Equal changes in the perceived hue of an incandescent body are not matched by equal changes in colour temperature. With its reciprocal there is a much closer match. As this figure is very small, in order to bring it up to whole numbers we multiply it by a million. The resultant figure is in units called *mireks*, from micro reciprocal kelvins ($10^6 \, K^{-1} \div$ colour temperature).

In the days when kelvins were degrees Kelvin, mireks were mireds, and many textbooks still refer to them as such. 'Hue' is a more exact term than 'colour' here, because 'colour' also involves lightness and saturation (see Chapter 6).

Because hue shift expressed in mireks is more or less constant within the visible spectrum, colour filters to convert the colour temperature of various black body sources to match daylight or tungsten light emulsions are specified in mireks, negative for a blue shift and positive for a red shift. For example, a blue filter designed to match a 3200 K tungsten floodlamp (312 mireks) to 5500 K daylight film (182 mireks) has a value of (182 − 312) = 130 mireks. There is more

Photometry, lighting and light filters

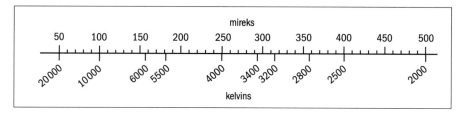

Figure 2.10 Colour temperature equivalents in mirek and kelvin scales.

information on colour temperature conversion filters in Chapter 12. Figure 2.10 relates colour temperature to its mirek equivalent.

Note: Some types of light source that are not incandescent, and therefore have an irregular or discontinuous spectrum (e.g. electronic flash, colour-balanced fluorescent tubes), have an overall spectrum close enough to a Planckian curve within the visible range to be allotted an honorary value, known as a *correlated colour temperature*.

Types of light source

Light sources for imaging can be continuous, as in daylight and floodlamps, or transient, as in flash, and their spectra can be quasi-black body or irregular. The main light sources used in photography are given below.

- *Daylight* This is a very variable source. Around sunrise and sunset the colour temperature is as low as 3000 K, but it rises to more than 5000 K at midday, and in summer the presence of blue sky can raise it to 6000 K and more. On an overcast day the colour temperature is higher still, up to 8000 K. On a clear day the colour temperature of the sky itself (which is the illuminant for shadows) may be as high as 16 000 K.
- *Tungsten filament lamps* Photographic floodlamps have a colour temperature of 3200 K; the halogen types have a colour temperature of 3400 K (Figure 2.9). Household lamps are 2800–3000 K.
- *Fluorescent lamps* These have an irregular spectrum, which is the sum of the spectra of the various phosphors used in the tube, plus the line spectrum of low-pressure mercury vapour broadened into bands (Figure 2.8c). Lamps intended for colour matching are balanced close to daylight. Ordinary fluorescent lamps do not render colours correctly in a colour transparency, and show a greenish cast; film manufacturers recommend combinations of weak magenta and yellow CC (colour correction) filters. Fluorescent lamps have a 100 Hz flicker (120 Hz in the USA), which may cause strobing effects with TV or digital cameras.
- *Discharge lamps* Pulsed xenon tubes are popular for cine and video work. Their overall hue is fairly close to that of daylight; the better examples incorporate various metal halides, which produce a near-exact match. Any flicker effect is avoided by converting the applied voltage to a very high frequency a.c.
- *Flashbulbs* Now seldom seen, these were at one time the mainstay of press photographers. They operate by the electrical ignition of a bunch of very fine zirconium or magnesium ribbon in an atmosphere of oxygen at reduced

The kelvin (K) is the SI unit of thermodynamic temperature. It has the same magnitude as the degree Celsius (°C) but starts at absolute zero (−273.15 °C), the point at which all molecular activity has ceased, so that 0 °C ≅ 273 K.

William Thomson Kelvin, 1st Baron Kelvin, (1824–1907) was Professor of Natural Philosophy at Glasgow University for 53 years. His researches made thermodynamics a respectable branch of science, and he was knighted for his part in the design and laying of the first Atlantic cable.

The Science of Imaging

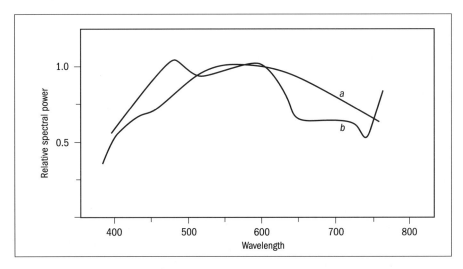

Figure 2.11 Spectral power distributions of (a) sunlight; (b) electronic flash.

pressure. Their inherent colour temperature is 3800 K, but the protective coating usually contains a blue dye, which raises the colour temperature to 5500 K. The duration of a flash may be between 5 and 50 milliseconds (ms) according to the type of bulb. The shape of the intensity curve is much the same as that of Figure 2.7, but with a much-expanded time scale.

- *Electronic flash*   Xenon gas in a tube at low pressure is ionised and made conducting by a trigger pulse of high voltage. This allows a bank of capacitors to discharge very rapidly through the tube, generating an intense flash of white light, typically of 1 ms duration, but which can be quenched to give durations as short as 20 microseconds (µs). Older flash heads produce a spectrum that is slightly deficient in green, and may have a pale yellow-green tint to the protective screen to compensate. Modern flash tubes contain small amounts of other gases added to produce a spectrum that is a close match to daylight. (Figure 2.11b.)

Photographic light filters

When photographers talk of filters they almost always mean light filters. A light filter is an optical device that modifies the colour, quality or intensity of the light it transmits. There are nine categories of light filter, each with a different purpose.

- *Emulsion balancing filters*   These are now chiefly of historical interest, though older photographers will remember the yellow filter they used in order to make the colour rendering of orthochromatic (red-blind) emulsions more believable, especially in the rendering of skies and flesh tones. More recently, a pale yellow-green filter has been used to balance the tone rendering of high-speed panchromatic films with enhanced red and blue sensitivity. Dufaycolor films (see Chapter 6) included a gelatin emulsion-balancing filter with each film. An oddity among these filters was the panchromatic vision (PV) filter, which was a deep purple and was intended to be looked through to give a visual impression of the way the monochrome image would appear. Nowadays this category of filters applies chiefly to the colour correction (CC) filters necessary when making very long exposures on colour transparency film, and to so-called haze-cutting filters, which are UV-absorbing filters with a very slight pinkish-yellow tinge.

- *Contrast filters*   These belong specifically to black-and-white photography. They function by lightening parts of the image that are the same colour as (or similar to) that of the filter, and darkening parts that are of colours different from its own, particularly the complementary (opposite) colour. (See Figure 2.12.)

Thus a red filter lightens red, and to a lesser extent yellows and greens, but darkens cyan, and to a lesser extent blues and greens. It is the effect on tone quality that is important. If you were photographing, say, a red mahogany surface and you wanted to bring out the detail of the grain, you could use a green filter to increase the contrast of the reds. This works because a pale red reflects a good deal of green light, whereas a more intense (saturated) red reflects scarcely any. The tone scale of reds is thus expanded. A yellow or orange filter darkens the rendering of blue sky, in the same manner. On the other hand, if you were photographing a green landscape in bright contrasty sunlight, a green or yellow filter would lower the contrast to give a more delicate range of tones.

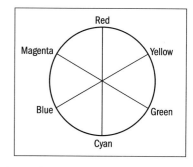

Figure 2.12 The colour wheel: a colour filter will lighten the tone of its own and adjacent colours and darken those diametrically opposite to it.

However, to say that a filter lightens its own colour is an over-simplification. Of course, a filter can only absorb light, so it cannot actually make a colour lighter. In practice you achieve the effect by increasing the exposure. This introduces the concept of a *filter factor*. This is the factor by which you have to multiply the exposure in order to obtain the same rendering of a neutral grey with the filter on the lens as you would without the filter. Typically, a medium yellow filter has a factor of ×2, and a deep red a factor of ×5. Figure 2.13 shows the spectral transmittances of three grades of yellow filter.

- *Colour separation filters*   The production of some types of colour print, and most colour photomechanical processes, require the making of *colour separations*. Briefly, you need to make three negatives (or the equivalent in other imaging techniques), recording respectively the red, green and blue content of the subject matter. The transmittances of the three filters (often termed a *tricolour set*) are shown in Figure 2.14a. They are such that the combined

The principles of colour separation are described more fully in Chapter 6.

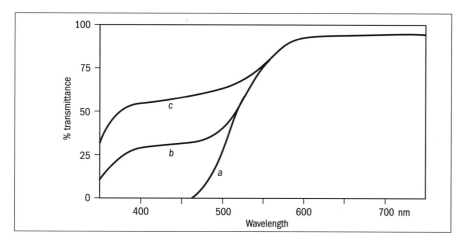

Figure 2.13 Spectral transmittance of three grades of yellow filter: (a) deep yellow (minus blue); (b) medium yellow; (c) light yellow.

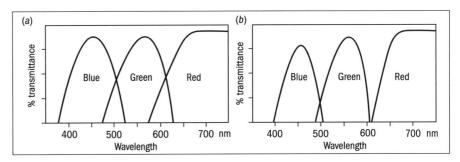

Figure 2.14 Tricolour separation filters: (a) broad band; (b) narrow band.

transmittances cover the whole visible spectrum, with the overlaps arranged such that the aggregate transmittance is uniform.

The separation filters used in the printing process (in processes where these are used) have a different set of transmittances. They are described as *narrow-band filters*, and there is little or no overlap (Figure 2.14b). You have to use this type of filter when you want to make colour separations from transparencies or printed matter, otherwise the result will be desaturated.

- *Lighting balancing filters*   These apply to colour films, and are of two general types. Both types of filter are calibrated in mireks. The first group compensates for the spectral mismatch between daylight-balanced films and tungsten lighting and the converse. The second group has a series of small colour balance shifts (5–15 mireks), and pictorial photographers use them for making subtle alterations in colour temperature to enhance the lighting quality of a scene in a colour transparency.
- *Colour print (CP) filters*   Because of small variations in colour balance in both negative and print materials, you may have to use these in the enlarger light path to obtain a satisfactory colour balance in the print. They are available in graded densities in cyan, magenta and yellow, and in the better enlargers are built into the head and adjusted by rotating wheels. Their application is dealt with more fully in Chapter 6.
- *Neutral density (ND) filters*   These reduce the intensity of light reaching the emulsion (or other sensitive medium) without affecting its spectral distribution. In practical photography you need ND filters only when you need to cut down the exposure and there is no possibility of adjusting either the shutter or the lens aperture. ND filters are used mainly in laboratory research and in sensitometric testing (Chapter 10).
- *Infrared (IR) and ultraviolet (UV) filters*   IR films are also sensitive to red and blue light, so an IR film has to be shielded from blue light and at least some of the red. IR filters typically pass a range of wavelengths from 680 to 900 nm. UV filter material transmits radiation between 300 and 400 nm, but also allows some red light through, so UV filters in practice also contain blue dye. To photograph fluorescence you need to have the UV filter on the illuminating lamp. You also need a UV-absorbing filter on the camera lens, so that only the visible fluorescence is recorded.
- *Special effects filters (SFX)*   These include graduated density filters, multiple-image filters, soft-focus and starburst filters and filters giving various diffraction effects. These do not lie within the remit of this book, but see 'Digging deeper' below.

Polarising filters

A polarising filter transmits only linearly polarised light. Ordinary unpolarised light consists of waves vibrating in all planes, and of those vibrations the polarising filter transmits only the vector components that are parallel to its polarising axis. This means theoretically that only 50 per cent of the light incident on the filter actually gets through it. So the filter factor ought to be ×2. In fact it is nearer ×$3\frac{1}{2}$, as the filter material is not completely transparent. Two polarising filters with their axes orthogonal will block off virtually all light (Figure 2.15).

The earliest polarising filters were called *Nicol prisms*. William Nicol (1768–1851) was a Scottish geologist who used polarised light in the microscopic examination of rock specimens. A Nicol prism is a single crystal of the doubly refracting material calcite, cut so that one of the rays is suppressed (the rays are orthogonally polarised). Modern polarising filters are sheets of plastics material containing a layer of transparent polymer with its molecular axes all aligned.

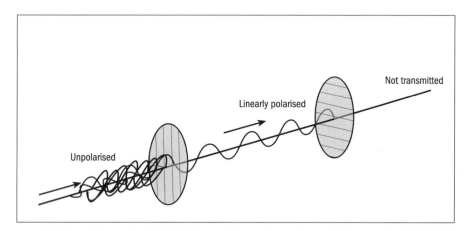

Figure 2.15 Action of a polarising filter.

Applications of polarising filters

There are two main areas of application for polarising filters. The first is in the suppression of specular reflections, as in the photography of objects behind glass or under water. Light reflected from a shiny non-metallic surface at an angle is partially polarised; at angles of reflection around 56–57° (the *Brewster* angle) it is totally polarised, in a plane that is orthogonal to the plane containing the incident and reflected rays and the normal.

This is called *s*-polarisation (the *s* is the initial of the German word *senkrecht*, which means perpendicular). Polarisation parallel to this plane (i.e. at right angles to the surface) is called *p*-polarisation (the German word *parallel* is the same as in English). Setting a polarising filter on the camera lens with its axis perpendicular to the surface will eliminate surface reflections. The Brewster angle for water is less, about 53°.

The Brewster angle is defined as the angle of incidence for which the reflected and refracted rays are orthogonal (Figure 2.16). At this angle the refracted ray is partially *p*-polarised and the reflected ray is totally *s*-polarised. The Brewster angle is the angle whose tangent is equal to the refractive index of the material.

Sir David Brewster (1781–1868) was awarded the Rumford Medal by the Royal Society for his discoveries in polarisation. He is better known for his invention of the kaleidoscope. He also had a hand in the development of the stereoscope. He was a friend and mentor to Henry Talbot, and introduced David Octavius Hill to Robert Adamson.

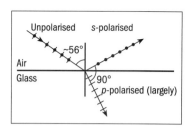

Figure 2.16 Reflection at the Brewster angle.

An extension of the polarisation principle is to illuminate the subject with light that is already polarised, and to use a polarising filter on the camera lens with its axis crossed with respect to that of the illuminant. Specular (mirror-like) highlight reflections are suppressed, but diffuse reflected light, which is depolarised, passes through the second filter.

The Science of Imaging

The maximum degree of polarisation occurs along an arc at right angles to the axis joining the sun to the camera.

The second use is in landscape photography. Skylight away from the sun is polarised to varying extents. If you fit a polarising filter with its axis directed at the sun you will darken the sky without changing its hue. Using the filter with its axis vertical will also improve the colour saturation of fields and horizontal spaces by minimising the effect of reflected skylight.

Circular polarisation

As I explained in Chapter 1, circular polarisation is the condition where the polarisation vector rotates one complete revolution for each cycle of the light wave. A circularly polarising filter (also called a *quarter-wave plate*) is a sheet of doubly refracting material such as mica, of a thickness such that one of the two rays is delayed by one quarter of a wavelength. This is the condition for circular polarisation.

Why do we need circular polarisation? There is a problem in most camera metering systems when the light entering the camera is linearly polarised. A polarising beamsplitter directs part of the incoming light to the exposure and autofocus sensors, and if you use a polarising filter, rotating it will vary the division of the light, resulting in incorrect settings and possible loss of focus. This problem is overcome by mounting a quarter-wave plate on the camera-facing side of the polarising filter. It has no effect on the behaviour of the filter in eliminating reflections, etc., but it allows the sensors to operate correctly for any orientation of the filter axis.

Digging deeper

Except that it still gives the old obsolete SI definition of the candela.

Many textbooks contain chapters on photometry that are less than reliable. Some otherwise excellent texts, especially from the USA, use obsolete and confusing units such as the erg (CGS) and the foot-lambert (FPS), sometimes together. One thoroughly reliable source is the *Manual of Photography* (9th edition, ed. Ralph Jacobson, Focal Press, 2000). The most comprehensive coverage of the subject is *Lamps and Lighting* (4th edition, ed. J R Coaton and A M Marsden, Arnold, 1996), which is updated at regular intervals. It consists of sections on light, vision, colour and measurement; lamps and control equipment; and lighting fittings and techniques. Eastman Kodak publishes *Filters Handbook*, with a list of all the filters available from Kodak, their characteristics and their Wratten numbers. The Kodak Workshop series includes *Using Filters* (1995). If your interests are largely pictorial, Joseph Meehan's *The Photographer's Guide to Using Filters* (Watson-Guptill, 1998) is excellent; and although it is now long out of print, Clyde Reynolds's *Focal Guide to Filters* (Focal Press, 1974) packs a lot of useful information into a small space. Your local library should be able to find a copy.

Chapter 3 Visual perception

The eye and evolution

The eye is such a complicated organ that creationists have used it for many years as evidence against the theory of evolution, saying 'What use to an animal would an only half-developed eye be?' In his book *The Blind Watchmaker* Richard Dawkins demolishes such arguments, showing that at every stage of the building of the mammalian eye, from a simple light-sensitive patch to its present complexity, the organ became steadily more useful to its owner, and, moreover, that every stage in that development is still extant, in creatures ranging from the earthworm to the eagle.

Optics of the eye

Older textbooks on light and optics always contained a section on the eye, and it was often the weakest section in the book. The authors usually treated the eye as a kind of camera, and got the optics wrong too, showing all the refraction taking place at the lens. Discussion of the visual process has now largely moved to textbooks on the psychology of perception, where the eye is treated more in terms of a computer input. Both these views contain a measure of truth, and both may be helpful in an appreciation of what really goes on in the process of visual perception. Certainly, the eye is an optical device that focuses an image on a light-sensitive surface, so that is probably the best place to begin.

You can demonstrate the optical workings of the eye quite easily, though unless you have some rather old-fashioned equipment such as a large goldfish bowl and the condenser lens from an old photographic enlarger (or a large and powerful magnifying glass) it will have to be a thought experiment. If you fill the goldfish bowl with water you will have made a lens – a very thick one, but a lens none the less. If you shine a beam of collimated (parallel) light such as sunlight or the beam from a focusing torch through it, it will bring the light to a focus about one-third of its diameter behind its rear surface. (Many a table runner was set on fire this way in the 1930s.) Water has a refractive index of 1.33, and glass around 1.5, so if you suspend your lens in the water near the front of the bowl, it will shorten the focus so that it is more or less at the rear surface of the bowl. If you use a large glass ball (like a fortune-teller's crystal) instead of the lens, it acts as a more powerful lens, and by moving it backwards or forwards in the water you can bring nearer or farther objects into focus. In fact, this is the way a fish's eye works. You can see the focusing effect if you hold a thin sheet of paper against the rear surface of the bowl. But to simulate the human eye more exactly you need a flexible lens, say one made of gelatin. You change the focus not by moving it, but by squeezing it to make it fatter. But the image still won't be very sharp. You can sharpen up the focus by placing a sheet of opaque material with a medium-sized hole in the middle between the lens and the front of the bowl. You have now created a model of the optical system of the human eye. Figure 3.1 shows a simplified anatomical model of the eye, seen from above.

The optics of the eye are simple. As with the goldfish bowl, the front surface, the *cornea*, does most of the refracting. The medium in front of the lens is called the

The Science of Imaging

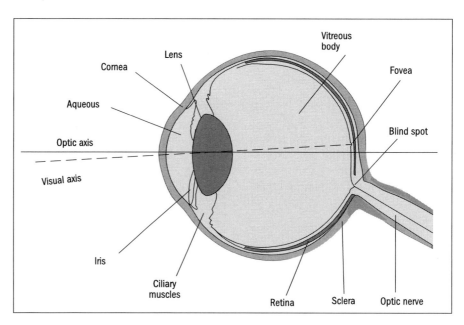

Figure 3.1 Basic structure of the human eye.

Graded refractive index lenses have been around for only about twenty years. Their use solves many of the problems associated with lens aberrations (see Chapter 4). It seems that in this respect, as in many others, Nature got there first.

aqueous humour (or, more often, just the *aqueous*, 'humour' being an antiquated word meaning simply 'fluid'). Its refractive index is 1.34, the same as seawater – which says something about our fishy ancestry. Behind the lens is the jelly-like *vitreous body*, also with a refractive index of 1.34. The lens is built in layers like a shallot, and has a refractive index that varies from 1.38 on the outside to 1.41 at its centre.

The lens is kept in tension by the ciliary muscles, which relax for close focusing, allowing its curvature to increase. As we get older the lens becomes less flexible, so that it becomes difficult to focus on close objects. Sometimes the lens slowly turns yellowish and cloudy – a pickled shallot. Its refractive index increases and the eye becomes short sighted. This condition is called cataract, and it can be cured by a simple surgical procedure, which breaks up and removes the lens and replaces it with a soft silicone implant.

Some animals do much better than this. Domestic cats have a slit pupil that can vary its aperture between $f/1.5$ and around $f/100$, a useful capability in an animal that hunts by both day and night.

The *iris* is an opaque muscular diaphragm with a circular aperture, the *pupil*. The iris controls the level of illumination that reaches the light-sensitive tissue that lines the back of the eye, called the *retina*; when you move from bright light into dimmer surroundings it opens wide until your retina has adjusted to the new conditions. Its range of apertures, in photographic terms, is from about $f/3$ to $f/16$ – a ratio of about 28 to 1. (f-numbers are explained in Chapter 4.)

Short and long sight

Other eye defects not associated with ageing are short and long sight. In short sight the eyeball is too long or the curvature of the cornea too high, resulting in the image being focused in front of the retina; in long sight the eyeball is too short or the curvature of the cornea too low, so that the image is focused behind the retina (Figure 3.2). Astigmatism is a combination of these defects: the cornea has too high a curvature in one direction and too low in the other (i.e., instead of being

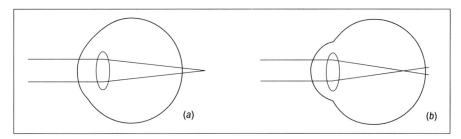

Figure 3.2 (a) Hypermetropia (long sight); (b) myopia (short sight).

Note that this description of ocular astigmatism has nothing whatever to do with the lens aberration also called astigmatism. It is somewhat unfortunate that two totally different optical phenomena happen to have been given the same name – one by the medical profession and the other by physicists. Chapter 4 discusses lens aberrations.

part of a sphere, like a soccer ball, it is like a rugby ball). The eye is short sighted for verticals and long sighted for horizontals (or some other pair of orthogonal axes).

The retina

The embryonic eye does not grow as an entity. The cornea and the frontal part of the eye, including the eyeball itself, are outgrowths of the skin, but the retina is an outgrowth of the brain. The way the retina grows has one unfortunate consequence: it is effectively inside out. The layer of light-sensitive cells is overlaid by relay cells, nerves and blood vessels. There is also a sizeable blind spot (some 3° across) about 15° out from the centre of your vision; this is the spot where the individual nerve fibres disappear into the optic nerve. You cannot usually see the nerves and blood vessels, as they are stationary with respect to the retina, but when an optician examines your eyes with a bright light it will sometimes surprise your retina into perceiving the shadows of this tree-like network. Figure 3.3 is a schematic of the structure of the retina.

The optical image produced at the back of the eye is not a good one by photographic standards, though it is rather better than you might imagine from the description above.

Only a tiny area at the centre of vision, the *fovea*, with a field of about 1°, picks up a really sharp image. The visual receptor cells become fatigued very quickly, and turn off altogether after a few seconds unless there is a change in the stimulus. These two factors are alleviated by constant small involuntary oscillations of the eye, which allow the stimulation to continue, and spread the effective angle of sharp vision to several degrees.

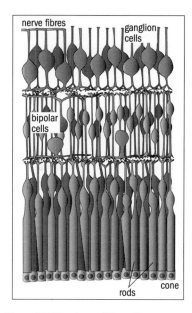

Figure 3.3 Structure of the retina.

Rods and cones

The primary light receptors of the retina are of two types, *rods*, which function only in dim light, and *cones*, which respond only to bright light. There is a twilight area (literally) where both types are operating, though neither operates very efficiently, as anyone who has to drive a car around dusk can confirm. Cones operate efficiently at light levels above $10\,\text{cd}\,\text{m}^{-2}$, and cease functioning altogether at about $0.1\,\text{cd}\,\text{m}^{-2}$. Rods, on the other hand, continue to function down almost to single photon level. They contain a purple dye called *rhodopsin* that is destroyed by light, and this provides the operating mechanism. It takes up to half an hour to

The Science of Imaging

build up a full complement of rhodopsin, and this is why it takes so long to become fully dark-adapted.

Cones operate in a similar manner to rods, but the dye is different. It is called *iodopsin*, and comes in three different forms. These have sensitivity peaks in the red-orange, green and blue regions of the spectrum, so the cone system can detect colour, much in the manner of the three-colour additive process of television. The blue-sensitive cones are much less numerous than the others, but make up for this by being more sensitive. Cones are much more numerous than rods in the central region of the retina. About 6000 cones are packed very tightly in the fovea, where there are no rods at all. This is why, on a starry night, you can see dim stars better if you look slightly away from them.

Additive colour systems are discussed in more detail in Chapters 6 and 15.

Cone vision, called *photopic vision*, has a maximum sensitivity at about 555 nm. This represents a yellow-green hue, and it closely matches the peak spectral output of the sun. Rod vision, called *scotopic vision*, has a sensitivity peak around 507 nm, well into the bluish-green region, and is totally insensitive to red light. As there is only one type of rod, there is no colour discrimination with scotopic vision. By moonlight everything looks grey. The shift in maximum spectral sensitivity from yellow-green to blue-green is known as the *Purkinje shift*.

You can test this by watching an electric fire in the dark, when you switch it off. The element goes a deeper red as it cools, but just before it goes out it turns grey. As rods are insensitive to red, you can dark-adapt your eyes in full sunlight by wearing red goggles, a fact I discovered many years ago when working as a photographer in the Far East.

Because of the huge number of cones in the foveal area, humans have excellent visual acuity in bright light. In dim light, however, the resolution of the human retina (though not its sensitivity) is poor. Rod cells are connected together in parallel, sometimes in quite large numbers, which increases their efficiency at detecting very low light levels. The foveal area is turned off. The iris is fully open, increasing the effect of optical aberrations. So although everything *looks* bright and sharp on a moonlit night, you can't read a newspaper, apart from the largest headlines. About 45° out from the fovea the cones peter out altogether: there is no colour vision in your peripheral retina. Near the edges of the retina the rods are much larger, and can only detect movement.

Jan Evangelista Purkinje (properly spelt Purkyně) (1787–1869) was a Czech histologist of considerable distinction. Apart from his work on the structure and functions of the eye, he investigated neural structures in the brain and heart and in embryos (several structures in each of these bear his name); he discovered the sweat glands in skin, founded the techniques of fingerprint identification and was the first to use a microtome in preparing specimens for microscopy. He also translated the works of Goethe and Schiller.

Sensitivity range

The sensitivity range of the human eye is enormous. From full sunlight to starlight represents an illuminance ratio of more than 10 million to 1. It's just as well, then, that we don't perceive equal increments of luminance as equal increments of brightness. Instead we perceive them on something close to a logarithmic scale, that is, equal *multiples* of luminance (1, 2, 4, 8, 16, etc.) are perceived as equal increments of brightness (1, 2, 3, 4, 5, etc.). This is an important characteristic of visual perception, and it appears more than once in later chapters. This principle also applies to other senses, and is equally important in hearing. It is called the *Weber–Fechner Law*.

Visual pathways

Most of the operations of the right-hand side of the body are processed and controlled by the left hemisphere of the brain, and vice versa. This also applies to visual processes, though humans are wired up differently from many other animals. The terminus of the neural pathway from the retina is at the rear of the outermost part of the brain, and is known as the *visual cortex*. In most animals all the nerves

The Weber–Fechner Law

This consists in fact of two laws. The earlier one, due to the German physiologist Ernst Heinrich Weber (1793–1878), states that the 'just noticeable differences' in stimuli are proportional to the magnitudes of the original stimuli. For example, if you can just distinguish between two weights of 10 grams and 11 grams, then you will just be able to distinguish between weights of 100 grams and 110 grams, *not* 101 grams, as you might have supposed. The German psychophysicist Gustav Theodor Fechner (1801–1887) carried this principle further. His law states that the intensity of subjective sensation increases as the logarithm of the stimulus intensity. Thus it is Fechner's Law that we are concerned with here.

from the right retina go to the left visual cortex and those from the left retina go to the right visual cortex. However, in humans and some other mammals (particularly primates) it is the left *halves* of the retinal fields that both go to the left visual cortex and both the right halves to the right visual cortex. The foveal areas cross over entirely. The important thing about this is not that the field is split down the middle (it is not: there is some overlap), but that both retinas send their messages to the cortex as it were superimposed. (We shall see the importance of this later.) This sorting out of the visual pathways takes place at a ganglion called the *optic chiasma*. From the optic chiasma the nerve bundles travel to two locations right and left within the midbrain called the *lateral geniculate nucleus* (LGN). From there they fan out to the visual cortex (Figure 3.4).

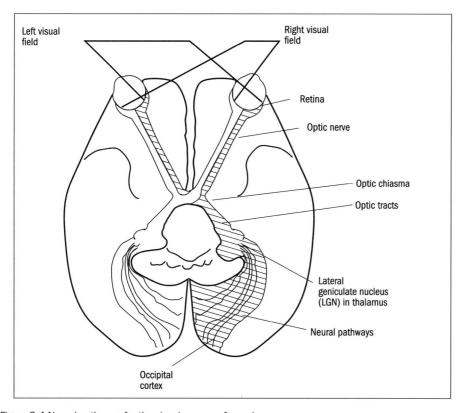

Figure 3.4 Neural pathways for the visual process, from above.

The Science of Imaging

David Hubel (b. 1926) was until his retirement professor of neurobiology at Harvard University, where he met Torsten Wiesel (b. 1924). For many years they worked together on the correlation of the anatomical structure of the visual cortex with the physiological responses to visual stimuli. In 1981 they were awarded the Nobel Prize for Physiology for their work.

Neural processing of the visual signal

The details of neural processing in the visual pathways are complicated. Most of the pioneering work on this was done by Hubel and Wiesel. What follows is a much simplified version.

There are more than 100 million light receptors in each retina, but the two optic nerves, with diameters less than that of a pencil, contain only about one million fibres each, so plainly a good deal of processing has to have gone on in the retina itself. When a receptor cell in the retina fires, its signal is passed first to a so-called bipolar cell, then to a further cell called a ganglion cell, which may be connected to a large number of adjacent bipolar cells. It passes a (modified) signal down the optic nerve to the LGN. The connections between retinal cells are intricate, and appear to collect and modify the stimulus in such a way as to compensate for optical deficiencies in the eye. For example, where there is a stimulus such as a point of light, the signals from the cells surrounding the area are inhibited. This makes the image clearer and sharper, in much the same way as adjacency effects operate in a photographic developer (Chapter 12).

The optic nerve can carry some 10 million bits of information per second. It passes this information directly to the LGN, which acts as a kind of relay station, sorting out types of signal: colour information, shapes, movements, contrast etc., and routeing them to the appropriate areas of the visual cortex. Here, different areas register lines and edges at various angles, movement in various directions, differences between the two retinal images, colour information and so on, and assemble everything into a coherent visual appreciation by matching the perceived stimuli to known objects and performing an interpretation.

Visual fields and binocular vision

Vertebrates have evolved with broadly bilateral symmetry, so that in general our organs are either symmetrical or duplicated. In the process of evolution our auditory and visual processes have taken advantage of this duplication, in the former case awarding us binaural hearing and the ability to locate sounds accurately. Binocular perception seems to have gone in two different directions, depending on whether the species in question belongs to the hunted (pigeons, antelopes) or the hunters (falcons, leopards). The hunted have their eyes at the sides and take in an all-round view, often nearly a full sphere (useful for spotting something sneaking up behind you or about to dive on you), each eye contributing half the visual field with only a small overlap at the front (helpful in avoiding hitting a tree). The hunters have both eyes at the front, giving them excellent binocular judgement of distance over a wide field of view. Humans, too, stem from a line of predators and tree-dwellers, and also have wide binocular vision. Each eye sees a slightly different view, and part of our visual cortex is dedicated to recording this difference and deducing the three-dimensional nature of what we are observing. The field of human binocular vision is about 125° vertically and about 120° horizontally (this depends to some extent on the size of your nose!). Outside this area we have some 40° each side of poorish monocular vision dedicated mainly to the detection of movement (our ancestors, after all, were sometimes hunted themselves). This is illustrated in Figure 3.5.

Visual perception

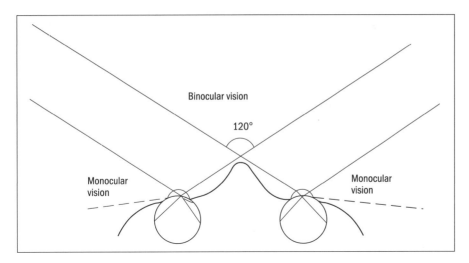

Figure 3.5 Horizontal limits of binocular vision in humans.

Stereoscopic perception or stereopsis, as it is called, is a consequence of the differences between the left-eye and right-eye images because of their different viewpoints; the bifurcated signal at the optic chiasma may be a sign of the evolutionary importance of the rapid and accurate distance judgement afforded by efficient stereoscopic perception. Be that as it may, most of us scarcely notice its presence, and it is only the surprise on viewing a stereoscopic pair of photographs for the first time, and suddenly seeing what looks like real depth in a photograph, that makes us aware of it. Stereoscopic perception is not a very powerful process in humans. Indeed, something like one person in four has poor stereoscopic vision and finds difficulty in 'fusing' traditional stereoscopic pairs. Roughly one person in twenty has no stereoscopic perception at all (usually the result of an uncorrected squint in early childhood). Other, largely unconscious, clues to depth are collectively more useful: apparent size of objects, obscuration of one object by another, parallax (change of appearance of a scene when you move your head), bluishness and lowered contrast in distant objects (called 'aerial perspective' by photographers), and the need for eye convergence and accommodation (change of focus) for near objects. Chapter 16 is concerned with three-dimensional images, and treats the subject in greater depth.

Colour perception

Few animals seem to possess as fully developed a perception of colour as we do, though some birds, fish, insects, and even advanced molluscs such as octopus and cuttlefish, may run us close. The cones of the human eye, as we saw earlier, are of three types, with sensitivity peaks in the red-orange, green and blue regions of the visible spectrum. The sensitivities overlap, especially those of the red-sensitive and green-sensitive cones. Figure 3.6 shows the (normalised) spectral sensitivity ranges of the three types of cone.

Roughly one person in ten has some deficiency in one type of cone (usually red or green). Such people, who are almost always male, find difficulty in distinguishing certain colours – usually red and green, which they tend to register as various shades of brown. Anomalous colour vision can be diagnosed by Ishihara test

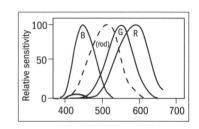

Figure 3.6 Spectral sensitivities of the three types of cone (normalised). Rod sensitivity is shown as a broken line.

cards. These are a set of patterns of coloured spots, which contain a number printed in a different hue. People with colour anomalies see either a different number or no number at all (see Figure 3.7, colour plate).

Seeing a range of colours

With only three types of cone, how is it that we can manage to see such a huge range of colours? It really is huge: Judd and Wyszecki in the 1970s carried out reliable research which indicated that people with normal colour vision could distinguish ten million colours. The answer lies in the way a coloured surface stimulates the three types of cone differentially. Physiologists refer to the three types by the Greek letters ρ (rho) for red, γ (gamma) for green and β (beta) for blue. If a certain combination of wavelengths stimulates the ρ and γ cones equally, we perceive the hue halfway along the spectrum between red and green, namely yellow. If the ρ cones are stimulated more than the γ cones, we see orange, and if the γ cones are stimulated more than the ρ cones we see yellow-green. Similarly, with equal stimulation of the γ and β cones we see blue-green (cyan), and with varying stimulations of the two types of cone we see the spectrum running from green through bluish green, cyan and greenish blue to full blue. When the β and ρ cones are stimulated together in various proportions we see the range of purples, magentas and crimsons, hues that are not in the spectrum at all, but are real colours none the less. This may seem odd, until you realise that light waves do not themselves possess colour: it is their effect on the retinal cells that produces the sensation of colour.

It must be clear now that there is more than one way of producing a particular colour sensation such as yellow. As long as you stimulate the γ and ρ cones equally you will get yellow, so you can use monochromatic sodium light with a wavelength of 589 nm, or two bands of wavelengths centred around 549 and 649 nm: the result will be the same colour sensation.

But what about brown, slate grey, white or pink? These colours certainly exist. Up till now I have been discussing only saturated (pure) colours, such as those in the spectrum. A desaturated colour is like a pure colour with grey added. Now, you get grey if you stimulate all three types of cone equally, to a low level. Add a little extra stimulus to the ρ and γ cones and you have brown, which is nothing more than a desaturated yellow or orange. Add a little stimulation to the β cones instead, and you see a bluish or slate grey. Increase the stimulus all round and grey becomes white; and take away a little from the β and γ stimulation and you are left with pink, which is simply a light desaturated red.

This description has been a simplified one. For one thing, you cannot produce a fully saturated cyan with green and blue light, because the ρ sensitivity curve overlaps the γ-sensitivity curve over the whole 500–600 nm wavelength range. But it is broadly correct, and is the basis of colour photography as described in Chapter 6.

On being introduced to the three-colour additive principle (Chapter 6), many people find the summation of red and green to produce yellow hard to accept. Yellow seems to be a primary hue, not related to either red or green. Yet when you project a red light and a green light on to a white screen, superimposed, the result is indisputably yellow – you have already seen why this should be so. Part of the reason may be cultural: yellow had a name long before cyan or magenta. Again, one can visualise a reddish blue and a greenish blue, but a reddish green? Indeed, red can readily be seen as a kind of opposite to green, just as blue seems opposite to yellow. This intuitive concept led Ewald Hering to propose an 'opponent-colour' theory of colour perception in opposition to Thomas Young's three-colour principle. It is interesting that although there is no doubt that the retina analyses colour according to the three-colour principle, it has recently become clear that the LGN does operate in a kind of opponent-colour mode, with some neurons firing or inhibiting firing for red versus green and for yellow versus blue.

Constancy

Our eyes are continually moving, and are mounted on an unstable platform. Why, then, do we see objects and landscapes as stationary? If you were to take a video camera and move it around in the same manner, the resulting effect would be vertiginous for the viewer. Again, if you stand at the foot of a tall building and turn your gaze upwards to the top, the building does not appear to move. But if you do a similar upward pan with a video camera, the verticals converge and the building seems to be toppling over backwards. Yet this doesn't seem to happen in the visual world. It seems that when we are looking at a scene, all the movements of our head and eyes are being recorded too; and the brain, which receives all this information in addition to the visual input, integrates the information and discounts the effect of the motion. The result is termed *constancy of position*. You can disrupt this if you move your eye without the cooperation of your eye muscles. With one eye closed, push gently with your finger on the side of the other eye, through your lower eyelid. You will see the whole field of vision moving. Position constancy is essential if we are to be able to make sense of the visual world – and to be able to look upwards without falling over.

It seems that some birds, particularly chickens and pigeons, have a different way of dealing with the problem. They keep their heads stationary with respect to the ground while taking a step forward, jerking their head forward at the beginning of each step to catch up.

Colour constancy is nearly as important. As a sunny day draws to a close the colour temperature of the sunlight may fall from more than 5000 K to less than 3000 K. But the trees and houses don't seem to change colour. Nor do the objects in your room appear to change colour when you switch on the light (although the fading daylight in the window seems to have suddenly gone blue). As a rule you don't notice much (if any) change in the colour of your surroundings when the colour of the illumination changes. Yet if you take a photograph on colour transparency material balanced for daylight in a room lit by filament lamps the result will have a deep orange cast. Photographers who make their own colour prints need to check the colour balance of test exposures under standard white illumination, and quickly learn not to examine a print for more than a few seconds without looking away. If you gaze at a print that is poorly colour-balanced for half a minute or so it will look closer and closer to being correctly balanced.

In the early days of High Street cheap'n'cheerful fast colour prints, it was a standard ploy to get customers who were unhappy with their results to examine the prints carefully, while they were assured meantime that the colour balance was fine. The traders usually got away with it, too.

Constancy of shape and size are also important. When an object moves away from you, you do not perceive it as shrinking. In fact, if it *does* shrink (as with a slowly deflating balloon), it is hard to resist the impression that it is in retreat. Again, a round plate that is tilted does not look any less round, even though the image on your retina has become elliptical.

Before the laws of perspective were properly established, painters invariably painted distant figures too large and tilted plates too round. A naïve person, asked to draw a plate presented almost edge-on, will invariably draw it insufficiently foreshortened.

Visual illusions

All of us have at some time been fooled by a visual illusion. Some of these are very powerful. Visual illusions can often reveal much about our visual and perceptual

The Science of Imaging

processes. Some illusions plainly originate in the retina, some almost certainly in the LGN and some in the visual cortex. Others appear to be purely cultural: we have learnt to expect a particular visual experience, and this colours our perception.

The most common retinal illusions are associated with neural fatigue. The best known of these is the colour-reversed United States flag, with black stars on a yellow ground, and black and cyan stripes (Figure 3.8, colour plate). When you stare at this in a good light for about half a minute, then transfer your gaze to a plain white surface, you see a phantom flag in its correct colours – an after-image. The yellow has fatigued your ρ and γ cones, and the cyan bars have fatigued your γ and β cones, while the black stripes and stars have left the cones in these areas of the image fresh. You can do the same trick with a photographic negative, even with a colour negative if it is contrasty (this works best with a colour transparency film 'cross-processed' to give a colour negative).

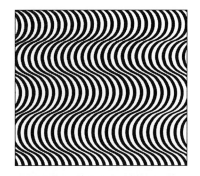

Figure 3.9 A typical 'op art' pattern. As you look at this it begins to flicker, because of the small involuntary movements of your eyes.

Another striking retinal illusion happens when you study a pattern of closely spaced circles, spokes or wavy black lines (Figure 3.9). After a few seconds you begin to see dancing patterns. The involuntary movements of your eyes cause the after-images to jump about and construct a set of dancing moiré patterns. The artist Bridget Riley exploited this illusion in her Op Art series of paintings in the 1960s.

A set of illusions probably associated with the LGN is a family of geometrical patterns which appear to be distorted, though they are not (Figure 3.10). Illusions of this general type can sometimes appear in real situations, such as the bathroom tiles of Figure 3.11.

Illusions associated with motion are also connected with the LGN. If you place a spiral on a record player turntable and watch it turning for about half a minute, then stop the rotation, the spirals will appear to be moving in the opposite direction (e.g. outwards instead of inwards). You may be familiar with the sensation of moving backwards when you have been driving a car at a steady speed and you have to stop suddenly. This is because your sensory system has become accustomed to seeing the scene in front steadily expanding. It has a similar origin to the spiral illusion.

Figure 3.10 Zöllner and Orbison illusions. It is almost impossible to see the two lines as parallel or the circles as true.

Most of the illusions originating in the visual cortex are concerned with either conservation or perspective. The best known is probably the Necker cube and its derivatives, beloved of the artist Maurits Escher (Figure 3.12). You can invert the

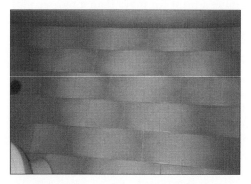

Figure 3.11 Bathroom tiles in a Greek hotel. It is hard to believe that they are perfectly aligned, rectangular and flat.

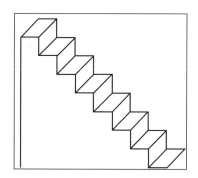

Figure 3.12 Reversing perspective; one of the derivatives of the Necker cube illusion.

Visual perception

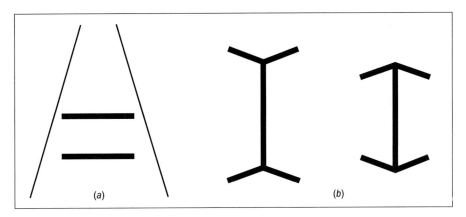

Figure 3.13 Perspective illusions: (a) Ponzo; (b) Müller–Lyer.

perspective of this type of drawing just by thinking. The Ponzo 'railway lines' illusion and the Müller–Lyer 'arrowheads' illusion are examples of misperception of the length of a straight line because of distractions which, though irrelevant, are impossible to ignore (see Figure 3.13). The converging lines of the Ponzo illusion create an overwhelming feeling of perspective, so that the upper line is perceived as farther away than the lower line, and therefore larger (conservation of size). The Müller–Lyer illusion is almost certainly cultural in its origins. One figure looks like the corner of a building, and the other like the corner of a room. This illusion is, it seems, not perceived by people unfamiliar with rectangular buildings, such as Bushmen and some Inuits.

There are large numbers of three-dimensional and moving illusions which, unfortunately, cannot be depicted in a book. (But see Digging deeper, below.) One of the most striking (and disturbing) is a hollow cast of a human face. The inside of a Hallowe'en mask will do. If you light it from below, it is almost impossible to believe that it is not convex, like a normal human face; and if you move around it, it seems to rotate in the same direction as you are moving in, but at twice the speed. This is a totally culture-driven illusion: we are so used to seeing human faces as convex, that the belief that this must also be convex overrides any genuine depth perception.

There are many more visual illusions, some of which are dynamic. Many are not at all easy to explain. But they are all indications of the vast amount of processing that goes on for us to make sense of the images that fall on our retinas. They also remind us that the process of visual perception is subtle, complex and altogether remarkable.

Perception and imaging

What, you may ask, has all this to do with the science of imaging? The answer is that every image, however technically perfect, has to go through the filter of human visual perception, which, as you have seen, is by no means straightforward. Some of it can be quantified and fitted into the matter dealt with in Chapter 5, but sometimes an image may have to be modified in order to fit what we *think* we see. The ancient Greeks knew this when they built the pillars of the Parthenon and other temples with a slight bulge in the middle so that they appeared to be truly parallel sided. Modern architects are no less crafty. In later chapters we shall see, from time to time, how the nature of images is affected by the way we look at them.

Digging deeper

For a book that takes the subject of visual perception about as far as a non-biologist would want to go, I strongly recommend *Eye and Brain* by Richard L Gregory (5th edition, OUP, 1998). It has a really comprehensive bibliography including such research papers as Hubel and Wiesel's original publications. *Perception*, by John Rock (Scientific American Library, 1984) is another excellent introduction to some extent complementing Gregory's book, but omitting colour perception. *The Art and Science of Visual Illusions* by Nicholas Wade (Routledge, 1982) is a comprehensive study, and includes a number of transparent moiré pattern generators to play with, some rather naughty. *Visual Perception*, by Nicholas Wade and Michael Swanston (Routledge, 1991) covers the same ground as this chapter, but in much greater detail. *The Ishihara Tests for Colour Blindness* are published by Arnold. They are available in three degrees of complexity, but as the simplest costs nearly £100 I don't suggest you should get yourself a set. Your local eye hospital would probably be pleased to show you a copy. Hubel and Wiesel wrote up their researches in the *Journal of Neurophysiology* in 1965, but the report is strictly for the specialist. David Hubel wrote a much more approachable article for *Scientific American*, November 1963, and this was reprinted in the Open University's reader for their early course *The Biological Bases of Behaviour* (Harper and Row, 1971). This reader also contains an article by Alfred Sherwood Romer on the structure of the eye, and a comparative study of the octopus eye and mammalian eye by E J W Barrington, showing how two very different evolutionary lines converged to develop a remarkably similar organ (another nail in the coffin of creationism). This book is out of print, but copies are easy to find second-hand or from a public library. One of the best collections of visual illusions is available on the Internet at www.illusionworks.com. In addition to all the best-known illusions it has many interactive and animated examples; and detailed explanations of the illusions.

Chapter 4 Camera lenses

A model for the geometry of camera lenses

Although, as we have seen, there are a number of models describing the behaviour of light, the one used by lens designers is the simplest, namely the ray model. But designing a new lens is still a complex business. Before there were computers, it meant months of laborious ray tracing. Today, designers can optimise a lens design in a few minutes using a computer program. The variables they need to juggle are the refractive indices of the components, the colour dispersions of the glasses, the number of components, their curvature, thickness and separation, and the position of the stop. These represented a daunting set of variables for the old-time designer equipped with only a slide-rule and a set of tables of glass types, but to a modern computer they are no more than a light breakfast. The result still doesn't predict everything about the performance of the final lens; this demands a more sophisticated model. However, the simple ray model is sufficient to describe how a lens forms an image, and why it is necessary to have more than one glass element in a camera lens. But to begin with we will look at the properties of a single simple convex lens.

The simple lens

The most basic ray model for a lens makes a number of assumptions:

- The thickness of the lens can be ignored.
- The lens aperture is small compared with the focal length.
- The refractive index of the material is the same for all wavelengths.

The geometrical optics of a simple or 'thin' lens was established by Isaac Newton, and for this reason is usually known as *Newtonian optics*.

The lens laws

The Newtonian lens laws consist of three basic principles concerning a simple lens. The first concerns the relationship between the distance of the object and its optical image from the lens. The position of the object and the image (which is inverted) are together termed *conjugate foci* (Figure 4.1). They are connected by the *focal length* of the lens, which is the distance of the image from the lens when the object is at infinity. If the distance of the object from the lens is u, the image distance is v and the focal length is f, the relationship is given by the formula

$$1/u + 1/v = 1/f$$

The second principle concerns the scale or *magnification* (M) of the image (usually less than 1). This is given by the formula

$$M = v/u$$

and this can be substituted into the formula for conjugate foci to give

$$v = f(1 + M) \quad \text{and} \quad u = f(1 + 1/M)$$

The Science of Imaging

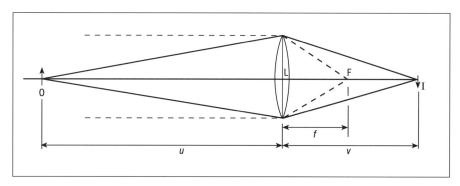

Figure 4.1 Conjugate foci. F is the rear principal focus and *f* the focal length. The object–lens distance OL is *u* and the image–lens distance IL is *v*. As *u* increases, I moves nearer to F.

The third principle is the effect of combining two or more lenses. If two lenses have focal lengths f_1 and f_2 and are mounted close together, their combined focal length *f* is given by

$$1/f = 1/f_1 + 1/f_2 \quad \text{or} \quad f = f_1 f_2 / (f_1 + f_2)$$

If the lenses are separated by a distance *d*, the formula becomes

$$1/f = 1/f_1 + 1/f_2 - d/f_1 f_2 \quad \text{or} \quad f = f_1 f_2 / (f_1 + f_2 - d)$$

For two convex lenses, separation shortens the focal length.

These formulae also apply to negative (concave) lenses, which are considered to have a negative focal length.

> Throughout this book I am using a sign convention known as 'real is positive'. This means that all distances to objects and real images are considered positive, and all distances to virtual images, as well as their magnifications, are negative. The focal length of a negative lens is also negative. Some optics textbooks use a convention that puts the lens at the origin of a Cartesian system. This gives *u* and *v* opposite signs, and in my view complicates matters unnecessarily.

Real and virtual images

If you are using the lens formulae, and a distance and/or a magnification comes out negative, the image is a *virtual image*, that is, the rays do not actually pass through it, but instead appear to have come from it. This happens with the image formed by a single negative (concave) lens, which is virtual, erect (not inverted) and smaller than the object.

A positive lens can form a virtual image too, if *u* is less than *f*. In this case the image is also erect, but is enlarged. This is the image you see with a magnifying glass (Figure 4.2a). A concave (negative) lens gives a virtual image for any value of *u* (Figure 4.2b). An image that the rays do pass through is called a *real image*, and you can catch it on a screen or on a photographic film.

Depth of field

In real life the object (i.e., the subject matter) usually has some depth, so that there is a u_{max} and a u_{min}, the farthest and nearest visible points on the subject matter. These give correspondingly different values for *v*. The question in practice is: how 'unsharp' can we allow the image to be before it becomes unacceptable? This begs the question of what we mean by 'sharp'. The answer is less simple than it appears;

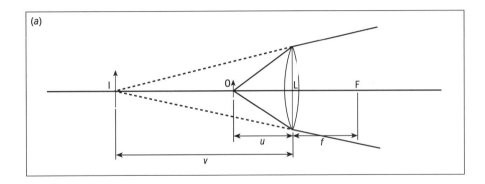
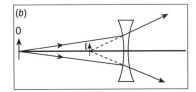

Figure 4.2 (a) Formation of a virtual image I by a positive lens with the object O at a distance less than f from the lens. The distance v is negative, and the rays do not pass through the image, but appear to have originated from it. This is a magnifying-glass configuration: $|v/u| > 1$; (b) formation of a virtual image by a negative lens.

it is discussed in more detail in Chapter 5. But for the moment let us go by the most obvious definition: an image is sharp if it *looks* sharp.

Disc of confusion   Any point on the object that is not at the correct u-distance for the film plane setting will produce an image that is a disc rather than a point. This is called – or should be called – the *disc of confusion*. (Photographers whose grasp of mathematical nomenclature is weak call it a 'circle of confusion'.) If, when you view the final print, you can't distinguish this disc from a true point, then it is, by definition, sharp. The statistical norm for the resolution of the human eye is taken as the perception of 6 mm detail at a distance of 6 m, hence '6/6 vision' ('20/20' in the USA). So if the disc of confusion on an A4 size print (which would typically be viewed from a distance of about 40 cm) is no larger than 0.4 mm in diameter, the image is 'sharp'.

Relative aperture   At this point we need to introduce the concept of *relative aperture*. This is denoted by a number called the *f*-number (or *f*/no), and is the focal length of the lens divided by the aperture diameter (but see marginal note). This latter is controlled by an iris diaphragm, a circle of metal leaves which in camera lenses is within the lens and can be adjusted by a ring on the lens barrel. At a given *f*/no the illuminance on the film for a given subject is constant, no matter what the focal length of the lens may be. It follows that for a given *f*/no, the actual aperture diameter varies in direct proportion to the focal length.

A print intended for direct viewing, or a projected image on a screen, is usually much larger than the original negative or transparency, so the calculated diameter of the acceptable disc of confusion for the final image has to be divided by the scale of enlargement to give the required resolution on the film. This will give a diameter for the disc of confusion of (typically) 0.04 mm for a 35 mm format, 0.03 mm for the APS format, 0.02 mm for digital cameras and 0.15 mm for a 4×5 in film.

Because '6/6' is a statistical mean, half the population has better eyesight than this, and would be able to detect unsharpness in such an image. However, 6 mm at 6 m amounts to 1 milliradian (mrad), which is a conveniently round figure for general purposes, and suffices for pictorial images. Some technical applications require a higher standard than this, though.

Strictly, the *exit pupil*, which is the apparent diameter of the image of the aperture as seen from the rear principal focus. There is also an *entrance pupil*, which is its apparent diameter as seen from the front. These are not necessarily the same size as the aperture itself or each other.

I am using the terms 'film' and 'film plane' for the sake of brevity. All the material in this chapter applies equally to digital and TV camera lenses.

The Science of Imaging

Hyperfocal distance

If a camera lens is focused on infinity (the film plane is at the principal focus of the lens) the distance to the nearest object that will be rendered acceptably sharp is called the *hyperfocal distance* (h). It is calculated from the formula

$$h = f^2/Nc$$

where N is the f/no and c the diameter of the acceptable disc of confusion. When the camera is focused on a distance u, the farthest and nearest object distances that produce tolerably sharp images are respectively given by

$$D_{near} = hu/(h+u) \quad \text{and} \quad D_{far} = hu/(h-u)$$

From this we can show that the full depth of field D is given by

$$D = 2hu^2/h^2 - u^2$$

If you focus the lens on the hyperfocal distance, D is a maximum, extending from half the hyperfocal distance to infinity. This is the way cheap fixed-focus (so-called focus-free) cameras are constructed.

By the way, if you feel somewhat overwhelmed by all these formulae, don't worry. All 35 mm, APS and medium-format camera lenses (apart from compacts) have depth of field scales on them. And if you are operating a large studio camera you can examine the image on the ground glass screen with a magnifier.

Depth of focus

Depth of focus is the tolerance in the film plane. It becomes important in two situations: when you are having to use a film holder that is not properly matched to your camera; and when you are making ultraclose-up images and need to focus by adjusting the back of the camera rather than the lens. The formula for depth of focus is comparatively simple. When the magnification is small, the depth of focus t for a disc of confusion of diameter c is given by the formula

$$t = 2cN$$

In passing, you should remember that the optical image is not completely flat: it does have some depth. In fact, at a magnification of 1 it has the same depth as the object. However, the magnification along the axis is the square of the lateral magnification, so except in extreme close-ups this three-dimensional optical image is very much flattened.

Karl Friedrich Gauss (1777–1855) was one of the greatest mathematicians ever. He made important innovations in the theory of almost every branch of science, and even developed a non-Euclidean geometry, which, in the climate of his time, he prudently withheld from publication. He also made useful contributions to practical lens design: one of these is the model for most large-aperture camera lenses in use today.

Gaussian optics

One of Gauss's most important contributions to optics was a simplification of the geometry of multi-element ('thick') lenses. He showed that it was possible to apply Newtonian optics to any lens system by specifying six *cardinal points* on the principal axis of the lens.

One of these is familiar already: the *rear principal focus*. There is also a *front principal focus*, located by sending parallel rays into the lens in reverse. The four other points are called the *principal* and *nodal points* of entrance and emergence (or front and rear). In any lens wholly in air, the principal and nodal points coincide, though they are defined differently. The planes through these points perpendicular to the principal axis are called the front and rear principal (or nodal) planes.

Gauss's main thesis was that any ray entering the lens, when produced to cut the front principal plane, would, on emergence, appear to have originated at the corresponding point on the rear principal plane (Figure 4.3). The space in between could be ignored for geometrical purposes. So the *u*-distance is measured from the front principal (or nodal) plane, and the *v*-distance and the focal length are measured from the rear principal (or nodal) plane (Figure 4.3).

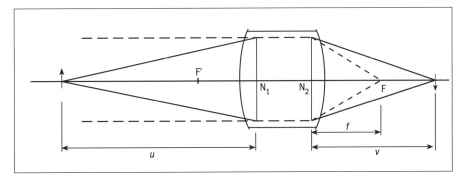

Figure 4.3 Gauss points in a thick (compound) lens. The focal length *f* and image distance *v* are measured from the rear nodal point N_2 and the object distance *u* is measured from the front nodal point N_1. F is the rear and F′ the front principal focus, and $F'N_1 = FN_2$. Two other points, the principal points, coincide with N_1 and N_2 if the lens is wholly in air.

In general-purpose fixed-focal-length (so-called prime) lenses, the nodal planes are situated about one-third of the way inside the lens, but for special purposes they may be designed to be in quite different positions. The important point is that wherever the rear nodal plane is located, you could in theory substitute an 'equivalent thin lens' that would do exactly the same job.

Telephoto lenses

Going back to the lens laws, it is plain that when *u*, the object distance, is large, *v*, the image distance, will be approximately equal to *f*, the focal length. The formula for magnification then simplifies to

$$M = f/u$$

i.e., the image scale is directly proportional to the focal length of the lens. So if you wanted to impress your friends with a full-frame shot of a charging rhinoceros, you would use a lens of very long focal length (a 'long-focus' lens), say 400 mm instead of the usual 35 or 50 mm. Now, a 400 mm lens of conventional design is a cumbersome object, not the kind of thing one would want to carry on safari. But the focal length is measured, not from the back of the lens, but from the rear nodal plane. If the lens designer can coax this plane out in front of the lens, you will be able to have a short lens with a long focal length. This is precisely what a telephoto lens does.

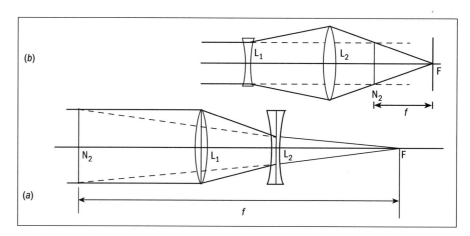

Figure 4.4 (a) Telephoto principle: the insertion of a weak negative lens behind the main lens increases the focal length, moving the rear nodal point (i.e., the position of the equivalent thin lens) well in front of the combination. *f* is the equivalent focal length. (b) In a reverse telephoto (retrofocus) configuration the rear nodal point is behind the combination, giving clearance for a reflex mirror.

The basic principle of the telephoto lens is that of the Galilean telescope. Figure 4.4a shows how the addition of a negative lens increases the focal length of the combination. If you produce the final convergent rays back until they cut the initial parallel rays entering the lens, you will have marked the position of the equivalent thin lens, the hypothetical simple lens that would have done the same job. The plane of the equivalent thin lens is the rear nodal plane of the combination, and its distance from the film plane is its focal length (strictly, *equivalent focal length*).

Retrofocus lenses

If you reverse the order of the two lenses (like turning a pair of opera glasses around) you also reverse the position of the rear nodal point. It is now *behind* the lens (Figure 4.4b). This is called a *retrofocus* configuration, and it is most often used in wide-angle lenses, especially those designed for single-lens reflex cameras where there would otherwise be insufficient clearance for the reflex mirror to swing out of the way for the exposure.

Varifocal and zoom lenses

Some early lens systems were designed so that they could have their focal length changed by removing the front cell, or by changing the separation of the elements; but in general such crude methods gave poor-quality images. A better way of producing a *varifocal lens* is to divide the positive focusing element into two, with a negative element in between (Figure 4.5). The negative element has a focusing power greater than that of either of the outer elements but less than their sum. If you shift this central element from close to the front element towards the rear element, you will move the rear nodal plane from behind the combination to its front, changing the configuration smoothly from retrofocus to telephoto.

This system is often employed in slide projector lenses, but has the disadvantage for cameras that you need to re-focus each time you move the middle element.

Camera lenses

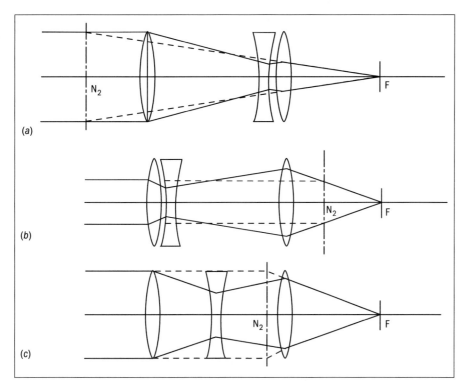

Figure 4.5 Principle of varifocal lens system. The negative element is slightly stronger than either of the positive elements, and when moved close to either of them it gives a negative combination, resulting in either (a) a telephoto, or (b) a retrofocus combination. (c) When the negative element is in the halfway position, the rear nodal point is in its conventional position.

By adding a further moving negative element at the front or rear it is possible to hold the focal plane very nearly constant, and this is the basis of the design of the modern *zoom lens*, which does not need re-focusing when zoomed. Other factors, some of which we shall consider below, complicate the design considerably, and some zoom lenses have fourteen or more elements.

Angle of field

The *angle of field* is the angle subtended at the rear nodal point by the diagonal of the image format, and in a lens giving correct drawing, i.e., giving an image free from distortion, is the same as the *angle of view*. A 'normal' angle of field is around 50°–60°, which is roughly the angle taken in by the human eye when viewing a scene. (This has been confirmed by observations of the distances from which people habitually look at pictures.) Typically, this is about the same as the length of the diagonal of the picture. Now, the diagonal of a 35 mm camera frame is about 43 mm, so a lens of focal length between about 35 and 50 mm is considered 'normal angle'. Any lens with a focal length much beyond 50 mm is considered 'long focus' (narrow angle), and any lens with a focal length much below 35 mm is considered 'wide angle'. The focal length of the lens used can affect the perceived perspective of a print when it is examined from the normal distance: this is discussed at the end of this chapter. As lenses are designed for specific formats, the *covering power* (disc of sharp imaging) of a lens is seldom much beyond the corners

The Science of Imaging

of the format, except for special 'shift' lenses, which are also discussed later. Thus it is in general unprofitable to try to adapt a lens for a format for which it was not designed.

Lens aberrations

Magnifying glasses are specified by their magnification when used in a standard manner; spectacle lenses are specified in dioptres. These are units of focusing power: dioptres are the reciprocal of the focal length in metres. Hence a 4 dioptre lens has a focal length of 0.25 m or 250 mm.

If you take a simple lens such as a magnifying glass or a 4 dioptre spectacle lens, and try to focus an image of a distant light bulb on a wall, as you change the distance of the lens from the wall the image becomes sharper or less sharp. On a close examination of the sharpest image you can get, you see that the image has a somewhat blurred surround, and that this is coloured, too. If you twist the lens out of alignment the image takes on a peculiar shape that changes with the distance of the lens from the wall. These effects are the manifestations of what are called *lens aberrations*. The name comes from a Latin word meaning 'to wander away', which is just what the light rays do. Apart from the colour fringes there are five other main aberrations. These were first described mathematically by L von Seidel, so are called Seidel aberrations, or sometimes 'third-order aberrations', from the form of the Seidel equation. They can be described individually in terms of their effects. One of the Seidel aberrations occurs uniformly over the focal field; the others occur only in the outer parts of the field.

- *Chromatic aberration* This is not a Seidel aberration. It is the result of dispersion, the lens deviating long wavelengths less than short wavelengths. The focal length of a simple lens is thus greater for red light than for blue. The image of a point of white light on the axis appears fringed with colour (longitudinal chromatic aberration), and an image away from the axis appears as a tiny spectrum with the red outermost (lateral chromatic aberration). The effect can be minimised by adding to the focusing element a weak negative lens of a material with a high dispersion (Figure 4.6). Such a combination is termed an *achromatic doublet*.

- *Spherical aberration* Lenses are traditionally manufactured in batches by a process that produces surfaces that are parts of spheres; but the curvature of a

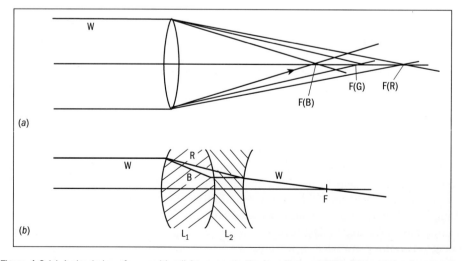

Figure 4.6 (a) A simple lens focuses blue light nearer to the lens than red light; (b) correction by a weak lens with high dispersion. (The thicknesses of the glasses have been exaggerated for clarity.)

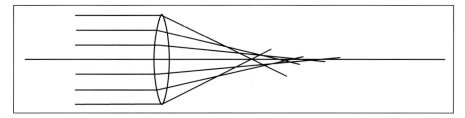

Figure 4.7 Spherical aberration: the outer zones of the lens have a shorter focal length that the inner zones.

sphere is not the theoretically correct curvature. It is too great towards the periphery of the lens, so that the outer zones have a shorter focal length than the inner zones (Figure 4.7).

Spherical aberration occurs uniformly over the whole field. In a single lens it is minimised by sharing the refraction equally between the two surfaces. The result (for parallel light) is very nearly a plano-convex configuration (i.e., one side of the lens is an optical flat). This accounts for the ubiquity of this type of lens in simple focusing devices such as the condenser lenses in enlargers and slide projectors. Adding a weak negative lens with high (negative) spherical aberration to the focusing lens can further reduce spherical aberration. The negative component of an achromatic doublet can do duty for this.

- *Coma* In a simple lens the image of an off-axis point is focused in a different position for each small area on the lens surface. The image from the peripheral zone of the lens is a circle displaced outwards from the true geometrical position (Figure 4.8). The intermediate concentric zones of the lens form circles that are smaller and nearer to the geometrical image. The overlapping of these gives rise to a *coma patch* (*coma* is the Latin for a comet). Coma can be minimised in a single lens by using a meniscus shape (i.e., like a watch-glass) with the aperture stop at a short distance from its concave side (Figure 4.9). This is a dodge used in cheap cameras with a lens aperture around $f/11$.

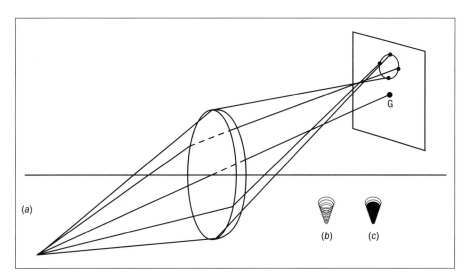

Figure 4.8 Coma: (a) The outermost zone of the lens produces an image of a point object away from the axis as a circle displaced from the geometric position G produced by the primary (central) ray. Intermediate zones produce similar, smaller circles (b), which overlap to produce a comet-shaped image (c) with the tail radially outwards.

The Science of Imaging

Figure 4.9 Two ways of minimising coma in a single lens by employing an aperture stop with a meniscus configuration.

Josef Max Petzval (1807–1891) was a Hungarian mathematician who designed the first photographic lens using a rigorous mathematical method. His lens had a relative aperture of f/3.4, bringing exposures down from minutes to seconds and revolutionising portrait photography.

Two functions are said to be *convolved* when the whole of one function is dealt out to every point on the other function. The noun is *convolution*.

Yes, it is a double negative, and Dallmeyer pedantically called its designs 'Stigmats', though this name didn't catch on. Ernst Abbe (1840–1905) was responsible for quantifying the dispersion of optical glass in terms of refractive index versus wavelength (a high *Abbe number* indicates low dispersion) as well as designing many optical devices. In particular, he found a rigorous method of defining the resolving power of a microscope.

- *Curvature of field* The 'focal plane' of a simple lens is actually part of a sphere, the concave side facing the lens. This means that when the centre of the field is in focus, the periphery is out of focus, and vice versa. This can be controlled by closing down the aperture to increase the depth of focus. However, if the lens is split into several components, their respective surface curvatures can be adjusted to give a flat field. The reciprocal of the field curvature is called the *Petzval sum*, and when this is zero the field is flat.

- *Astigmatism* This term comes from the Greek for 'not a point'. Even when all the foregoing aberrations have been minimised, the image of an off-axis point is still not a point. In one plane it is a short line radial from the axis, and in another it is a short line that is part of a circle drawn around the axis (Figure 4.10). In a simple lens this effect is convolved with the coma patch and is difficult to see clearly; but in any camera lens made before 1886 it is plain enough. In that year, the glassmaker Schott, working from theoretical material by Abbe, developed new types of optical glass, which, when incorporated into new lens designs, produced lenses substantially free from astigmatism. These became known as *anastigmats*.

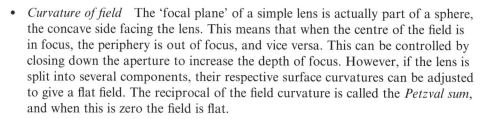

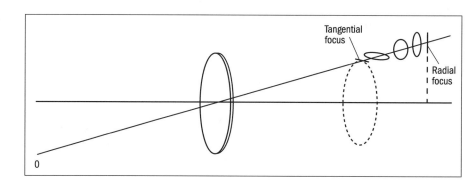

Figure 4.10 Astigmatism: rays from an off-axis point object pass through different thicknesses of glass, and are incident at different angles. The image is not a point but a line that is part of a circle drawn round the optic axis at one distance, and a line radial from the axis at a different distance. In between is a 'disc of least confusion', which is reduced in diameter by the use of a small lens aperture.

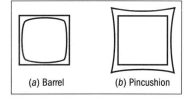

Figure 4.11 Distortion: (a) in barrel distortion the magnification decreases with increasing angle of field; (b) in pincushion distortion it increases.

- *Distortion* If the aperture stop is positioned in the centre of a simple lens (split in two for the purpose) all the rays will pass through the central area, and over the entire field the image points will be in their correct geometrical position, i.e., the lens has correct drawing. However, if the stop is situated in front of or behind the lens, it will select rays that are deviated too much (barrel distortion) or too little (pincushion distortion): the magnification changes with angle of field (Figure 4.11). Distortion can be eliminated most easily by designing lenses to have a quasi-symmetrical configuration of components. By a careful choice of stop position it can be minimised in telephoto and retrofocus lenses too, but it is very difficult to eliminate in zoom lenses.

Aspheric surfaces

The introduction of a single aspheric surface can correct a multitude of residual aberrations, even in a single lens. Many simple cameras are now made with a single aspheric lens that is substantially free both from spherical aberration and coma, the most serious aberrations in a normal-angle lens. In more sophisticated lenses such as the multi-element zoom lens of Figure 4.12, a single aspheric surface takes care of almost all residual aberrations.

Fall-off

In any camera, the optical image is sharper at the centre of the field than at the corners, as the off-axis aberrations increase as the angle of field increases. Closing down the aperture makes the image sharper up to the point where diffraction takes over, but the required exposure duration can become excessive, and both lateral chromatic aberration and distortion become more obvious at small apertures. The best modern lenses show little fall-off over the field they were designed for, but they all show some fall-off in illuminance away from the centre of the field. This is called *cos⁴ fall-off*. With a simple lens the fall-off in illuminance is indeed proportional to the fourth power of the cosine of the semi-angle of field. This is for the following reasons:

- The distance of the image plane from the lens is proportional to the cosine of the angle made with the optic axis, so the inverse square fall-off is inversely proportional to $\cos^2 \theta$, where θ is the semi-angle of field.
- Owing to the obliquity of the beam, which is spread out in the film plane over an area proportional to $1/\cos \theta$, there is a further fall-off proportional to $\cos \theta$ (Lambert's Law).
- The projected area of the aperture itself is reduced proportional to a further $\cos \theta$.

Hence 'cos⁴ fall-off'. This amounts, for example, to a 75 per cent fall-off in illuminance in a 90° wide-angle lens.

Lens designers tackle this problem by redirecting the rays in the front lens component so that they pass more directly through the iris diaphragm (Figure 4.13). In a quasi-symmetrical lens corrected for distortion, this can reduce the fall-off to something like $\cos^2 \theta$.

You can observe this effect in a modern wide-angle lens (or a zoom lens set to its shortest focus). If you turn the lens so that you are looking at the front element almost edge on, the iris diaphragm, which you would expect to be reduced to a narrow ellipse, seems to have risen up to face you – a very peculiar effect (Figure 4.14). This is sometimes termed the *Slussarev effect*, from the designer who originated the idea of introducing coma into the front component and cancelling it out with opposite coma in the rear component.

By allowing a measure of barrel distortion, as in a fisheye lens, fall-off can be almost eliminated, even with a 180° angle of view (Figure 4.15).

Aspheric means simply 'not a sphere'. The simplest aspheric surface is the surface of an ellipsoid, paraboloid or hyperboloid of revolution. Other aspheric surfaces include cosinusoids, catenaries and other mathematical curves and combinations of them. They cannot be made using conventional methods, and must be either diamond-turned (glass) or moulded (plastics).

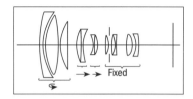

Figure 4.12 Arrangement of elements in a typical modern zoom lens (Vivitar). The arrows indicate the direction in which floating groups of elements travel when zooming from 35 to 85 mm.

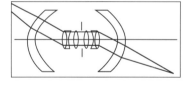

Figure 4.13 To reduce cos⁴ fall-off a lens can be designed to deviate the rays to pass less obliquely through the iris diaphragm.

Figure 4.14 The Slussarev effect: fall-off reduction in a modern wide-angle lens.

The Science of Imaging

Figure 4.15 A fisheye lens gives an increase in peripheral illuminance at the expense of severe barrel distortion. (Photographs by Sidney Ray.)

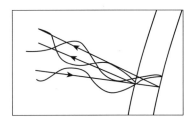

Figure 4.16 When a lens has a low refractive index coating one-quarter of a wavelength thick, the light waves reflected from the two surfaces interfere destructively, suppressing the reflection.

Lens coating

Although the basic designs of almost all of our present-day lenses had been evolved by the end of the nineteenth century, many potentially good designs could not be used because internal reflections caused flare and 'ghost' images of lights and of the iris diaphragm. Losses in a lens with eight glass–air surfaces (such as the Cooke Aviar aerial reconnaissance lens) amounted to 30 per cent or more, much of this light reaching the emulsion as contrast-reducing glare.

The proportion of light R reflected directly back from a glass surface at normal incidence is given by Fresnel's law, which states that

$$R = (n_2 - n_1)^2 / (n_2 + n_1)^2$$

where n_1 and n_2 are the refractive indices of the first and second medium respectively. For air and optical glass the figure comes out at around 1/25 or 4 per cent. For oblique incidence the amount is somewhat greater. If we coat the glass with a material of intermediate refractive index and do the sums for both interfaces, the total reflectance is lowered to about 2 per cent. But we can do better than this. By making the coating just one quarter of a wavelength thick, the two reflected wavefronts will be one half wavelength out of phase, and will interfere destructively (Figure 4.16).

Multi-coatings matched to three well-chosen wavelengths can reduce reflections to well below 1 per cent. This makes it possible to construct lenses with a dozen or more glass–air surfaces, retaining excellent image contrast and freedom from glare. (Though you may still see a row of 'ghosts' when a careless TV camera operator pans too close to the Sun. This can happen with simpler lenses, too (Figure 4.17, colour plate).)

Specialised lenses

Besides the zoom lens, which has become almost a universal adjunct to the prime lens in general cameras as well as standard in TV cameras and camcorders, there are a number of specialised lenses (to be strict, objectives) that have been evolved comparatively recently.

- *Catadioptric (mirror) systems* The optics of these systems is a hybrid of Cassegrain and Schmidt telescope objectives. The principle is that of a telephoto lens, but the part of the main lens components is played by optical mirrors. As the optical path is folded, such objectives are very short, but have a very long focal length. Another advantage of using mirrors is that they do not suffer from chromatic aberration, though they do have all the Seidel aberrations. These are taken care of by aspheric glass elements called 'corrector plates' and field-flattening lenses (Figure 4.18).

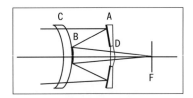

Figure 4.18 A simplified catadioptric system. A and B are the primary and secondary mirrors, corresponding to the positive and negative lens elements in a telephoto lens. C is a correcting element for aberrations and D is a field flattening lens.

- *Telecentric lens systems* These consist of two lenses positioned so that the rear focal plane of the first coincides with the front focal plane of the second. The stop is placed at the common focal plane. The effect is that the magnification is independent of the subject depth. The subject matter is at the front focus of the first lens, and the image is located at the rear focus of the second lens. The angle of field is very small, and the chief use of these systems is in scanning devices and microscopy (Figure 4.19).

- *GRIN lenses* These are a kind of longitudinal lens, a rod with graded refractive index. The refractive index is lower at the periphery than at the centre of the rod, and light passing into the rod is turned inwards towards the centre. Light passing into the rod is converged to a focus, which may be at or beyond the end of the rod. Long GRIN rods may have many foci along their length, and can be used as image relay systems in probes (Figure 4.20). It is now possible to make conventionally shaped lens elements with graded index. Such lenses can correct spherical aberration without having to be ground aspherical.

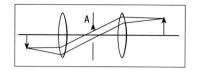

Figure 4.19 A telecentric lens system. The aperture A is at the common principal focus of the two lenses, and the object and image are at the other two focal points.

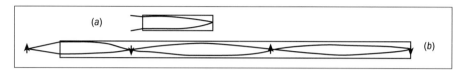

Figure 4.20 (a) GRIN rod focusing at one end; (b) long GRIN rod acting as a series of relay lenses.

- *Diffractive optical elements (DOEs)* As the name implies, these devices operate by diffraction, not refraction. They can be manufactured holographically, or by computer-controlled engraving. Many optical elements can occupy the same space, so that beams can be combined or split as well as focused. DOEs have considerable potential in optical computing systems; but owing to their high dispersion they are in general suitable only for monochromatic (laser) light. However, as the dispersion is in the opposite sense to that of optical glass, a DOE can be used to counteract chromatic aberration in a conventional optical system. DOEs are discussed further in Chapter 15.

Perspective

In general, the term 'perspective' refers to the pictorial representation of three-dimensional objects and scenes on a two-dimensional surface. Perspective was well understood by the ancient Greeks and Romans and used in their paintings, but the knowledge was subsequently lost, and was only rediscovered in the Renaissance years by Italian painters.

In photography, perspective is best thought of as being concerned with the relative sizes of the images of objects at different distances, and the projected proportions of large deep objects such as buildings. Although your camera lens records precisely what it sees, and thereby avoids the difficulties facing painters, it can still produce effects that seem somehow wrong. Even professional photography students often don't really understand what is going on in photographic perspective. There is just one simple fact to remember: *in any scene, the perspective depends only on the viewpoint*. It does not depend on the focal length of your lens, nor on the film format.

Let me give an illustration. Suppose you are photographing a building from a distance of 100 m from its nearest point, and the farthest part visible is 20 m farther away (Figure 4.21). The farthest part will be 120 m away, and from the laws of optics (remember $M = u/f$) it will appear on the image as 100/120 of the height of

The Science of Imaging

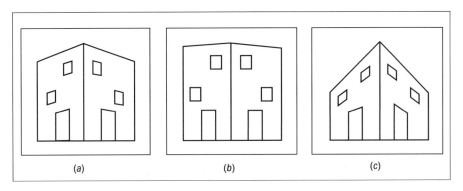

Figure 4.21 Apparent distortion of perspective with change of distance: (a) normal distance; (b) far distance; (c) close up.

the front, i.e., 5/6 times the size. If you move your viewpoint back to 1000 m, the proportions will now be 1000/1020 or 50/51, and the angle of the corner will appear flattened. If you change to a long-focus lens to bring the image to the original size, this will not result in any difference in the proportions: the perspective is unchanged. If you now move up close, say 20 m from the nearest point on the building, the proportions will now become 20/40 or $\frac{1}{2}$, and the steepness of the corner will be exaggerated. If you substitute a wide-angle lens to get the whole image in, again the perspective stays the same. You can stand in one position with a zoom lens, and make the image any size you like, but you cannot change its proportions. Next time you watch a TV play or film, watch what happens when a shot of a person moves in from mid-distance to close-up. If the background enlarges at the same rate as the subject, the camera operator is zooming the lens; if the background goes only slightly larger, and partly disappears behind the subject, the camera is tracking in physically.

Long and short focus perspective There is another important effect in perspective that I have already hinted at. It is connected with the way we habitually view a picture. We tend to hold a photograph so that it subtends an angle of around 50° to 60°, which matches what we consciously assimilate visually when we view an actual scene. This represents a distance from the picture roughly equal to its diagonal. A standard (prime) camera lens is designed to cover about the same angle, so that when we look at a print made from the full frame the perspective looks natural.

It is a different matter when the photograph has been taken with a long-focus lens. The objects within the field of view seem to be flattened and piled up one on top of the other (Figure 4.22a). You can notice this especially in TV sports programmes where the cameras have to be far away from the action, as in motor racing or cricket, especially in head-on views. It looks as if the cars are very close together, and batsmen seem to be almost running on the spot. That is, if you watch your TV from the usual distance. If you look at it from the bottom of a long garden, the perspective becomes believable.

The opposite applies to short-focus (wide-angle) shots. Here, the perspective seems exaggerated. Near objects seem huge and far objects tiny, and interiors may appear enormous (a dodge often employed by brochures for small holiday hotels with smaller restaurants and even smaller swimming pools). In addition, objects towards the edge of the frame seem stretched out. However, when you view the picture from much closer than usual, the apparent distortions disappear.

You can think of it like this: if you were to hold the picture up in front of the scene itself, at your normal viewing distance, everything in the picture would match, size for size, with what you see in the scene. If someone then whipped away the print you would, so to speak, see no difference.

Camera lenses

Figure 4.22 The effect of using (a) a very short and (b) a very long focal length lens with the same format. (Photographs by Sidney Ray.)

Photographers call these effects 'wide-angle distortion', though the distortion only appears when you view the print from an inappropriate distance (Figure 4.22b).

Converging verticals   A disconcerting effect often seen in amateur photographs of buildings is 'converging verticals'. If you have to tilt a camera in order to get the whole of a building into the frame, the top of the building will be farther from the plane of the film than its base, and will appear smaller because the magnification (f/u again) is less. Looking at a real building, we don't notice this, because our visual interpretation mechanism takes account of the position of our head. In fact, if you take a print with converging verticals and hold it some way above your head to view it, the effect will disappear. Architectural photographers use cameras in which the lens panel can be raised while the lens and film remain upright (Figure 4.23). In this way the film plane remains parallel to the subject, and the verticals do not converge. The same facility has been introduced for 35 mm and medium-format cameras, in the shape of what are called 'perspective control' (or simply 'shift') lenses, which can be moved in the same way as the lens boards of larger cameras.

Diverging verticals are possible, too. You get them when you take a photograph of the neighbourhood from the top of a tall building or from the air, with the camera tilted downwards. They are not often seen except in low-level aerial shots of buildings, but we have all seen the unfortunate effect when a society photographer takes a shot of a celebrity standing up at a party, using a wide-angle lens at eye level.

It could be argued that the converging verticals represent correct perspective and that 'correcting' by using a rising front actually distorts the perspective; I wouldn't want to enter into an argument about this. What matters is what *looks* right on a print. In practice many professional photographers feel that a small amount of convergence often gives a more realistic image under some circumstances.

The Scheimpflug rule

Both converging and diverging verticals are part of a phenomenon known as *keystoning*. A keystone is the trapezoidal shaped stone at the top of an arch, and it matches the shape of the image of a rectangle photographed with the camera axis not perpendicular to the plane of the rectangle. Sometimes an existing image with

The Science of Imaging

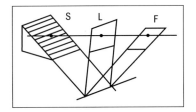

Figure 4.24 Scheimpflug's Rule: for overall sharp focus of a tilted subject plane S, this plane must meet both the lens plane L and the film plane F in a single line.

keystoning needs to be corrected or 'rectified'. This happens not only with shots of buildings, but also with aerial survey photography, where the camera axis may not be truly vertical for every shot. A rather different problem arises when a professional photographer needs to have a particular item in the image (such as a carpet, or a table laid with food) all in focus simultaneously. The solution to all these problems lies in what is known as the *Scheimpflug Rule*, after the person who first worked out its geometry. The rule states that when a camera is used to photograph a tilted plane, that plane will appear in sharp focus only when the planes of the subject, the lens panel and the film all meet in a single line (Figure 4.24). The proof is an extension of the Newtonian lens laws.

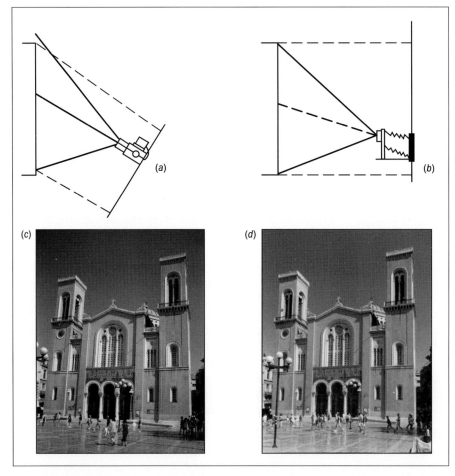

Figure 4.23 (a) When a camera is tilted, the top of a building is farther from the film plane than its base; (b) use of a rising front keeps the camera back parallel to the wall; (c) if you need to tilt the camera to get the top of a building into the picture, the verticals will converge and the building will appear as if tilted backwards; (d) use of a rising front counteracts this effect, but should not be overdone. It is usually advisable to leave a very small amount of convergence for a realistic effect.

Digging deeper

In this chapter I have been conscious of having had to skim over a great deal of material. If you feel frustrated by the lack of derivations and proofs for the various formulae, you can find them all (including the Scheimpflug rule) in *The Manual of*

Photography (9th edition, ed. R E Jacobson, Focal Press, 2000). There is also a fuller account of lens aberrations. Sidney Ray, who wrote the optics section in the *Manual*, has written a number of books on photographic optics for Focal Press, the most recent being *Applied Photographic Optics* (2nd edition, 1988). One of the best books aimed at a more general readership level is *Optics in Photography* by Rudolf Kingslake (SPIE Press, 1992). For a really good history of the photographic lens from its beginnings to about 1936, it is worth hunting for a copy of Keith Henney and Beverly Dudley's classic *Handbook of Photography* (McGraw-Hill, 1939).

Chapter 5 Resolution in optical systems

In the previous chapter I asked 'How sharp is "sharp"?', and suggested that for practical purposes if an image looked sharp, then it *was* sharp. If you felt that this was an unsatisfactory answer you were right. But even if your image does look sharp, if you keep on enlarging it far enough, you will eventually reach a point where further enlargement will not show any more detail, but will simply enlarge up the blur. Microscopists call this condition 'empty magnification'. The point where this begins to happen is called the *limit of resolution* of the optical system.

In the real world, that is. I once saw a courtroom drama film where the crucial piece of evidence was a snapshot of a street scene, blown up to the point where the date on a newspaper held by a man some 50 metres away was clearly readable. In Hollywood anything is possible.

You can obtain a rough estimate of the limit of resolution of a camera system by photographing a grid of parallel lines that become closer together with distance across the negative, such as a set of park railings seen obliquely (Figure 5.1). The spacing at which you can no longer distinguish the separate rails at any magnification defines the limit of resolution. (Because park railings are not a very scientific kind of object, there is a standard test object of a similar type, called a Sayce test chart (Figure 5.4).)

Figure 5.1 Park railings seen in perspective represent a one-dimensional object with spatial frequency that increases across the negative.

The *resolving power* of an optical system is usually expressed in *line pairs per millimetre* ($lp\,mm^{-1}$), a line pair being a black and a white bar.

Testing for resolving power

The Sayce chart is not well suited to practical applications. Real subject matter does not often include long parallel bars of high contrast. Mostly, objects are chunky and of fairly low contrast. This is especially true of aerial and satellite photography, an area where resolution is very important. The standard test object for optical resolution was developed by the US Air Force during the Second World War, and consists of three parallel bars with an aspect ratio (i.e., the ratio of width to height) of 1:5, equally spaced. These are arranged in blocks of four, in a spiral pattern of steadily reducing size (Figure 5.2). They may be of high or low contrast as required, and the limit of resolution is taken as the line pair figure for the smallest block in which the direction of the bars can be seen.

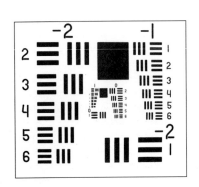

Figure 5.2 US Air Force test target.

Resolution in optical systems

Diffraction limitation

For large lens apertures the limit of resolution of a lens is governed by the residual aberrations, but for well-corrected lenses at small apertures the resolution is limited by diffraction. If the aperture is circular, the diffraction pattern produced by a point object is an Airy pattern (see Chapter 1). More than 80 per cent of the light intensity is in the central patch, the *Airy disc*.

Camera lenses are designed, as a rule, to let as much image-forming light through as possible, so at the maximum aperture there will be some residual aberrations. With most high-quality prime lenses these will have all but disappeared by two stops down from maximum aperture, by which point the Airy disc will have increased in size to approximately match the diameter of the blur disc resulting from aberrations. This is the point at which the lens is said to be *aperture limited*. At smaller apertures the Airy disc takes over as the main limitation to resolution.

There are some variations. Very large aperture lenses ($f/1.2$ or so) may be aperture limited at three or even four stops down, and process lenses give their best performance at around $f/22$ (this is to allow the halftone screen optics to operate correctly). The older types of wide-angle lens for large-format cameras also need to be used at a small stop; but this kind of performance is unacceptable in short-focus lenses for 35 mm cameras. High-quality long-focus lenses often give their best performance at maximum aperture. The intensity profile of the image of a point is called the *point spread function* (PSF), and its shape changes as the lens is stopped down, from the aberrated image at full aperture to an Airy pattern at small apertures (Figure 5.3).

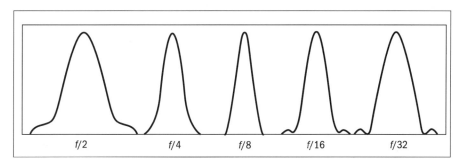

Figure 5.3 Profile of a point spread function with decreasing aperture in a camera lens.

The Rayleigh criterion

When observing double stars through a telescope, Rayleigh found that two stars could just be distinguished as separate if the centre of the Airy pattern formed by the image of one star fell on the first zero of the Airy pattern of the other. You will remember from Chapter 1 that the radius of the Airy pattern to its first zero is $1.22\lambda N$, where λ is the wavelength of the light and N is the f/no of the optical system. This is the *Rayleigh criterion* for the resolution of a diffraction-limited optical system.

In the early days of photogrammetry it became apparent that the figures found in practice for optical resolution didn't always fit the perceived sharpness of an image, particularly at the edges of details. The concept of *acutance* evolved with reference to the sharpness of image edges (such as the image of a roof seen against a bright sky). This could (with some difficulty) be quantified, but was generally treated simply as high, medium or low.

John William Strutt, third Baron Rayleigh (1842–1919) was an outstanding figure in both astronomy and physics. He succeeded Maxwell as Cavendish Professor of Experimental Physics at Cambridge University, and was elected President of the Royal Society in 1905. He discovered the element argon, for which he received the Nobel Prize for Physics in 1904, and gave a scientific explanation of the blue colour of the sky. He was the first person to point out the inconsistency in the (then) laws of radiation that led to Planck's quantum model. His work on diffraction was only a small part of his contribution to the advancement of science.

The Science of Imaging

In silver halide photography, film processing methods play a large part in the acutance of a photographic image. The human retina also possesses an edge-enhancing mechanism (see Chapter 3).

A trained photo-interpreter was usually aware of these optical idiosyncrasies, and could readily compensate for them mentally. The successful identification by Constance Babington-Smith ('Babs') of the first V-1 flying bomb, from a tiny smudge on a high-level reconnaissance photograph, was an outstanding example of the photo-interpreter's craft.

Selwyn's work was not published at the time as his wartime research was covered by the Official Secrets Act, and the credit has usually gone to Duffieux, although his book *L'Intégrale de Fourier et ses applications à l'Optique* actually acknowledged Rayleigh as the true originator of the Fourier approach. In turn, Rayleigh's published work refers back to Airy and Helmholtz, and beyond... Shoulders of giants, again?

The inadequacy of resolving power measurements

Aerial reconnaissance played a vital part in the conduct of the Second World War. A group of specialists, known as photo-interpreters, examined the details of reconnaissance photographs under high magnification, and it quickly became clear that the resolving power criterion for an optical system was only one factor in the successful interpretation of detail. Indeed, some lenses produced more readily interpretable material than others of higher measured resolving power. Some lens types even produced distortions in the fine detail.

Lens designers, laboriously tracing the paths of light rays, and making allowance for the effects of diffraction as well as estimating engineering tolerances, were able to predict the shape and size of the image of a point of light in various parts of the field, and thereby to foretell something of the behaviour of the optical system under realistic conditions. This 'star image', which eventually became known as the point spread function, was, however, difficult to quantify mathematically.

The modern approach to image quality

The solution arrived in the early 1950s, following the pioneer work by E W H Selwyn in England and, independently, P M Duffieux in France. Both noticed that in the image of a Sayce-type chart, as the spatial frequency increased the intensity profile became markedly sinusoidal and the contrast fell progressively, becoming zero at the limit of resolution (Figure 5.4).

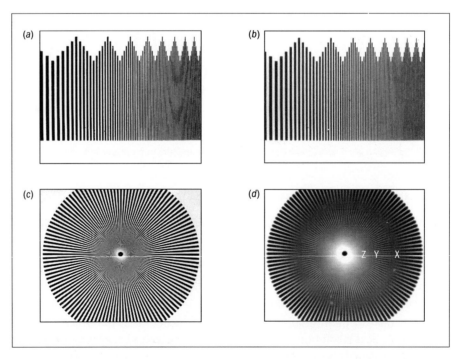

Figure 5.4 (a) Sayce chart. (b) The same chart photographed at a very small lens aperture. Note the way the density profile of the bars becomes increasingly sinusoidal and the contrast decreases as the spatial frequency increases. (c) Siemens star test object. (d) The same object photographed with an out-of-focus lens. Note the spurious resolution beginning at X, and further reversals at Y and Z.

Resolution in optical systems

In some cases the image appeared to be resolved again at a still higher spatial frequency, but with the positions of the light and dark bars interchanged. This is typical of lenses that are slightly out of focus, and is called *spurious resolution*.

A test chart with bars having sinusoidally varying density showed no change in intensity profile, i.e., they remained sinusoidal, but the contrast fell in the same way. It thus became clear that there was something fundamental about a sinusoidal bar pattern.

Analysis of a square wave

If you connect the output of an audio signal generator set to 'square wave' to a loudspeaker you will hear, not the bland tone of a pure sinewave, but a somewhat raspy timbre reminiscent of a badly played clarinet. If you listen carefully you will hear a distinct note an octave and a half higher than the basic (fundamental) note, and if you have a keen ear and the fundamental note has a low pitch you may hear further tones at still higher pitches. Unlike the eye, which sees a combination of several discrete wavelengths as their mean, the ear is an analytical organ that is capable of separating out individual frequencies. A square wave can be assembled from a series of sinewaves of frequencies 3, 5, 7, 9, etc. times the fundamental frequency, with amplitudes that decrease successively in the same proportions (Figure 5.5).

This is known as *Fourier synthesis*. Its converse, the teasing out of the frequency components of a complex wave, is known as *Fourier analysis*. By putting your signal generator output through a variable low-pass filter and suppressing the higher frequency components one by one, you will finally be left with a pure sinewave at the fundamental frequency.

(Appendix 3 looks further into the principles underlying the Fourier approach to imaging.)

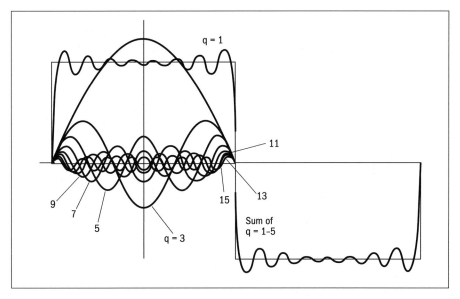

Figure 5.5 Synthesis of a square wave from sinusoidal components. q is the relative frequency of the component (and the order of the harmonic).

The Science of Imaging

The breakthrough that led to a revolution in optical thinking, and a new era in optics technology, embraced two major insights:

- Any linear optical system (i.e., a lens free from flare and a sensitive material with a linear response) will always produce a sinusoidally varying image from a sinusoidally varying object grating.
- Any scene can be considered as the sum of a spectrum of sinusoidal gratings of varying spatial frequency, amplitude, phase and orientation.

This means that if you set up an optical system using an object consisting of a sinusoidal grating of constant amplitude (i.e., constant maximum density) and progressively increasing spatial frequency with distance (sinusoidal Sayce chart), and record the amplitude and spatial phase (positional shift) of the resultant, you will be able to deduce the response of the system to both coarse detail and fine detail, and will also be able to predict the extent of any image distortion within the fine detail. In practice this information is plotted graphically as a function of the spatial frequency of the image of the test object.

Modulation

Contrast is a difficult concept to put into figures. The most obvious way would seem to be to define it as the ratio of the luminances of the brightest and darkest areas of the subject matter. This is misleading, however, as a contrast of 1000:1 looks very little different from a contrast of 100:1. We obtain a more realistic result by taking the logarithm of the luminance ratio instead. (I discuss this further in Chapter 8.) But this rule does not hold accurately once you get to contrasts higher than about 10:1 (a logarithmic contrast of 1). A closer fit to perceived contrast comes from the use of *modulation* as a quantifier for contrast. The modulation M of a subject is given by the expression

$$M = (L_{max} - L_{min}) \div (L_{max} + L_{min})$$

where L_{max} and L_{min} are the respective luminances of the brightest and darkest areas of the subject matter. For a transparency luminance is replaced by transmittance, and for a print it is replaced by reflectance (fractions or decimals, not percentages).

At low values (up to about 0.7) the numerical scale for modulation does not differ much from logarithmic contrast, but above this the two values begin to diverge, as the logarithmic value can in theory go right to infinity, whereas the modulation cannot go higher than unity. (See Table 5.1.)

Modulation does provide a closer approximation to the way we see contrast than does logarithmic contrast. It is a comparatively recent concept in imaging, though you will have come across it if you have studied communications technology.

Table 5.1 Numerical scales of contrast compared.

Arithmetic	1:1	1.5:1	2:1	3:1	4:1	6:1	10:1	100:1	1000:1	∞
Logarithmic	0.00	0.18	0.30	0.48	0.60	0.78	1.00	2.00	3.00	∞
Modulation	0.00	0.20	0.33	0.50	0.60	0.71	0.82	0.98	0.998	1.00

Resolution in optical systems

The optical transfer function

The *optical transfer function* (OTF) is a combination of the *modulation transfer function* (MTF) and the *phase transfer function* (PTF). These are defined as follows:

- The MTF is a graph of modulation transfer (image modulation ÷ object modulation) plotted against image spatial frequency in cycles per millimetre ($c\,mm^{-1}$).

- The PTF is a graph of the displacement of the image compared with its geometrically correct position, in radians (rad) or milliradians (mrad), plotted against the image spatial frequency ($c\,mm^{-1}$). A displacement of one whole cycle is $\pm 2\pi$ radians. Figure 5.6 shows a typical OTF for a camera lens. The MTF shows us how the system performs with reference not only to the finest detail (i.e., the limit of resolution), but also to the coarser detail and (indirectly) to the edges of objects. The PTF is less useful for general purposes, but it does predict any distortion of detail, in particular whether there is likely to be a problem with spurious resolution.

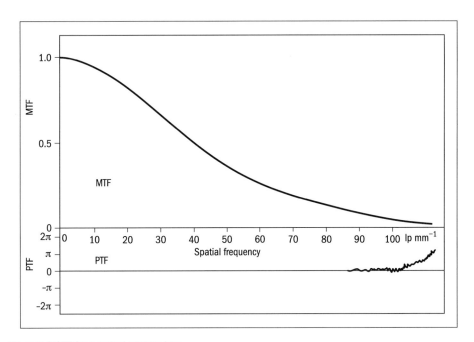

Figure 5.6 OTF for a typical camera lens.

As a grating of parallel bars is effectively a one-dimensional object and a photographic image is two-dimensional, the practical measurement of the MTF for a camera lens involves two sets of data made with orthogonal set-ups. The directions chosen are with the bars of the test object respectively radial and tangential to the optical centre of the image field. To evaluate the MTF over the whole image field you have to make measurements at a sufficient number of points between the optical centre and the periphery of the field.

The Science of Imaging

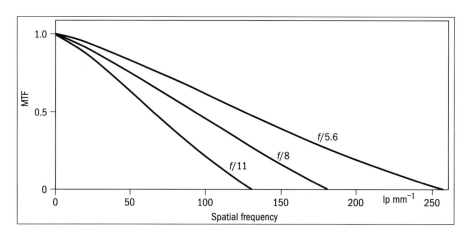

Figure 5.7 The MTF of an 'ideal' (diffraction-limited) lens.

The MTF of an 'ideal' lens

The MTF for an 'ideal' or totally non-aberrated lens can be calculated from the diffraction equations, as diffraction is the only limitation on image quality. It looks like Figure 5.7. Plainly, the MTF for any real lens must lie below this curve at all points.

> **The OTF is the Fourier transform of the LSF (and vice versa). It may have been the discovery of this that led indirectly to the adoption of the Fourier model for optical imaging, though they are not otherwise connected.**

OTF and spread functions related

As with the OTF, the spread function is usually measured in radial and tangential directions separately, as *line spread functions* (LSFs), using a line instead of a point as test object. The tangential LSF will always be symmetrical (provided the lens has not somehow become decentred), whereas the radial LSF will be symmetrical only if the lens is completely free from distortion and coma. Quantitative analysis of the LSF is not at all easy, but it does not need to be measured in practice as its shape can be readily derived from the OTF. Nevertheless, when the concept of the OTF first appeared, OTFs had to be computed from the shape of the LSF before equipment for measuring the OTF directly was in existence.

There is a further function, namely the edge spread function or ESF. This is a plot of luminance across the image of a sharp edge. This needs to be measured both inwards and outwards for the tangential figures. The ESF is related mathematically to the LSF: it is the first derivative of the LSF. Its shape gives us information about acutance, and it is therefore more generally useful in relation to the performance of photographic films. In practice none of the spread functions is used much, as the OTF is so much more directly informative; but the PSF often appears (much enlarged) in ray-tracing computer program displays.

Cascading of transfer functions

One of the major advantages of the OTF over the various spread functions is that the OTFs of a complex system (such as camera lens, film, printer optics and print material characteristics) can be readily combined – something that is difficult with spread functions. The final OTF is found by 'cascading' the function: multiplying the MTFs together, and adding the PTFs. Some imaging systems (such as the one instanced above) contain a number of components, each with its own OTF. In practice there can be many of these. For the complicated and exacting imagery of aerial or satellite survey you would need to include OTFs for haze, lens flare, camera vibration, recording medium, viewing system and even the perceptual processes of the image interpreter! Some of these factors may add to the complexity by having a response that is not completely linear, but that takes us beyond the scope of this text.

Spread functions are combined by convolution, i.e., every point on one of the functions is 'dealt out' to every point on the other. This is not an easy task.

Granularity and pixel size

In a developed silver halide film, the photographic image is made up of millions of tiny grains of opaque silver. Under a jeweller's loupe these look like small pieces of coke, and under a microscope they look like spaghetti bolognese. The reasons for this are explained in Chapter 9. Although silver metal is shiny, the texture of the grains causes the light to be reflected back and forth until it has all been absorbed, so the grains look black. On a microscopic scale, therefore, a photographic image is a binary device: light either passes unimpeded or is blocked completely. However, as the grains are in general too small to see with the unaided eye, the image appears in varying shades of grey, depending on the number of grains in a given area – in fact, somewhat similarly to the halftone image in a newspaper, which is made up of black dots, and is also binary. The silver grains vary in size, and we obtain the figure for *granularity* by statistical methods, using a plot made by a travelling microdensitometer, a light meter measuring through a tiny pinhole. As the spatial period of the image of the test object (the inverse of the spatial frequency) approaches the mean grain size, the MTF falls rapidly to zero; for the optimum resolving power for the system, this should occur at the point where the lens MTF also falls to zero. But in practice this does not necessarily result in optimum final image quality, as the film MTF may be a completely different shape from the lens MTF (Figure 5.8).

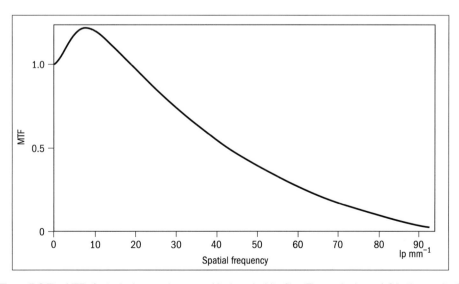

Figure 5.8 The MTF of a typical general-purpose black-and white film. The peak above 1.0 is the result of adjacency effects during development (see Chapter 12).

The Science of Imaging

In an ideal lens, using light of a single wavelength (such as laser light) the MTF does not fall off at all until the cut-off point (which is determined by the wavelength and lens aperture) is reached.

This also applies to halftone reproductions, images from digital cameras and any image (such as that from a liquid crystal display) that is made up of pixels. Because the dots form a regular pattern and are not random as in the silver grains of a photographic image, the cut-off point is very sharply defined. In a pixel image, by its nature, the MTF can remain at unity right up to the cut-off point.

Conclusion

Although quasi-realistic test objects such as the USAF test target may give a good idea of the performance of an existing optical system, it is the OTF that gives us the greatest insight into the nature of the information that can be extracted from the image; and knowing its required characteristics in advance can help the system designer enormously. The concept of the OTF brought about a revolution in optics. It is sad that although the principle was initially developed in Britain it was left to other nations to take full advantage of it, most notably Japan. As a result Japan has dominated the camera lens market for nearly fifty years.

Digging deeper

It has not always been easy to find good, readable textbooks on image quality and optical resolution. Most of the accounts are buried in more general texts, and may vary from sketchy to downright misleading. There is an excellent introduction to the subject in Rudolf Kingslake's *Optics in Photography* (SPIE Press, 1992). There is a good account of practical field-testing of aerial survey cameras in G C Brock's *Physical Aspects of Air Photography* (Dover, 1967), which can be applied to other types of camera; you should be able to borrow a copy through your local library. You will also have to pay it a visit if you are to find *Evidence in Camera*, by Constance Babington-Smith, which is an absorbing account of wartime photographic reconnaissance. It sounds dull, but it isn't.

There is an excellent account of the effects of haze in Ron Graham and Roger Read's *Manual of Aerial Photography* (Focal Press, 1986). Sidney Ray's *Applied Photographic Optics* (2nd edition, Focal Press, 1988) contains a chapter on the physical optics of lens systems including a vigorous if somewhat terse explanation of the Fourier approach to the OTF concept. Both granularity and OTFs are covered in some detail in *The Manual of Photography* (9th edition, ed. R E Jacobson, Focal Press, 2000), and a chapter on Fourier optics is thrown in for good measure. But if you really want to know everything there is to know about OTFs in theory and practice, there is only one choice: Tom Williams's *The Optical Transfer Function of Imaging Systems* (Institute of Physics Publishing, 1999).

Chapter 6 Images in colour

Early attempts

When photography first appeared early in the nineteenth century, its images were in monochrome; but it was immediately accepted as a promising newcomer to the visual arts. As far as we can tell from contemporary writings, nobody seemed to be concerned about the absence of colour from the image, at least in landscape photography. It is perhaps surprising that the lack of colour went unremarked, as very few finished paintings have been made in a single colour (though Picasso's 'Guernica' does spring to mind). It worried many of the pioneer photographers, though. There were stories of daguerrotypes that showed colour when they were viewed from a particular angle, but these are probably apocryphal: certainly no example exists now.

> **Family portraiture was another matter: Victorian portraits were often hand-coloured.**

Today, almost all amateur and commercial photographs are in colour. Monochrome keeps a tenuous foothold in fine-art photography, and (for its impact) in photo-reportage.

> **I am using the term 'monochrome' here rather than the more usual 'black-and-white', because a majority of exhibition photographs of this type are toned, or are on warm-tone print material.**

Lippmann photography

If not quite the earliest attempt to produce a photographic image in colour, Gabriel Lippmann's was certainly the boldest. Lippmann was a physicist who thought of colour in terms of wavelengths, and his method used the interference of light waves to produce colours in a similar manner to the formation of colours in soap bubbles or oil films. In the 1890s, using some of the very first panchromatic materials (i.e., materials sensitive to the whole visible spectrum), Lippmann succeeded in producing colour photographs of such high quality that he was awarded the Nobel Prize for Physics in 1908.

The problem Lippmann faced was that if you want an image of (say) a rose to look red and its leaves to look green when illuminated with white light, you have to arrange for the image of the flower to reflect only red light and for the image of the leaves to reflect only green (Figure 6.1).

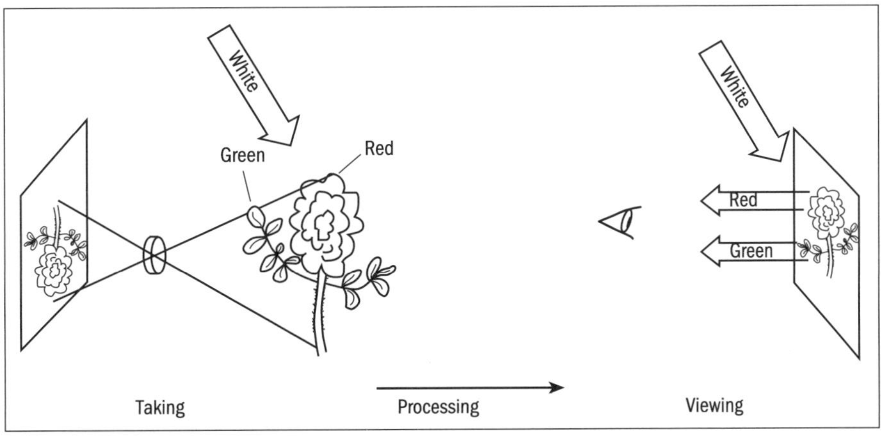

Figure 6.1 In a colour photograph, the image must reflect only the colours in the subject.

The Science of Imaging

Figure 6.2 Lippmann's special mercury-containing plateholder made by the Penrose Company.

Bragg reflection The type of interference described here is called *Bragg reflection* (or, more correctly, *diffraction*). It is named after the Braggs, father and son, who successfully investigated the principle (Figure 6.3). William Henry Bragg (1862–1942) and his son Lawrence Bragg (1890–1971) pioneered and perfected the techniques of X-ray diffraction that led to the establishment of the atomic structure of crystals. They were jointly awarded the Nobel Prize for Physics in 1915 for this work. Both men were knighted, and both became in turn Director of the Royal Institution. Sir William Bragg is remembered particularly for inaugurating the famous series of Christmas lectures at the Institution. There is more about Bragg reflection in Chapter 18.

A more rigorous explanation invokes the Fourier model. Lippmann was familiar with the work of Joseph Fourier, and was almost certainly aware of its application to his own work. It is interesting to speculate that had he seen fit to publish this aspect, it might have advanced the 'revolution in optics' by more than half a century.

Lippmann found an ingenious way to achieve this. He installed the photographic plate in his camera with the glass side towards the lens and the emulsion in contact with a mirror, actually a bath of mercury. Red light forming the image of the flower and green light forming the image of the petals would then be reflected back from the mirror through the emulsion (Figure 6.2), interfering with the incoming waves, producing standing waves and a stationary interference pattern which on development would produce sheets of semi-transparent silver parallel to the surface of the emulsion and spaced exactly one half-wavelength apart. Such a system would reflect strongly only light of the appropriate wavelength, by constructive interference.

But what, you may ask, about desaturated colours: pinks, browns, greys? It would appear as if only saturated colours with a narrow bandwidth could be reproduced. One would expect that a wide spread of wavelengths would wipe out the delicate structure of interference planes within the thickness of the emulsion. But the pinks, browns and greys *were* all there, as well as startlingly accurate flesh tones. Lippmann's scientific papers do not explain this, although historical research suggests that he knew precisely what was going on. For nearly a century various theories were proposed, but only recently has the correct explanation resurfaced (see Box).

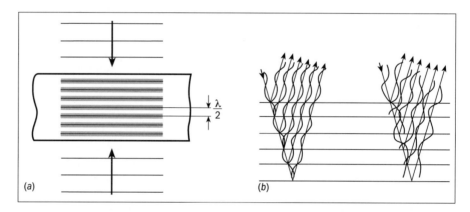

Figure 6.3 (a) Two light beams travelling in opposite directions within the emulsion form stationary interference planes one half-wavelength apart; (b) light of the original wavelength is reflected from adjacent layers in phase and is reinforced; light of other wavelengths interferes destructively with adjacent reflections and is suppressed.

Lippmann's desaturated colours

Light of a single frequency is said to be *temporally coherent*. Such light forms stationary interference patterns when it re-crosses its own path. In Lippmann's photographs the image-forming light was reflected directly back through the emulsion, forming interference planes one half-wavelength apart. In this situation, the distance over which the interference planes can occur (i.e., the number of planes that can be formed) depends on the *coherence length*, which is the distance over which the waves remain in phase (in step). The coherence length is inversely proportional to the frequency bandwidth (Figure 6.4).

In some laser beams the coherence length can be many metres. Most light sources, though, have a very low coherence length. Even sodium light has less than a millimetre. Light reflected from natural objects has a coherence length of only a few wavelengths; and in the Lippmann process this light would form at the most about half a dozen interference planes, so that when the image was illuminated with white light the wavelength selection would be less precise. In the extreme – white light – only one reflecting plane would be formed, and there would be no wavelength selection: all the light would be reflected.

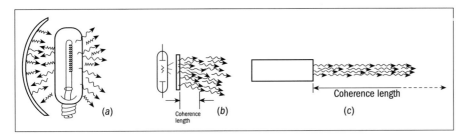

Figure 6.4 Coherence length: (a) incoherent radiation, bandwidth ~60 THz or 1 µm; (b) partly coherent, bandwidth ~300 GHz or 1 mm; (c) highly coherent, bandwidth ~2 MHz or 150 m.

A number of Lippmann's contemporaries, including the Lumière brothers Louis and Auguste (the inventors of cine photography) and Frederic Ives (an American pioneer of colour photography), followed up his work; but the process was so exacting that few photographers were prepared to attempt it, and the technique languished for nearly a century. Figure 6.5 (colour plate) is a reproduction of a Lippmann photograph.

There are still plenty of examples of Lippmann photographs around, mostly in French museums. Some of these are very beautiful, and most are in nearly perfect condition, owing to the method of mounting (behind a thin glass wedge, to make it possible to view the photograph at the correct angle without one's head getting in the way of the viewing light). A few are in the Science Museum, London. With the renewed availability of ultrafine-grain panchromatic emulsions (mean grain diameter less than 20 nm), interest in the process has begun to revive (see Digging Deeper).

A certain Dr Spirna of Norway was *nct* put off, though. He took literally hundreds of Lippmann photographs of a stuffed parrot, many of which (unlike the parrot) are still extant.

According to the late Jack Coote, the finest collection of Lippmann photographs is in the Preus Fotomuseum at Horten in Norway.

The Young–Helmholtz theory of colour perception

At the beginning of the nineteenth century Thomas Young discovered that almost every possible colour could be matched by superposing no more than three coloured lights, specifically red, green and blue, in various intensities, directed at a white screen. You can easily demonstrate this with three slide projectors loaded with plain slides bearing respectively red, green and blue filters. If you have not seen a demonstration of this, you may be surprised (most people are) to see that where the red and green patches overlap the colour is a brilliant primrose yellow (this was discussed in Chapter 3). Less surprising is the combination of blue and green to form the colour we call cyan, a sort of kingfisher blue, and of blue and red to form magenta. The superposition of all three beams gives white (Figure 6.6a, colour plate).

By varying the relative intensities of the red and green lights you can obtain all the range of colours from pure red through orange, yellow and lime green to pure green. By doing the same with green and blue you will get a similar shift from pure green through emerald green, cyan, 'petrol' blue and finally pure blue. Mixing blue

The Science of Imaging

Hermann von Helmholtz (1821–1894) was a polymath who dominated German science during the mid-nineteenth century. He was a qualified surgeon, and invented the ophthalmoscope. He made the first precise formulation of the principle of conservation of energy. He developed thermodynamics in physical chemistry, and apart from his extension of Thomas Young's work to cover after-images and colour blindness, made contributions to auditory perception, acoustics, hydrodynamics and electromagnetism.

Maxwell was lucky to have had any success at all. His plates were completely insensitive to red. As it happened, his red filter was transparent to UV, and his red ribbon reflected UV strongly. His green filter needed to be very pale for anything to record. Hence his separations were not red, green and blue, but UV, blue and violet. Maxwell was a theoretical physicist of genius, but he knew very little about photography. Perhaps this explains the apathy with which the Photographic Society of London (now the Royal Photographic Society) received his presentation of his findings.

Lumière colour transparencies have stood the test of time well. There are some excellent examples in the archives of the Royal Photographic Society.

The reason for the high density of additive mosaic transparencies is that the maximum transmittance of any tricolour filter cannot exceed 33 per cent, and every cell is covered by such a filter. However, in the 1980s additive colour photography surfaced again, with an instant-slide material from Polaroid. While still somewhat dark, the resolution is better than with the older films, as the pitch of the colour filter mosaic, which is a raster of plain horizontal rulings, is much finer, and invisible even in a large projected image viewed from a normal distance (Figure 6.6b, colour plate).

with red similarly takes you through violet, purple and magenta to crimson and pure red. All the colours from red to blue approximate to pure spectral colours. In colorimetric terms they are 100 per cent saturated, as is the magenta range (which is not part of the spectrum). To obtain desaturated (pastel) colours such as pinks and blue-greys you have to add the third patch of light and adjust the relative intensities of all three.

In order to explain these results, Young suggested that there were three types of colour receptor in the eye, sensitive respectively to red, green and blue, and that simultaneous stimulation of pairs of these gave the sensation of the three intermediate colours cyan, magenta and yellow. Stimulation of all three types of colour receptor would result in the perception of desaturated colours – pink, brown, grey etc. Nearly half a century later, Helmholtz was to take up Young's hypothesis and add a sound physiological basis.

Additive colour synthesis

The Young–Helmholtz model for colour perception was taken up in the 1850s by James Clerk Maxwell, who further extended it in terms of physiology. (Like Helmholtz and Young, he was a qualified medical practitioner.) He constructed various models to demonstrate additive effects, including a spinning disc with adjustable red, green and blue sectors. In 1861 Maxwell took three separate photographs of a bow of coloured ribbon through a red, a green and a blue filter respectively. He made positive slides from his negatives and projected these in register through red, green and blue filters. The result was an image that approached natural colour – at least within the limits of his material.

It was fully fifty years before patents began to appear for colour processes based on the additive principle. In 1907 Louis Lumière patented a method of coating an emulsion with starch grains dyed red, green and blue. The film was reversal developed to a positive, with the randomly distributed filter grains *in situ*. Others such as Paget and Dufay used ruled colour mosaics in a similar way. Dufaycolor materials continued to be available up to around 1950, and gave excellent colour rendering, particularly of flesh tones; but the material could not be printed successfully, and the transparencies were not very bright. And by this date new, much brighter, materials suitable for home processing, such as Kodak Ektachrome, had appeared.

The principle of additive colour imaging underwent a rebirth with the advent of colour television. Colour TV screens carry a mosaic of red, green and blue phosphor spots (in some cases vertical bars), and the red, green and blue elements of the scene are delivered by three electron guns. A precisely aligned perforated screen allows the beam from each gun to 'see' only its appropriate colour phosphor spots. (I shall be discussing colour television in more detail in Chapter 15.)

Quantifying colour: the CIE chromaticity diagram

If, to an artist, it seems vaguely indecent to reduce the mysteries of colour to mere physiology, to a physicist it may seem frustratingly difficult to remove the subjective element from the physics of colour. But the mechanism of colour perception became clearer in the early years of the twentieth century, and the Young–Helmholtz theory was eventually confirmed by the identification of red, green and blue sensitive cone cells in the retina. In 1931 the *Commission Internationale de l'Éclairage* (CIE) began to establish colour criteria using extensive tests carried out on human subjects to obtain statistically sound quantitative figures for the perception of colour. In 1947 Robert Hunt and his team published the CIE chromaticity diagram (Figure 6.7a, colour plate). In its more recent (1978) form (Figure 6.7b) it represents equal perceived changes in hue and saturation by equal distances within the area of the curve.

It is sometimes difficult to believe that when you are watching colour TV there really *are* only three colours, red, green and blue, present, especially when you are looking at something that appears brilliant yellow or purple. But go up close, and you will see that the yellow is just red and green dots, and purple is just red and blue. And if you are wondering how you get brown, look in close again. You will see that it is just red with some green, but less bright – in fact, a dark yellow or orange, as I pointed out in Chapter 3.

Measurement systems for colour

In the 1947 CIE chromaticity diagram the axes represented percentages of red (horizontal) and green (vertical). The blue content could be deduced from the missing percentage (e.g., the coordinates 30R, 50G implied 20B). All possible hues and saturations were to be found within the closed curve. The reason the curve does not completely fill the triangle is that the sensitivity ranges of the cone cells of the retina overlap, and a pure spectral green and the deepest blue both provide some stimulation to the ρ (red) cones. In 1978 it was agreed to modify the curve by a transformation which rotated it anticlockwise through about 40° and then shrank it by about 30 per cent horizontally, replacing the axes x and y by u' and v' respectively. The exact relationship is

$$u' = 4x/(-2x + 12y + 3), \qquad v' = 9y/(-2x + 12y + 3)$$

Equal distances on this later diagram represent equal perceived changes in hue and/or saturation (lightness does not appear). Complementary hues are at the ends of straight lines drawn through the neutral-colour point. Saturation decreases with distance away from the boundary of the curve.

Other scales of colour measurement

There have been a number of colour scales; those that were most flawed were eventually dropped, but six or seven are still in use. Of these, the most important are probably the Munsell and DIN systems.

- *The Munsell System*   This was originated by the artist A H Munsell in 1905, and has been continuously refined ever since. It is the system most used by artists. There are five principal hues, red, yellow, green, blue and purple, and these hues are spaced equally in terms of perceived hue shift. There are finer divisions of hue between these, designated by numbers and letters. For each hue there is a chart with colour patches, which have a constant lightness (called *value* in this system) from right to left but decreasing saturation (called *chroma*). The vertical scale of patches runs from lightest at the top to darkest at the bottom. All adjacent patches have the same perceived difference. An example is shown in Figure 6.8 (colour plate). The *Munsell Book* is an atlas of these patches, with a progressively different hue for each page.

- *The DIN System*   (DIN stands for *Deutsches Institut für Normung*) was developed in 1986. It replaces 'lightness' by 'darkness', and is based on a sphere

The Science of Imaging

with white at the upper pole and black at the lower. Hues are represented by successive radial sections, and saturation by angular distance out from the grey-scale axis. As for the Munsell system, the DIN scales are obtainable as an atlas, the *DIN Colour Chart*.

Subtractive colour synthesis

Additive colour synthesis is fine as long as we are dealing with coloured light, as in additive projection systems and colour television. But additive transparencies using réseau or mosaic filtering are always going to need a great deal of projector light; and additive reflection prints are out of the question. If we are to obtain transparencies and prints where the whites really are white, i.e., transmit or reflect *all* the light falling on them, we need to use a radically different technique. The way ahead was pointed out by a Frenchman, du Hauron, who, in a paper published in 1869, *Les Couleurs en Photographie*, formulated the principles of both additive and subtractive colour synthesis.

> Louis DuCos du Hauron was born in 1837. His novel ideas, which he never exploited commercially, were recorded in a vast number of scientific papers. In the words of D A Spencer (see Digging Deeper): 'There is scarcely a principle or process in colour photography that his genius did not foreshadow... Rarely has any inventor shown such imaginative foresight, or received so little encouragement. He died, greatly honoured, in abject poverty in 1920.'

When painters mix colours they name the 'primaries' as red, yellow and blue, whereas colorimetrists (and photographers) think of 'primary hues' as red, green and blue. There is a confusion of terminology here. The photographer's 'blue' is a royal blue, like the painter's ultramarine, whereas the painter's 'blue' is Prussian blue, nearer to the photographer's cyan. The photographer's 'red' is the painter's vermilion, but the painter's 'red' is crimson lake, a purplish red akin to the photographer's magenta. So the painter's 'primaries' are the photographer's 'secondaries' (or *complementaries*), cyan, yellow and magenta.

> I agree that these correspondences aren't exact; but the point is that the painter is mixing pigments, not lights, and is therefore working in subtractive, not additive, colour.

Now, in the additive process we obtained yellow by adding red light to green light. This is the same thing as starting with white light and removing blue. For this reason photographers often call a deep yellow filter 'minus-blue'. In a black-and-white photograph a minus-blue filter darkens a blue sky without affecting red flowers or green grass. Similarly, magenta, which is red light added to blue light, is white light minus green; and cyan (green plus blue) is the same as white minus red. You can test this by putting a cyan filter on top of a red filter, or a magenta on a green, or a yellow on a blue. In each case the result is black – no light. If you take the secondaries and superpose them in pairs, you get the three (photographer's) primaries. Thus magenta and yellow give red (white minus green and blue); yellow and cyan give green (white minus blue and red); and cyan and magenta give blue (white minus red and green). Again, all three give black (white minus red, green and blue). This is illustrated in Figure 6.9 (colour plate).

Colour separation negatives

Until the end of the 1940s the only way to make colour prints was to superpose three images in register, one recording all the red component of the image, one the green and one the blue. As in Maxwell's experiment, the first step was the making of separation negatives, one each through a red, a green and a blue filter respectively. These had to be correctly exposed, and developed to the same contrast – a difficult task, as with most emulsions both speed and contrast vary with wavelength. It was usual, therefore, to include a neutral grey scale card at the edge of the scene for control.

> These cards remain useful for ensuring colour accuracy, and are still obtainable from Kodak materials stockists. They include colour patches for identification of the separation negatives (Figure 6.10, colour plate).

Images in colour

To avoid the labour and delay in changing plates and filters, it was possible to obtain repeating backs, which were triple plateholders in which the tricolour separation filters were installed directly in front of the plates. They were coupled to the shutter release mechanism so that when you operated the shutter each plate was brought into position in turn and exposed, making the three exposures in rapid succession. The differing filter factors were adjusted by the incorporation of neutral density filters. A better, though much more expensive, solution was a specialised 'one-shot' camera, which exposed all three plates simultaneously, using two dichroic beamsplitters (Figure 6.11).

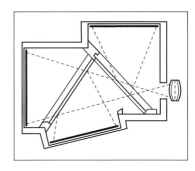

Figure 6.11 Schematic layout of a one-shot colour camera using dichroic beamsplitters. From *The Illustrated History of Colour Photography* (Coote), courtesy of Fountain Press.

Colour prints from separation negatives

In the subtractive process as visualised by du Hauron, red would be produced by removing green and blue from white light. The red content of a print could thus be controlled by a positive cyan image on white paper. (It would have no effect on the green or blue content of the reflected light.) Similarly, the green content would be controlled by a magenta image, and the blue content by a yellow image. When all the images were superposed in register, the whole colour content, red, green and blue, would be controlled, and the image would be in natural colours. Two methods emerged for achieving this: carbro and dye transfer. Both were complicated, messy, unreliable and expensive, but capable of producing outstandingly beautiful (and absolutely permanent) results. Because of the light these techniques throw on modern printing techniques, I have included a simplified description of both below, as well as the closely related Technicolor process.

- *Tricolour carbro*   The name is a conflation of 'carbon' and 'bromide' as it involves pigmented gelatin papers ('carbon tissues') and silver halide prints ('bromide prints'). The process was introduced by Thomas Manly in 1899 (hence the old names). When a silver halide print is bleached in a dichromate solution in contact with a pigmented gelatin paper, the gelatin is hardened to a depth proportional to the image density. When the soft gelatin is washed away with warm water a positive image in pigment remains. You begin by making matched prints from the separation negatives, and make gelatin positives from them using cyan, magenta and yellow pigmented papers. You strip the images on to transparent plastic sheet (originally waxed celluloid) and transfer the gelatins, one at a time, in register on to a final white paper base. The process was never easy, and only one company offered carbro printing as a service, using a variant of the technique named Vivex. Production of carbro materials ceased in the 1950s.

- *Dye transfer*   This evolved over the first years of the twentieth century from the Woodburytype principle, in which a relief image in gelatin was used in the manufacture of a printing plate. Eastman Kodak introduced the dye-transfer (then called 'wash-off relief') process for colour prints in 1946. In this process you made relief positives from the separation negatives using large films coated with very soft emulsion, using a hardening developer and bleach, and washing away the unwanted gelatin with warm water (later, dissolving it chemically). In order to prevent the hardened image also washing away, you exposed the emulsion through the back. These matrices were pin-registered. You then soaked them separately in cyan, magenta and yellow dye baths, the dye being taken up in proportion to the thickness of the gelatin. Finally, you squeegeed the matrices in turn in register on to mordanted paper, leaving a print in full

This is a much-simplified account. To say it was not an easy process is no understatement. Variables such as pH and temperature were critical; material varied from batch to batch; and it took up to ten hours to produce a finished print. The delicate gelatins often tore during registration and at worst slid off the substrate and went down the plughole. But when it *did* work the results seemed well worth all the hours of frustration.

The Science of Imaging

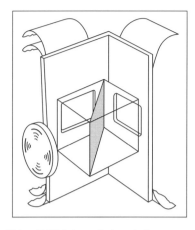

Figure 6.12 Schematic layout of a Technicolor camera (1948). From *The Illustrated History of Colour Photography* (Coote), courtesy of Fountain Press.

colour. Dye transfer printing was easier and more reliable than carbro, and had the advantage that once you had borne the expense of producing the matrices the prints themselves were very cheap to produce. The pin-registration equipment was expensive, and there were dodges for obtaining exact superposition without it; but modern printing processes capable of image modification by dodging and shading superseded the technique. Materials for dye transfer are still available, and a few dedicated workers continue to use it.

- *Technicolor*   From its inception in the 1920s, Technicolor motion picture technique went through a number of stages, the most successful being the exposure of three separate films. A blue/yellow dichroic filter reflected the blue component on to a blue-sensitive film. The remaining light fell on two films running in contact, one green sensitive (orthochromatic) and the other red sensitive. These films were then used as masters for the dye-transfer process on a clear film coated with mordanted gelatin. After 1950 standard colour film was used, the separations being made from this. A grey image was added to strengthen the shadows in the final print. Although the early Technicolor prints are now faded beyond hope, the original black-and-white separations are still in good condition, and it has been possible in many cases to resurrect old colour films in the kind of quality we expect from present-day cinema (Figure 6.12).

Tripack colour transparencies

Kodachrome   The first colour film based on the subtractive process was Kodachrome. It was invented, surprisingly, by two professional musicians, Leopold Mannes and Leopold Godowsky.

Mannes and Godowsky were not particularly outstanding musicians like their fathers (David Mannes was a distinguished violinist and conductor, and Leopold Godowsky Senior was arguably the most technically accomplished pianist who has ever lived), but both had been groomed for a career in music. They were fascinated by photography, and carried out their experiments in kitchens and bathrooms. They were recruited by the far-sighted Kenneth Mees, head of the Eastman Research Institute, and joined the research team, abandoning the musical profession – much to their distinguished parents' dismay. However, both men eventually returned to their musical careers.

The principle, foreshadowed by du Hauron, was to expose three emulsions through red, green and blue filters, and finish with three superposed positive transparencies in cyan, magenta and yellow respectively. In the Kodachrome process, launched in 1935, the three emulsions were coated on the same base, as what became known as a *tripack* emulsion. The outermost emulsion was blue sensitive only, and behind it was a yellow (minus blue) filter layer. The middle emulsion was orthochromatic, i.e., sensitive only to green and blue, and therefore recording only green. The innermost emulsion was sensitive to red and blue only, and thus recorded the red.

The processing of a Kodachrome film was a long and complicated business, at its outset requiring twenty-eight separate steps, all under precise control. Over the years the processing has become somewhat more simplified, though it is still too complicated for home processing. Popular demand, however, has kept the process alive. Processing begins by developing the film to a negative. It is then exposed to red light and developed in a developer containing a cyan dye, which couples only to the developing grains (which are in the red-sensitive emulsion). The film is next exposed to blue light and developed again, this time in a yellow-coupling developer. The remaining emulsion is now chemically fogged and developed in a magenta-coupling developer. The positive colour image is now complete, and the final stage is to remove the silver with a bleach-fix solution, leaving the colour image in pure dyes. As might be expected, only a few specialised units are licensed to process Kodachrome material.

Development is explained in full in Chapter 9.

Ektachrome   During the Second World War it became necessary to secure colour aerial photographs on bombing missions, to establish whether the targets had been found from the images of coloured flares dropped by 'pathfinder' aircraft.

These were to be spliced into the films that were used to photograph the target itself, using flash. These colour inserts would have to be processed under ordinary darkroom conditions.

Eastman Kodak responded with a new idea. Instead of the dyes being in the developers they would be formed in the emulsion itself. Colourless molecules anchored in the emulsion layers would react with the second developer to form the appropriate dyes wherever an image was formed. The first development was to a plain black-and-white negative, after which the film was exposed to white light, and then the remaining silver (the positive image) was developed in the special colour-forming developer. Finally, as in Kodachrome, all the silver was removed, leaving the positive image in dyes. This wartime film, called Kodak Aerial Color Reversal Film, was the forerunner of Ektachrome, which first appeared in 1946. The processing procedure has been steadily improved and the processing times shortened, but the principle remains the same today. The re-exposure step, which was somewhat unreliable, has been replaced by the inclusion of a fogging agent in the second developer.

Other colour transparency films Other film manufacturers followed Kodak as soon as they had found alternatives to Kodak's patents: Agfa, Ilford, Ferrania, Gevaert and Orwo were among the front-runners. To begin with, each make of film had its own processing chemistry, but eventually all the emulsions were made compatible with the Kodak processing routine. The principles and processing are illustrated in Figure 6.13 (colour plate).

The run of photographs included two colour photographs, a black-and-white flash shot (a four inch magnesium-powder 'bomb' of about 1 000 000 candelas, dropped with a parachute) and two black-and-white exposures without flash, one short and one longer, to record traces of fires and enable the photo-interpreters to deduce the precise location and path of the aircraft.

Casual photogaphers call these films 'slide films'. In professional photography the standard term for this type of processing is *reversal processing*, and the result is called a *transparency*, as except for audiovisual shows photographs are seldom used as slides. In other European languages the term is *diapositive* (or a variant).

Prints from transparencies

If you want to make a print from a transparency there are two routes you can adopt. The one most used is a direct positive print. There are two technologies available to you:

- *Direct positive* This can be a reversal material that works in exactly the same way as transparency film, but has a low inherent contrast to compensate for the high contrast of the original. This is also the usual way for making copies of transparencies. An alternative print material uses a *dye-bleach* process. Here, the print material contains latent dyes, which are destroyed progressively by exposure to light. The processing chemistry is completely different from the reversal processing system, and involves only one development. For both types of material you can correct any colour imbalance by using CP (colour print) filters, which are available in all six primary and secondary hues in gradations of 0.25 density.

- *Internegative* The second route is to make a colour 'internegative' from the transparency and a colour print from the internegative. There is a description of the process later in this chapter. Although the process involves two stages, it is theoretically possible to obtain an exact replica of the original, whereas in a 'positive–positive' system the inherent colour errors of the reversal process are doubled. (It is only fair to say, however, that with present-day materials such errors are virtually undetectable.)

The density is measured through a filter of complementary hue. The principle is explained fully in Chapter 10.

Infrared emulsions

In the early part of the twentieth century new dyes were discovered that could sensitise silver halide emulsions to infrared (IR) radiation in the region 700–

The Science of Imaging

Do not confuse optical IR with *far infrared*, which requires special thermal imaging cameras. I discuss this type of imaging in Chapter 20.

950 nm. This region is called the *optical infrared*, as the radiation can be focused by conventional camera lenses.

The important characteristics of IR radiation for photography are:

- It is invisible, so you can use it to shoot photographs in total darkness.
- It penetrates atmospheric haze, human skin and some opaque plastics materials and textiles.
- It is reflected strongly by broad-leaved plants.

The uses of the first characteristic are self-evident. It is fortunate that almost half of the radiant output of a xenon electronic flash is in the optical IR region of the spectrum; so by covering a standard electronic flash head with a 'black' IR filter you can take covert photographs in darkness.

It does not penetrate fog, though, because it is susceptible to refraction, as is visible light. (Nor nylon underwear, despite optimistic press claims.)

Black-and-white IR photographic emulsions have been in use for long-distance aerial photography since the First World War, because of the ability of the radiation to penetrate haze. The presence of atmospheric haze causes low contrast owing to Rayleigh scatter, which is inversely proportional to the fourth power of the wavelength of light. As the wavelength of optical IR radiation is more than twice that of blue light, it penetrates haze about 20 times more effectively.

Perhaps the most important property of IR radiation is that it is absorbed and reflected in a manner very different from visible radiation. Tarmac surfaces and water absorb it completely, and chlorophyll reflects it strongly. A black-and-white landscape taken on IR film shows roads as jet black and deciduous trees as snowy white. Pictorial photographers have made much use of these qualities. Figure 6.14 compares the tonal qualities of a garden in normal panchromatic rendering and in infrared.

Figure 6.14 (a) Visible spectrum and (b) infrared photographs of an ornamental garden. The yellow flowers in the foreground reflect IR radiation strongly and appear very light, as does the shrubby foliage, whereas the painted labels and the rocks absorb IR and appear unnaturally black. (Photographs by Adrian Davies.)

Images in colour

Infrared false colour This tripack material was, like Ektachrome, developed by Eastman Kodak during the Second World War for aerial reconnaissance, though it arrived too late to be useful. It has subsequently more than earned its keep in ecological, agricultural and demographic surveys. It is a false-colour system, with three emulsion layers sensitive respectively to IR, red and green. Blue is excluded by a yellow (minus blue) filter on the camera lens. The IR emulsion is reversal developed to cyan, the red emulsion to magenta, and the green emulsion to yellow. The result is that IR records as red, red records as green and green records as blue. Blue, of course, records as black. If you think of infrared as a fourth primary, along with red, green and blue, you can draw the four primaries as a colour quadrant, with a shift round of one quadrant between the subject hue and the image hue (Figure 6.15).

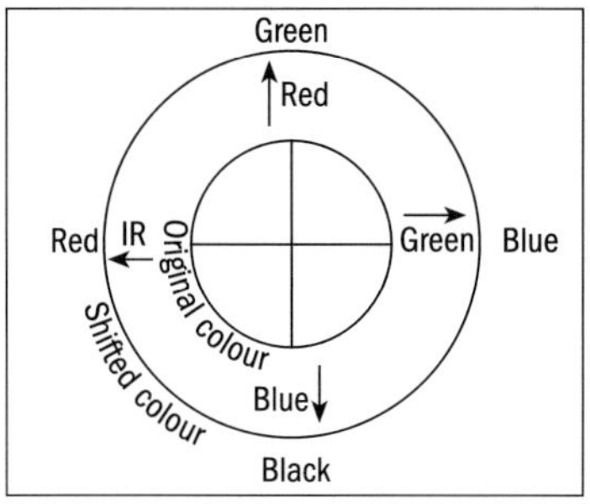

Figure 6.15 Colour shift diagram for infrared false-colour film.

The most important colour shifts in ecological terms are for live vegetation (IR + green → red + blue = magenta), and dead or dying vegetation (yellow → cyan, brown → grey-green). These differences show before anything is visible to the unaided eye. Silt and other contamination show up in sharp contrast to the black water of seas and rivers. Freshly ploughed fields show blue-grey, roads are black and towns a bluish patchwork. An example is shown in Figure 6.16 (colour plate). You can buy infrared colour roll film, and creative photographers occasionally exploit its anomalous characteristics, sometimes using filters other than plain yellow.

Polaroid colour

The Polaroid Corporation was originally formed by Edwin Land to manufacture and market a sheet material that would polarise light passing through it. Land was conducting a campaign to have polarising anti-glare screens fitted to all car headlights throughout the United States, and in order to finance this he invented the Land instant camera, which first reached the public in 1948. The campaign failed, but Polaroid photography went from strength to strength. There are two main markets: the casual photographer, who wants instant party and beach photographs; and the professional, who needs to check the set-up, lighting and composition before the all-important shot.

Polaroid photography has traditionally used a diffusion transfer process. In the black-and-white version, the action of ejecting the film and paper in an opaque package releases a developer solution. This contains chemicals that cause the undeveloped residue of the negative to diffuse into the paper, where a reducing agent blackens it. In some versions you can wash, dry and re-use the negative for making prints.

There are two versions of Polaroid colour. The first type, used chiefly by professional photographers, also uses a diffusion transfer principle. In the tripack emulsion, the cyan, magenta and yellow dyes are rendered immobile in the developing negative; in the unexposed regions the dyes migrate freely into the print paper, resulting in a positive colour image. The process takes about a minute and allows some manipulation, especially while the gelatin is still soft and easy to lift off and distort.

The Polaroid Corporation actually encourages this treatment. In recent years it has begun to make awards to creative Polaroid photographers.

The second type is aimed at casual photographers, and requires a special camera. It works on a broadly similar principle, but there is no separation of negative and positive after development. Instead, the dyes diffuse through an opaque white layer of titanium oxide, which becomes the substrate for the developing picture. The

The Science of Imaging

mysterious appearance of the image in broad daylight is an intriguing phenomenon (and not only to lay persons), and no doubt has helped sales. One consequence of there being no stripping procedure is that the optical image needs to be reversed right to left, and this is achieved in the camera by the use of an inverting mirror in the optical path.

Colour negative–positive systems

Most amateur photographers prefer to work with colour negatives rather than transparencies. The principle is simple: the colour film is developed in a single stage to a negative, using a colour-forming developer. The silver and unchanged silver halide are then both removed, leaving the negative image in dye. The light parts of the subject record dark and vice versa, just as in a black-and-white negative. But the red-recording layer produces its negative image in cyan (minus-red), the green-recording layer in magenta (minus-green) and the blue-recording layer in yellow (minus-blue). So the colours are negative too. As you might expect, saturation is unaffected: saturated colours remain saturated. The print material is coated with similar emulsions, so that after exposure to the negative and a similar development, the tones and hues are again reversed, and the final print has the correct scales of hue, lightness and saturation.

Colour print materials are usually balanced to give a print in approximately correct colour rendition when used with a normal enlarger filament lamp, but may need some correction using CP filters. Unlike positive–positive printing, the filters need to be the *same* hue as the error (e.g., a magenta cast in the test print demands a magenta CP filter). The filtration has a much stronger effect than with positive–positive prints, too, because of the much higher inherent contrast of negative–positive print paper.

The principles of negative–positive colour are shown in Figure 6.17 (colour plate).

The clever thing about the negative–positive process is that instead of doubling the errors inherent in colour recording as in the positive–positive process, they can be almost totally eliminated. As the colour negative is an intermediate and not for display, its physical appearance is immaterial. Its tones and colours can be, as it were, pre-distorted, to compensate for subsequent errors in the print material. You will no doubt have noticed that the backgrounds of colour negatives have an orange-yellow cast, and that their contrast is low. That is part of the pre-compensation mechanism.

The figures given here are not absolutely accurate; they vary from one type of film to another. There is no more than 5 per cent error in the yellow, so there is no real need for compensation here. Modern transparency materials incorporate subtle masking effects that depend on the inhibition of one layer of emulsion by the image as it forms in an adjacent emulsion.

Colour masking

The colours of dyes are dependent on complicated atomic vibrations within the dye molecule, and there are limits to what we can do to modify these. In our present state of knowledge we cannot produce a perfect magenta that absorbs all green and transmits all red and all blue light. The best magenta we can do still absorbs around 20 per cent of blue light, and thus behaves as if it had 20 per cent yellow dye mixed into it. Thus magenta shows a hue shift towards red. Cyan is an even poorer performer: it absorbs about 20 per cent of green and 15 per cent of blue. It can thus never demonstrate 100 per cent saturation, and behaves as if it had 15 per cent red, plus 5 per cent magenta, mixed into it. Fortunately, we hardly ever need to record 100 per cent magenta or cyan. So in principle all we need to do to get the colours correct is to remove some yellow dye wherever magenta occurs, and to remove some red wherever cyan occurs. This is achieved by incorporating 20 per cent yellow dye in the green-record emulsion and 20 per cent red dye in the red-record emulsion, both dyes being bleached wherever the image is formed in that emulsion. The result is that the magentas and cyans are adequately corrected, and the whole negative is effectively overlaid by a uniform layer of 20 per cent yellow and 20 per cent red. This overall orange cast is allowed for in the manufacture of the print paper by extra sensitivity to blue and green. By careful balancing of this integral masking, as it is called, the similar errors in the dyes of the print material can be allowed for in advance. Masking is dealt with more fully in Chapter 12.

Cross-processing

It is possible to process transparency material to a colour negative using standard colour negative processing techniques. This is called *cross-processing*, and a print made from such a negative shows high contrast and colour saturation, usually with some anomalous shifts in hue because of the lack of integral colour masking. Commercial and fine art photography sometimes makes use of the technique, and it can be useful in producing diagrams for projection, using the negative itself as a slide. However, you can usually obtain the result you want more easily by digital modification of the image (Chapter 12).

Although the photographic darkroom still holds its attractions, and in many circumstances has no substitute, many photographers are moving towards the use of scanning and printing systems that bypass the whole enlarging and printing stage. I shall be tackling this subject in Chapter 13.

Digging deeper

In this chapter I have cut a good many corners. In particular I have had to leave out any discussion of dye-coupler chemistry. For a comprehensive account of this and colour processes in general, *The Theory of the Photographic Process* (4th edition, ed. T James, Macmillan, 1977) is the standard work (earlier editions were by Kenneth Mees, then Mees and James). You can keep up to date with *The Imaging Science Journal*, published bi-monthly by the Royal Photographic Society. For a simple expansion of the material in this chapter, *The Manual of Photography* (9th edition, ed. R E Jacobson, Focal Press, 2000) is helpful, but the previous 8th edition has more of the chemistry background. If you are interested in the way colour photography developed, don't miss the late Jack Coote's *The History of Colour Photography* (Fountain Press, 1997), which is a fine exposition of the principles of colour photography from a historical viewpoint. A classic to hunt down in second-hand technical bookshops is D A Spencer's *Colour Photography in Practice* (3rd edition, Pitman, 1937), which is a gem. Spencer was Kodak's senior technologist and the managing director of Vivex, the only company to make carbro prints commercially. After his death his book was regularly updated by Focal Press's authors, but the original, with its beautiful illustrations printed separately, and individually tipped in, is a book to treasure. Spencer discusses both the carbro and dye-transfer processes in detail, as well as more recent techniques.

My marginal note on du Hauron is a quotation from the dedication of this book to du Hauron's memory.

If you were in any doubt about the validity of colorimetry as a scientific discipline, have a look at Robert Hunt's *The Reproduction of Colour* (5th edition, Focal Press, 1998) and his companion volume *The Measurement of Colour* (3rd edition, Focal Press, 1998), where there is enough mathematics to satisfy the most masochistic seeker after colorimetric enlightenment. (Robert Hunt was the begetter of the CIE chromaticity diagram.)

I have not said much about subjective colour effects. This is a large and controversial subject. Among those books I can safely recommend are *Eye and Brain*, by Richard Gregory (5th edition, OUP, 1998), *Visual Perception*, by Nicholas Wade and Michael Swanson (Routledge, 1991) and *Perception*, by Ian Rock (Scientific American Library, 1984).

The development of reflection holography (Chapter 17) has sparked off a renewed interest in Lippmann colour photography, with which it has some common elements. Some of the best recent research papers may be found in the *Proceedings of the 5th Lake Forest Symposium on Display Holography* (SPIE Press, Vol. 3358, 1997). In the UK, Dr Hans Bjelkhagen is currently leading a research team at De Montfort University, Leicester.

Infrared photography has become something of a cult among certain groups of pictorial photographers. Much of the available literature consists almost entirely of photographs, nearly all of them in monochrome. One book I can fully recommend is *The Art of Infrared Photography*, by Joseph Paduano (4th edition, Amherst Media, 1998). This includes a full account of infrared colour, with examples of the use of different filters. If you feel like working in this curious medium, this is the book to get.

Chapter 7 Still cameras

Early cameras

Although cameras come in a large variety of shapes and sizes, they all consist basically of a lightproof box with a lens at one end, a light-sensitive recording material at the other, and a shutter to control the exposures. A further necessity is a sighting device or viewfinder. The earliest cameras of Louis Daguerre in France and Henry Talbot in England were indeed little more than wooden boxes with a lens cap for a shutter. Talbot's wife called his cameras 'Henry's mousetraps'.

Talbot's full name was William Henry Fox Talbot. 'Fox' was a forename, not a surname, and he never used it himself.

Shutters

The earliest photographic materials were not very sensitive to light, and often demanded exposure times of many minutes; but with the advent in 1841 of Josef Petzval's wide aperture lens, with some fifteen times the transmittance of other current lenses, and of Frederick Scott Archer's wet collodion process ten years later, exposure durations of a second or less became common, and it became necessary to add shutters to cameras. At first these were crude roller blinds fitted to the front or rear of the lens. George Eastman's Kodak camera, introduced in 1888, had a rotating sector shutter driven by a spring. Rotating sector shutters are still used in modern cine cameras.

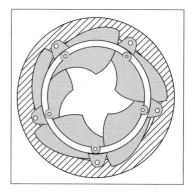

Figure 7.1 Action of a leaf shutter. The outer ring is fixed. Rotation of the inner ring causes the blades to open or close.

Today there are two different types of shutter used in cameras for still photography. The first is the *leaf shutter*. This is usually situated within the lens close to the iris diaphragm (the 'intralens position'), though in studio cameras it is often positioned behind the lens. It consists of a number of metal leaves (usually two to five) mounted on pivots on a ring which, when rotated, flicks the leaves rapidly outwards (Figure 7.1) and returns them at the end of the exposure time.

On non-automatic cameras the series of exposure settings is a geometrical progression of halving: 1 s, then $\frac{1}{2}$, $\frac{1}{4}$, $\frac{1}{8}$, $\frac{1}{15}$, $\frac{1}{30}$ s, down to $\frac{1}{500}$ s (sometimes $\frac{1}{800}$ too).

The figures are rounded off, by international agreement, though it is hard to see why, as '256' takes up no more room on a dial then '250'. It would have been more logical to have agreed to rate exposure durations in milliseconds rather then fractions of a second.

Blind shutters are situated immediately in front of the film, and for that reason are known as *focal-plane* (FP) shutters (Figure 7.2). In modern cameras FP shutters have two separate blinds (or curtains). The exposure duration is controlled by the length of time between the releasing of the leading or front blind and the releasing of the trailing or rear blind. In non-automatic cameras the series of exposure durations also follows a progression of halves, but the minimum figure can be as low as $\frac{1}{8000}$ s. To achieve the shortest timings the rear blind is released before the leading blind has completed its traverse, so that the two shutter blinds form a moving slit.

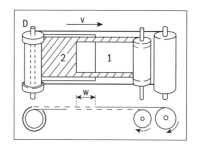

Figure 7.2 Action of a focal-plane shutter. The exposure duration is controlled by the interval between the release of the leading blind (1) and the trailing blind (2). For the shortest exposure durations the two blinds form a slit, width *w* in the diagram.

Shutter efficiency With both leaf and FP shutters it takes a finite length of time to uncover the whole lens aperture. The total time for which the lens is partly or wholly uncovered is greater then the *effective* exposure duration, and this gives rise to the concept of *shutter efficiency*. The effective exposure duration is approximately equal to the time from the shutter '25 per cent open' point to the '75 per cent closed' point, and dividing this by the time taken from the beginning of opening to the end of closing gives the approximate efficiency figure. More accurately, if you make a plot of the illuminance at the focal plane against time,

The Science of Imaging

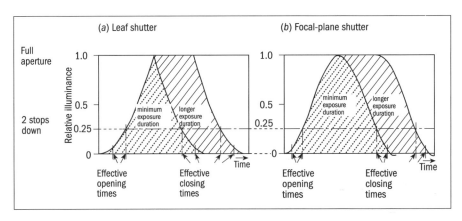

Figure 7.3 Shutter efficiency: At full aperture the shutter spends its entire time opening and closing, and efficiency is a minimum: 45 per cent for a leaf shutter and 50 per cent for a focal-plane shutter. At longer durations and/or smaller apertures the efficiency is increased; at smaller apertures the effective exposure duration is also increased. Note: these diagrams ignore shutter acceleration, which if present distorts the leaf shutter curve somewhat, but has a negligible effect on shutter efficiency.

the shutter efficiency is the area under this curve divided by the area of the rectangle enclosing it. This is illustrated in Figure 7.3.

From the plots in Figure 7.3(a) and (b) you can deduce two facts:

- Shutter efficiency is higher at small relative apertures.
- Shutter efficiency is lower at shorter exposure durations.

At the shortest durations the shutter spends its entire time opening and closing. The efficiency for the FP shutter is a little over 50 per cent and for the leaf shutter about 45 per cent, at full aperture. You may have to take this into account when trying to 'freeze' the subject in an action shot; the shutter may be at least partially open for up to twice the time set. FP shutters are generally more efficient than leaf shutters, as the efficiency depends chiefly on how close the shutter blinds are to the film plane, whereas the efficiency of a leaf shutter varies inversely with the diameter of the iris diaphragm.

Timing mechanisms Mechanical leaf shutters achieve their longer exposure times by a ratchet delaying mechanism, and shorter times by varying the spring tension on the movable ring that operates the leaves. Both of these mechanisms wear with time, and older cameras need checking occasionally. FP shutters also use a delaying mechanism for longer exposure times, but obtain the shorter times by varying the slit width. To check a timing mechanism for accuracy, you simply shoot off a series of exposures at a plain white wall, varying the duration but keeping the exposure nominally the same, e.g., $\frac{1}{500}$ at $f/2.8$, $\frac{1}{250}$ at $f/4$, $\frac{1}{125}$ at $f/5.6$ and so on. Place the developed film on a sheet of white paper and check the density (darkness) of the negatives. They should all be equally dark. If they are not, you will be able to see which way the exposure error goes, and compensate as necessary by adjusting the f/no. (You can usually take the $\frac{1}{60}$ setting as approximately correct.)

Remember, the f/no series goes in steps of $\sqrt{2}$, each step representing a halving of the light transmission of the lens.

Focal-plane shutter distortion There is another important property of FP shutters, namely that for short exposure durations the slit is narrower than the frame, and the time it takes to traverse the frame is much longer than the time it takes to expose a single point on the film. With a fast-moving object the image may have moved some distance in the frame between the start and the finish of the exposure.

For example, if you have a shutter blind that moves downwards (most of them do), it will photograph the bottom of the scene before the top (remember, the image is upside down), and if you are photographing, say, a racing car, it will appear to be leaning forward. With a horizontally moving blind the car will be compressed or elongated, depending on whether the shutter moves in the opposite direction to the (inverted) image of the car, or in the same direction (Figure 7.4).

Figure 7.4 (a) When a focal-plane shutter aperture moves in the same direction as a moving image, the result is elongated; (b) when it moves in the opposite direction, the result is compressed. (Photographs by Jon Tarrant.)

Other types of shutter There are other shutter types, such as louvre and contrarotating sector shutters, but these are used only in specialised cameras. There are also shutters operating on electro-optical principles, and these do not have any moving parts. The most important of these is the Pockels cell, a device consisting of an optically active crystal mounted between crossed polarisers. When an electric field is switched on the crystal rotates the polarisation through 90°. Such shutters are capable of controlling exposure durations measured in nanoseconds.

Types of camera

Simple cameras The simple camera has a long history. The first commercial model was marketed by George Eastman in 1888, and this led to the Brownie series, which was the archetypal model for the genre, with its roll film, sector shutter, and single $f/14$ meniscus lens set to the hyperfocal distance. Some later models had a range of fixed stops and a crude focusing arrangement. Folding models appeared around the turn of the century. So many variants of the simple camera have been marketed over the past hundred years that a historical account of them would fill a thick book.

Today the simple camera is 35 mm or APS format (16.7 × 30.2 mm on 24 mm film), the various other formats having fallen by the wayside. It is still a box with a fixed-focus lens ('focus-free' in marketing-speak), though it may have some extras such as a built-in flash.

Any volunteers?

Don't scorn the simple camera. Within its limits it can provide results as good as professional cameras costing thousands of pounds. The first photograph of mine to be hung in an international exhibition was taken with one. I always take a waterproof single-use camera on beach trips when on holiday (you can use them under water, and if you leave them around they are unlikely to be nicked).

The Science of Imaging

Single-use cameras These cost little more than the price of the film they contain. Manufacturers reclaim and reload them for resale, a throwback to George Eastman's original camera.

Compact cameras These were developed to fill a marketing niche for people who wanted a comparatively simple camera that was more versatile than the basic box. The present-day compact camera has automatic exposure control, automatic focus and a flash that operates when the light level is low. Some have zoom lenses. The most sophisticated compacts rival professional cameras, and professionals often have them as a second camera to use in the field. They are invariably 35 mm or APS format.

Rangefinder cameras Since the widespread adoption of automatic focusing systems, the amateur rangefinder camera has all but disappeared, but there are a number of professional cameras of this type. They stem from the old Leica and Contax designs and their offshoots, in which an optical rangefinder is coupled to the focusing mechanism. (I discuss rangefinder mechanisms in more detail later in this chapter.) Rangefinder cameras are particularly useful in poor lighting conditions where finding focus on a focusing screen is difficult; and they are a boon for a photographer cursed with poor eyesight, or in circumstances where automatic focus control is inappropriate. Medium-format rangefinder cameras using 120-size film are versatile and light compared with their single-lens reflex equivalents.

The forerunner of the medium-format rangefinder camera was built by Fairchild for the US Forces. It took 70 mm film in reloadable cassettes, and because of its large size it became known as the 'Texas Leica'.

Twin-lens reflex (TLR) cameras These have a viewfinder that is essentially a second camera, with a ground glass viewing and focusing screen and a lens coupled to the taking lens. The 'reflex' refers to the fact that the image in the finder is formed via a 45° front-surface mirror so that the viewing screen is horizontal. Most finders incorporate a Fresnel lens to improve the image brightness at the corners, and a magnifier to aid focusing. Because of the mechanical and optical excellence of the better models they were usually the choice of photojournalists and sports photographers during the 1940s and 1950s, but they had several serious drawbacks: the viewfinder image was laterally reversed; there were parallax problems in close-ups; they had to be operated from waist level; and (with one notable exception) the lenses were not interchangeable. They are now little used except by die-hards.

Single-lens reflex (SLR) cameras The earliest SLR cameras appeared in the 1890s, and were used chiefly for portraiture. The reflex mirror for the viewing and focusing screen was situated in the camera body, and the focusing lens was the actual camera lens, which was interchangeable. Before taking the photograph it was necessary to close down the lens aperture and raise the mirror, which blacked out the viewfinder, making the use of a tripod essential. (The TLR camera was designed to overcome these difficulties.) The first SLR camera to feature instant mirror return appeared soon after the end of the Second World War, and this innovation was quickly followed by the pentaprism viewfinder, which righted the image at eye level, and by a system that allowed focusing at full aperture, the iris diaphragm closing down only for the exposure itself.

SLR cameras have almost completely taken over from other cameras for photoreportage, action photography and most social events. The more advanced models contain computer programs that can be switched in to provide for all types of photographic situation, and complex systems for computing correct exposure and focusing. Nearly all such cameras are 35 mm, though at present there are a few medium-format SLR cameras, and APS format versions are gaining ground in the market for advanced amateurs.

Field and studio cameras   These are of two main types: *baseboard* and *monorail*. The baseboard type, usually called a *technical camera*, has a base plate with rails on which the lens assembly slides for focusing; the back rotates for portrait or landscape formats, and may have adjustments for tilt and swing (rotation about horizontal and vertical axes). The lens panel is connected to the back by bellows, and has sideways and vertical movements as well as tilt and swing. You can use a technical camera in the open for landscape and architectural work on a tripod, and can also operate it hand held. Fine focusing is by a ground glass viewing screen, but for hand held work most technical cameras have a coupled rangefinder adjustable for several focal length lenses. The cameras can be folded up for transport. Technical cameras, as you have probably guessed, are the descendants of the old wooden stand camera with a black cloth and of the big press camera with a baseboard and a huge flashgun. They are usually 4×5 in, though there are smaller formats. They can be used with sheet film, roll film, Polaroid material, plates and a digital back.

The monorail camera is not so much a single camera as a system. It is based on the optical bench principle, and all the components are interchangeable. The lens panel is mounted on lockable gimbals, usually (though unfortunately not invariably) pivoted so that the rear nodal point remains stationary and the image does not move when you operate the swing and tilt movements. Similarly, the swing and tilt axes of the camera back pass through the optical centre of the format. The bellows set is also interchangeable, with long bellows for close-up work and bag bellows for wide-angle photography with a short-focus lens. Tilt and swing angles are calibrated in degrees, and shift movements in millimetres. The format is usually either 4×5 in or 8×10 in.

Self-processing cameras   Polaroid and other *in situ* processing cameras differ from other cameras chiefly in their film transport system, which releases the processing chemicals and either delivers the film ready to separate from the print after an appropriate interval, or (in more recent models) simply delivers the developing print. The film-stripping cameras have normal camera optics, but the newer models have a folded imaging beam path, including a mirror, so that the image emerges the right way round.

Specialised cameras   There are scores of different types of specialised cameras, such as the fundus camera for photographing the inside of the eyeball, and the periphery camera for making a continuous record of the outside of a pipe (or any other 360° circumference). The more important specialised cameras turn up in the appropriate chapters of this book. Some of the others you may come across in a working environment are:

- *Underwater cameras*   These are sealed in watertight containers, or are themselves watertight. They have special controls that are easy to operate underwater. As a rule they have fixed lenses of short focal length, to compensate for the magnification due to the refractive index of water.

- *Aerial cameras*   Aerial survey cameras are carried on fixed anti-vibration mounts, and are operated by an *intervalometer*, which computes the time interval between exposures for correct overlap. Survey lenses are wide-angle (90°) and are designed to have less than 0.1 per cent distortion; each film register glass is individually calibrated. Cameras for aerial reconnaissance are no longer the long-focus monsters of a few decades ago (satellites have taken over their job), but small cameras using 70 mm film, with rates of exposure up to 12 pictures per second. Hand held aerial cameras are rugged box cameras with fixed focus lenses and handgrip operation.

The Science of Imaging

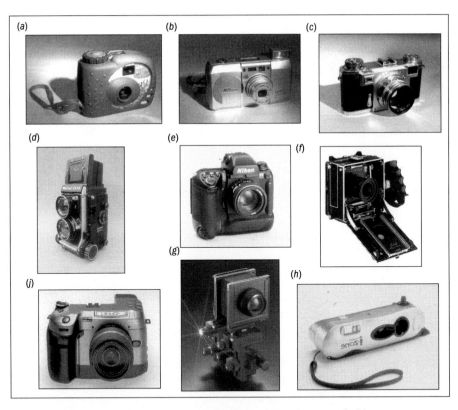

Figure 7.5 The main types of camera: (a) simple single-use camera (waterproof); (b) compact camera; (c) classic rangefinder camera; (d) twin-lens reflex camera; (e) single-lens reflex camera; (f) technical camera; (g) monorail camera; (h) instant colour print camera; (j) professional digital camera.

- *Wide-angle and panoramic cameras* These usually have a non-interchangeable lens with a rectilinear field of around 110°. The rotating type of panoramic camera has a normal-focus lens that rotates about its rear nodal point, coupled to a moving slit field aperture. Alternatively, the whole camera may rotate. In this case the slit is fixed and the film moves in synchronism with the image. There is more about panoramic cameras in Chapter 19.

- *Rostrum cameras* The rostrum camera has joined the scanner in replacing the enormous processing cameras that used to grace our press darkrooms. A rostrum camera is a sophisticated and versatile copying camera that can be used for copying drawings and transparencies and for animations.

- *High-speed cameras* These are important research tools, and are dealt with in Chapter 8.

Figure 7.5 shows examples of the major types of camera.

[Note: Digital cameras, once a short final paragraph in books of this kind, are now such an important item that they have a whole chapter to themselves (Chapter 13).]

Viewfinders

Focusing screens Studio cameras are not equipped with viewfinders, because you compose and focus the image on a ground glass screen. Such screens have always presented a problem: if they are coarsely textured the image is bright overall, but focusing is imprecise; if they are finely textured (acid etched), focusing can be more precise, but the image becomes very dim towards the corners of the frame. The situation is improved by positioning a Fresnel lens in contact with the ground glass to redirect the light into the viewer's eye. The lands of the Fresnel lens need to be closely packed so as not to be obtrusive, and the central area is usually left plain. New screens are being developed using microlithographic techniques that redirect the light by diffraction.

Frame and optical finders   Some technical cameras, particularly the larger ones formerly used as press cameras, have a wire frame finder; such finders are easy to use in bad light, and usually move with the lens panel when you use the rising or cross front, so as to keep the picture correctly framed. There are several patterns of optical finder, mostly based on the optics of a reversed Galilean telescope. Some, designed for cameras with interchangeable lenses, have variable framing to suit the different focal lengths. The *Albada* finder (Figure 7.6) has a rear component surrounded by a partially silvered rectangle that is projected into the view to indicate the edges of the field. This is useful in sports photography where the subject matter may be continually moving in and out of the frame. More recent versions have the framing rectangle generated within the camera and projected into the viewfinder field via a beam combiner.

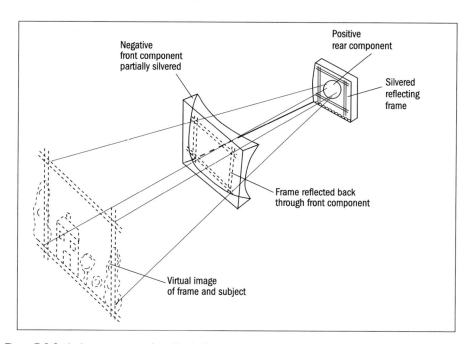

Figure 7.6 Optical arrangement of an Albada finder.

Twin-lens reflex finder   The twin-lens reflex (TLR) camera has a viewfinder that is optically a replica of the main camera, except that it has a 45° front-surface mirror to erect the image on a horizontal focusing screen. The two lenses are mounted on a single panel, so that focusing the finder also focuses the taking lens. One advantage over optical finders is that the image is full size; but there are several disadvantages, as we have seen. The fact that the viewfinder lens is displaced from the taking lens by up to 7 cm means that the actual frame is much lower than the viewfinder frame, and in close-up the top of the composition may be cut off.

In some TLRs a moving mask on the finder screen compensates for this, and supplementary close-up lenses come with an optical wedge for the finder lens to further compensate for parallax; but the viewpoints of finder and taking lens still differ, and there may even be an obstruction your finder does not see, for example if you are photographing through a grille.

Single-lens reflex finder The basic single-lens reflex (SLR) design goes back to the end of the nineteenth century. Early SLR cameras had a mirror that needed to be raised and lowered manually. The viewfinder image was reversed right to left as in TLR viewfinders, and disappeared when the mirror was raised. Some smaller SLR cameras with additional sports finders were made during the 1930s, but the breakthrough only came in the 1950s with the appearance of the instant-return mirror and the pentaprism finder. A pentaprism is a kind of roof prism with an extra reflecting surface (Figure 7.7). Early SLR cameras were difficult to focus at small lens apertures, and this led to the development of full-aperture focusing, with the iris diaphragm remaining wide open except during the actual exposure.

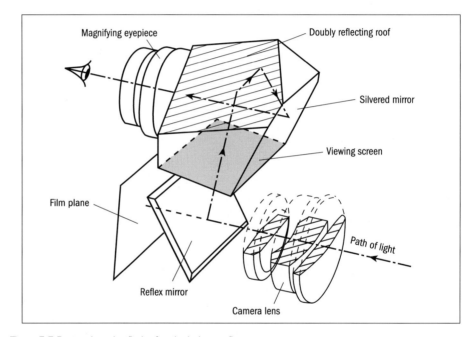

Figure 7.7 Pentaprism viewfinder for single-lens reflex camera.

Rangefinders and focus finders

The coupled rangefinder appeared at the same time as the 35 mm camera. The optical principle was an old one, used in the First World War by artillery regiments and naval vessels. Two 45° mirrors a metre or more apart produced images in a telescope sight via a beam combiner (Figure 7.8). One of the mirrors was rotated until the two images coincided, and the range was read from a scale attached to it.

The rangefinder in a camera is basically a smaller version of this. In the rangefinders of technical cameras a system of levers connects the mirror movement to the focusing system. In 35 mm and medium-format rangefinder cameras both mirrors are usually fixed, with a combination of moving wedges in one of the two optical paths; this raises the mechanical advantage and reduces possible backlash (Figure 7.9).

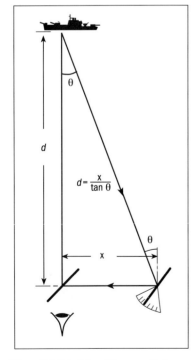

Figure 7.8 Principle of the rangefinder.

Still cameras

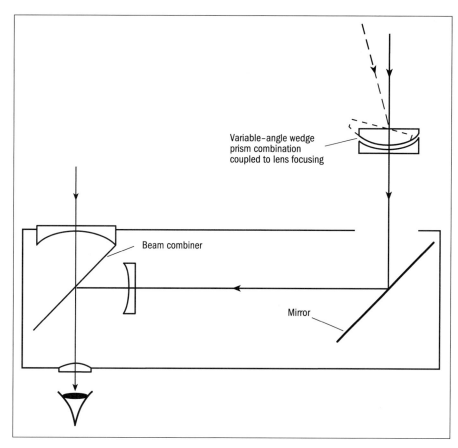

Figure 7.9 Rangefinder as used in the post-war Contax camera.

TLR and SLR cameras have a ground glass focusing finder and, as the image is small, it can be difficult to find an exact focus with the unaided eye. In the studio, of course, you can use a loupe, and most TLR cameras have a foldaway loupe hinged to the hood. Most SLR cameras now have an optical focus finder at the centre of the field. It consists of two small wedges, called (rather misleadingly) a 'biprism', one above the other, with equal and opposite slopes. The crossing point is in the plane of the focusing screen (Figure 7.10). Light entering the biprism is deflected in opposite directions to give a horizontally split image. The two halves coincide when the image is in the plane of the focusing screen.

Instead of a single biprism an array of 'microprisms' can be used. This gives a shimmering mosaic pattern that disappears when the image is in correct focus. Both systems depend on the use of a fairly large lens aperture. At apertures below about $f/11$, as well as with very long-focus lenses, you will simply see a black disc or half-disc at the centre of the field unless your eye is very carefully aligned.

Automatic focus control

There are three techniques for automatic focus control in use at present: ranging, contrast measurement and phase detection.

Ranging   Little used now, the sonar autofocusing principle was the earliest successful method of ranging. In this system the lens starts off set to its closest

The Science of Imaging

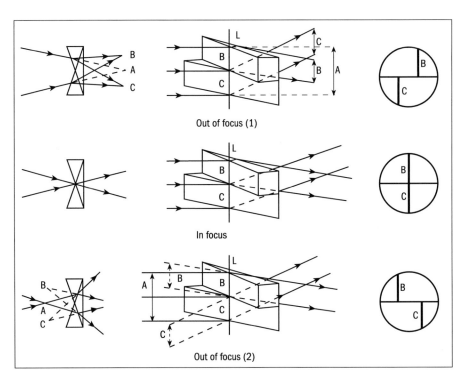

Figure 7.10 Principle of the SLR biprism focus finder. (From *Manual of Photography*, ed. Ralph E Jacobson. Reprinted by permission of Butterworth-Heinemann.)

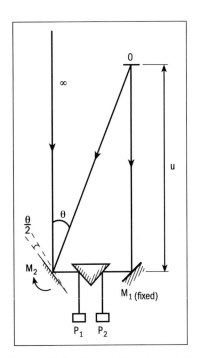

Figure 7.11 Ranging autofocus (scanning system). The swinging mirror M_2 scans the scene until the outputs of the photoreceptors P_1 and P_2 are correlated. From the mirror deflection $\theta/2$ the distance of the object O can now be computed and set.

focusing distance. A piezoelectric transducer emits a pulse of ultrasound, and at the same time the lens begins to move towards its infinity position; the returning echo is used to stop it. As the elapsed time is proportional to the subject distance, the rate of movement of the lens has to be suitably matched to the speed of sound. This system will operate in the dark, of course, but it is thrown by the presence of a glass window. Other ranging systems include a scanning system, where the output of a swinging mirror synchronised with the focusing movement is correlated with the output of a fixed mirror (Figure 7.11); when the correlation is a maximum the subject is in focus. This principle is similar to that of the optical rangefinder, and works best in bright light.

Image contrast When you focus an image on a focusing screen your main clue to correct focus is not so much the fine detail as the contrast of edges, which is a maximum when the image is in focus. Most autofocus compact cameras use this principle. A row of charge-coupled devices (CCDs) is set at the equivalent focus (that is, at an optical path distance equal to the focal length, using either a beamsplitter or a separate optical system) and its intensity profile analysed by a signal-processing unit for sharp/unsharp characteristics. A variant is to use two or even three sets of CCDs at different focusing distances (Figure 7.12). The signal-processing unit then selects from the outputs (equal output from two rows, sharpest output from three). This system is capable of very rapid response. The multiple-set CCD systems show unambiguously the direction in which the focusing is to move. The image contrast system does not work very effectively in poor light.

Phase detection focusing This is at present the preferred system for the better quality SLR cameras. The CCD arrays are mounted behind the equivalent focal plane, so that the pencils of rays have come to a focus and are diverging again.

Still cameras

They are re-focused on to the CCD array by small lenses (Figure 7.13); the distance apart of the images depends on the amount of divergence of the rays, being closer together for too far a focus and farther apart for too near a focus. This works very rapidly, and in poor light can be boosted by a dedicated flash head programmed to give a short burst of weak red light.

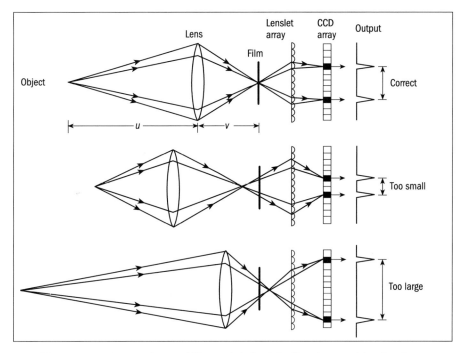

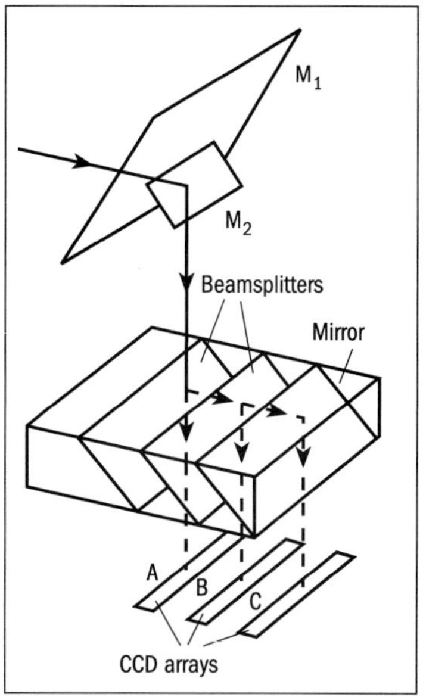

Figure 7.12 Image contrast autofocus (multi-array system). A small mirror M_2 on the back of the main reflex mirror M_1 directs part of the image-forming light on to three arrays of CCDs, the middle one of which (B) is in the equivalent film plane. The lens is driven until B shows the sharpest image edge characteristics.

Figure 7.13 Phase detection autofocus. ACCD array is behind the film plane, and the diverging rays are focused on it by a row of lenslets. The separation of the images depends on the divergence of the rays, being too large for too near a focus and too small for too far a focus. (From *Manual of Photography*, ed. Ralph E Jacobson. Reprinted by permission of Butterworth-Heinemann.)

Automatic exposure control

Exposure meters measure illuminance: light from the subject falls on a light-sensitive device, which may be photovoltaic (selenium), photoconductive (cadmium sulphide), or a photodiode or phototransistor (silicon). These last react many times faster than the first two types. They can even measure the luminous output of an electronic flash. With a separate meter, you can take a measurement at the subject with the meter pointed at the camera: this is called 'incident light measurement'. You will then get the same reading regardless of the brightness of the subject; and this is the way it should be, certainly for transparencies, where you want a black object to look black and a white object to look white. A meter that is built into the camera can only point forward at the subject matter, and it will give very different readings for a black object and a white one. If you believe the meter, you will finish with two images the same shade of grey. This is the reason incident light meters are de rigueur in professional studios. Out of doors, built-in meters come into their own. They are calibrated on the basis that the average subject has an overall reflectance of 18 per cent, which is, visually, a mid-grey. (You will see why this particular figure is chosen, in Chapter 11.) However, if your subject is a little unusual, such as a glade with dark shadows or a snow scene where almost

The Science of Imaging

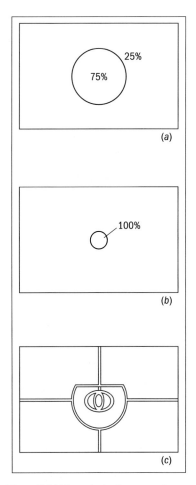

Figure 7.14 Through-the-lens metering options: (a) centre weighted; (b) spot metering; (c) matrix metering.

everything is white, you have a problem if you are working with transparencies. With a separate meter you can always go up close and measure a part of the subject that *is* a mid-grey (or take a reading on a standard 18 per cent grey card); but when, as is usual with modern amateur cameras, the meter actually controls the exposure, your glade will be overexposed and your snow scene will be grey. The better amateur cameras cope with this, after a fashion, by weighting the centre of the picture more heavily than the outer part (Figure 7.14a), though you still need to make allowances for non-standard subject matter. More sophisticated systems can be switched to measure only a very small spot which you choose yourself, holding the reading while you make the exposure (Figure 7.14b). The most advanced systems take simultaneous readings over five or so more areas of the field and ignore anomalous areas according to an inbuilt program that you can set according to the nature of the subject matter (Figure 7.14c).

How does the metering system control the exposure? With the cruder semi-automatic systems you have to match the meter needle, turning the iris diaphragm or shutter-setting ring until the needle coincides with a marker. With fully automatic operation, the signal from the meter is amplified and fed to servos that control the aperture setting and exposure duration. With through-the-lens metering in SLR cameras, a part of the area of the reflex mirror is made sufficiently transparent to allow a small amount of light through, and this is reflected on to the light-sensitive device(s). The simplest form of automatic exposure control is a program that starts off at the smallest aperture and shortest exposure duration for the brightest scene, then for dimmer scenes progressively opens to full aperture, thereafter progressively lengthening the exposure duration. Another program shares the exposure increase between the aperture and the exposure duration. Professional and semi-professional cameras offer a choice of *aperture priority*, where you set the aperture, and the meter controls the exposure duration, or *shutter priority*, where you set the shutter, and the meter looks after the aperture. These cameras also have a control that lets you uprate or downrate an exposure (bracketing) for unusual scenes.

Top-grade amateur and semi-professional cameras also contain specific exposing programs for such projects as sports shots, close-ups, silhouettes and so on.

Flash synchronisation

Until fairly recently, you could still come across cameras that employed expendable flashbulbs to illuminate dimly lit scenes or to fill in dark shadows. The flash was produced by the electrical ignition of a bundle of fine ribbon of zirconium or magnesium metal in oxygen. As the full power of the flash developed only after a delay of 5 to 40 milliseconds, depending on the type of bulb, it was necessary to delay the full opening of the shutter by a matching time interval, or at least to ensure the shutter was held open until the flash had passed its peak power. Flashbulbs are seldom seen today. Even the giant flash bombs used by the RAF in night reconnaissance during the Second World War have given way to electronic flash. Modern electronic flash heads have a flash duration of no more than one or two milliseconds, often much less, and synchronisation now is simply a matter of ensuring the shutter is wide open when the flash fires (Figure 7.15). The flashtube contains a mixture of gases at low pressure, mainly xenon, which, when ionised by a brief high-voltage pulse (the 'trigger'), conduct electricity freely, allowing a large current to flow briefly and generate a burst of light energy that is spectrally a close match to daylight. The power comes from a bank of capacitors

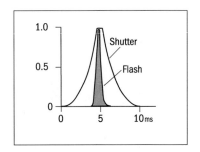

Figure 7.15 Synchronisation of electronic flash (leaf shutter).

that have been pre-charged. The procedure is quite safe as only the very small trigger current passes through the camera circuitry.

Most compact cameras have integral flash heads, which cut in when the light is inadequate for an ambient light exposure. The flash duration may be controlled in real time by the measured subject illumination in the better models. Flash heads for professional and semi-professional cameras are made to match specific cameras, and are said to be *dedicated*. They operate in a much more sophisticated way than general-purpose flashguns. Most of them measure the light actually falling on the film during the exposure, and quench the flash when sufficient exposure has been given. This is called *through-the-lens* (TTL) metering. It is not normally possible to use electronic flash with an FP shutter at very short exposure times, as the shutter is then only a narrow slit. But the more advanced dedicated flash heads can be programmed to break up the flash into a very rapid series of short flashes so that the illumination lasts for the whole of the shutter transit.

Most flash heads are mounted close to the lens axis, which is not the best place for the main light. However, most can be angled to reflect from a side wall or ceiling, to produce a more general illumination. They can also be used with extension leads, and have the capability of setting off slave flashes. No such facility exists for the cheaper compact cameras, and social photographs taken with them show the dreaded 'red-eye' effect caused by reflection of the flash from the subjects' retinas. Some compact cameras, and all professional equipment, now include the facility for a carefully timed pre-flash, which closes the subjects' pupils down sufficiently to avoid the effect.

In a night flash shot containing moving figures, any bright or self-luminous part of the subject will show a trail added to the sharp image. If the flash is synchronised to the opening of the front blind, and the shutter remains open for a short time after the flash, the trail will be in front of the moving subject, and the result looks odd. The newer dedicated flash heads have a 'rear curtain' option that fires the flash at the end of the exposure instead. Any trails are now behind the moving subject, and this looks more natural (Figure 7.16).

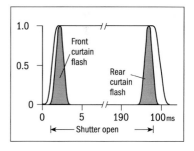

Figure 7.16 Front and rear curtain synchronisation (focal-plane shutter).

Guide numbers   A guide number is given to a flash head (not the dedicated kind) to relate the film speed, the f/no and the flash-to-subject distance. The relationship is

$$\text{guide no} \div \text{distance} = f/\text{no}$$

Guide numbers are usually given for distances measured in metres, and for various film speeds. For studio equipment they are usually also listed for half and quarter power.

Camera shake

In a static camera system with a stationary subject, the only kind of unsharpness you are likely to encounter will come from lens aberrations, incorrect focus and light scatter within the thickness of the film. But most amateur and many professional photographs are shot hand held, and here the major cause of unsharpness is what is known by the unlovely title of *camera shake*. It is such an important item in the photographer's tally of gremlins that it is worth examining in some detail. First, it is not shaking, of course, just movement. A hand held camera has six degrees of freedom, three translational (movement laterally, vertically and longitudinally) and three rotational (rotation about the camera axis, horizontal rotation and vertical rotation) (Figure 7.17)

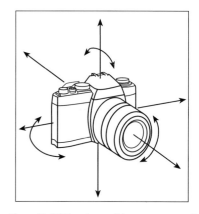

Figure 7.17 The six possible movements of a camera during a hand held exposure.

The Science of Imaging

The translational movements have very little effect on the image sharpness (except in close-ups): you would have to move some distance to get any unsharpness. But rotation is another matter. Rotation about the camera axis can be discounted, as it is very unlikely unless you do it deliberately. Most camera shake boils down to a combination of sideways and vertical rotation at the instant you press the shutter release. Of course, the effect is reduced by your giving the shortest exposure time you can (and not being afraid to use full aperture when you need it). But even that may not be enough. Suppose your camera swings at the rate of just one radian per second (a radian is about 53°). This is pretty slow: try it with your arm. With an exposure of $\frac{1}{1000}$ s (1 millisecond) you have swung the camera through 1 milliradian. And that is the resolution of an average human eye. Now, as I pointed out in Chapter 5, half the population has better visual acuity than that, so you can see that it doesn't take much movement to make your picture visibly unsharp. It is a nasty kind of unsharpness, too. Every point in the image is turned into a short straight line, usually diagonal, and there may be spurious double edges to fine details.

An analysis of the OTF for uniform camera rotation reveals a phase switch of 180° at high spatial frequencies, which accounts for this effect.

The use of long focus lenses increases the risk of camera shake in hand held photography. The rule of thumb is to use an exposure duration that is not more than the reciprocal of the focal length in millimetres, e.g., $\frac{1}{200}$ s for a 200 mm lens – and that is for a very steady pair of hands. One company has evolved a sophisticated device to cope with shaky hands: nothing less than a servo system that detects movement and immediately compensates by adjusting the position of a floating component in the lens. This system has also been applied to high-power binoculars.

Moving subjects are another aspect of unsharpness. Here, the simple answer is to follow the subject by 'panning' the camera. This is common practice in sports photography, where it doesn't matter if the spectators are blurred. It is best to use a tripod with a slightly slackened-off head for this, though camera shake does seem to be less of a problem in panned shots.

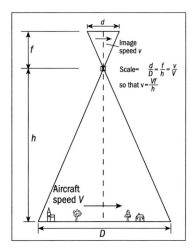

Figure 7.18 Forward movement compensation in aerial photography.

Image motion compensation

A special case of image movement occurs in aerial photography with a vertical camera, where the highest definition is essential. As the aircraft flies forward the image moves in the same direction, at a speed given by the relation Vf/h, where V is the forward speed of the aircraft, h is its height and f the focal length of the camera lens, all in the same units (Figure 7.18).

During an exposure the image moves a sufficient distance for the result to be unacceptably blurred. The film is therefore moved during exposure at synchronous speed. This technique is known as *forward motion compensation*, or *FMC*. The role of high-level reconnaissance aircraft has now been taken over by satellites, but the problem exists for those, too, and with present-day photographic sorties being made by ever faster aircraft from ever lower altitudes, even the small side-oblique cameras with their very short exposures need FMC.

Figure 7.19 Principle of photo finish camera.

An oddity among cameras – but an important one – is the photo finish camera. It has no shutter as such (apart from a capping shutter), but instead it has a narrow vertical slit in the focal plane, behind which the film can be moved at a controlled speed. The camera is aimed at the finishing line, and the film speed adjusted to match the speed of the horses, athletes etc. (Figure 7.19).

The slit panoramic camera, mentioned earlier, shares a somewhat similar arrangement. The periphery camera is also a slit camera that either moves round the object in a 360° circuit, or remains stationary while the subject is rotated.

Digging deeper

Much of this chapter seems to be more technology than science. But it is difficult, and unrealistic, to try to draw a hard and fast line between what is science and what is technology. You cannot have one without the other. The modern camera is a blend of roughly equal parts of science and technology, with plenty of craftsmanship thrown in. I have kept to generalities about construction and mechanical operation, partly because the machinery itself quickly becomes overtaken by next year's model, but mostly because this aspect of photography has been done, and continues to be done, much better by specialists writing in contemporary periodicals. Most photographic magazines devote regular articles to both new and classic photographic equipment. Of them the *British Journal of Photography* is perhaps the best. Among the textbooks, the *Manual of Photography* (9th edition, ed. R E Jacobson, Focal Press, 2000) is again a strong recommendation. Sidney Ray's monumental work *Scientific Photography and Applied Imaging* (Focal Press, 1999) is a mine of information about specialised cameras. One of the best books about flash photography is *Images Below*, by Chris Howes (Wild Places, 1997). Ostensibly about cave photography, it contains more hints and tips about flash than the average overground photographer might ever need, as well as some magnificent photography.

Chapter 8 Motion and high-speed photography

Persistence of vision

A photographic flashbulb has a total flash duration of about $\frac{1}{25}$ second, or 40 milliseconds (ms). An electronic flashtube produces a flash lasting about 1 ms. A pulse laser's flash lasts a mere 25 nanoseconds (ns). The energy they emit is similar, though the durations are very different. But to the eye the flash durations, too, look similar. This is because of a lag in our visual perception processes. We do not perceive the start of the flash until a fraction of a second after it has begun, and we continue to perceive it for a similar time after it has ceased. This is known as *persistence of vision*. And without it cinema and TV simply would not work. They would seem to flicker, just as a slowed-down cine picture does.

In fact, they do flicker, 50 times a second (60 for TV in the USA). Neon lights do, too, as do the LEDs on your hi-fi. The light seems steady when you look at them, but as you look away the image breaks up into a line of short dashes. If you deliberately look quickly away from a TV picture the result is more interesting: flick your gaze to the right and the chain of images appear as rhomboids tilted to the right; flick your gaze to the left and you will produce rhomboids tilted to the left. Flicking upwards stretches the image vertically, and flicking down compresses it.

The reason for these effects is that the picture is built up line by line successively. You can see flicker in the cinema, too, but as the whole image is displayed at once there is no distortion as you look away. A cine camera takes 25 pictures per second (pps) (24 on older models), but as 25 pps is only just above the rate at which the average person can see flicker in a bright image, the projector shutter runs at twice this speed, and shows each image twice. Home movies on 8 or 16 mm film are shot at 16 pps (as were the old silent films) and for these the projector shows each image three times.

The phi phenomenon When there is action during the filming, each successive image is a little different from its predecessor, and these successive changes give the illusion of movement. Even so, the more rapid movements would seem to be jerky, except for a quirk in visual perception going under the name of the *phi phenomenon*. If you watch a row of lights that are lighting up in succession, the impression that the light itself is moving is very strong. This effect is exploited in the 'moving' alphanumeric displays you see in public places. The phi phenomenon is so powerful that it will operate with as few as three lights. This effect, added to persistence of vision, and the slight blurring of fast-moving images, goes to confirm the impression of genuine movement in the image.

Early experiments

Hand-drawn moving pictures have been around at least since the eighteenth century. In Victorian times the *zoetrope* was a familiar toy (Figure 8.1). When it was rotated you could see the animated object moving as you looked through the slits at the inside of the cylinder.

A nanosecond is one thousand millionth (10^{-9}) of a second. It is a very short time indeed: in 1 ns a beam of light travels only about 30 cm. So in 25 ns a light beam would just about cross the average optics lab.

If you flick your eyes down rapidly you can squeeze the images to single bars; and if you do this fast enough (flick your head and neck too) you can actually succeed in inverting the picture. (But make sure nobody is watching you when you try this.)

Animated cartoons made before the computer age *do* seem jerky. This is because most of the drawn images were shown two or more times, thus saving a good deal of work in the draughtsmen's studio. Modern computer techniques can calculate how often a frame needs changing, and can blur edges of rapidly moving images.

Figure 8.1 The zoetrope.

Motion and high-speed photography

A simple version is shown in Figure 8.2 (colour plate). If you make a photocopy of this (copyright has been waived for this purpose), mount it on card, cutting out the slots, push a pencil or skewer through the centre, then hold it in front of a mirror while you rotate it looking through the slots with one eye, you will see the repetitive movement of the image.

The first moving pictures made by a photographic process were shot by Eadweard Muybridge (1830–1904), who began work on his system in 1872. He was commissioned to take photographs to settle a bet over whether a trotting horse had all four feet off the ground at some point. In order to do this he set up a line of twenty-four cameras, the shutters operated by threads stretched across the path of the horse, which broke them as it passed. The experiment was a success, and the bet was won. Muybridge subsequently made many more similar experiments, often of nude athletes (and himself), for the purpose of physiological research. These have several times been made into zoetrope images. The principle has recently been revived for making animated stereograms (see Chapter 14), using a bank of cameras fired under computer control.

> In the Science Museum, Kensington, there is a giant zoetrope with full-scale three-dimensional models of flying birds. The National Museum of Photography, Film and Television in Bradford also has a number of zoetropes on permanent display.
>
> His real name was Edward Muggridge. He spent the first part of his life in a large house in Kingston-on-Thames (which bears a blue plaque in his honour), but emigrated to California, where he became renowned for his photographs of the Yosemite Valley, shot with an enormous camera taking 20×24 in glass plates. He continued his experiments in the recording of motion into the 1890s before retiring and returning to England.

The modern cine camera

For many years the standard format for motion pictures has been 18×24 mm on 35 mm perforated film moving downwards through the projector gate. In order to fill a wider screen the optical image may be squeezed horizontally by an afocal cylindrical lens or by a prism combination, reversed on projection. The schematic layouts of a typical cine camera and projector are shown in Figure 8.3.

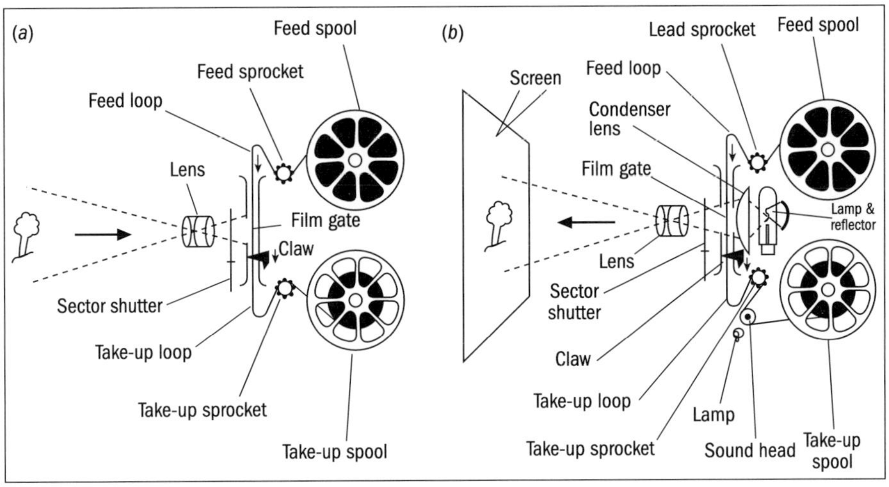

Figure 8.3 Schematic layout of a conventional cine system: (a) camera, (b) projector.

Smaller film widths include 16 mm, 8 mm and the nearly obsolete 9.5 mm. The normal framing rate is 16 pps, but some cameras have speeds of 32 and 64 pps for slow motion shots. Professional 16 mm cameras operate at 24 or 25 pps, with optical soundtrack. These are recorded on the side of the frame, and are usually generated by a light that illuminates a slit, which follows the sound waves by fluctuating in width; this is focused on the film. In the projector this track modulates a light beam passing through a fixed slit on to a photoreceptor, which passes the signal to an amplifier and speakers.

> 8 mm film can have a magnetic stripe coated outside the perforation, on which you can record commentary.

The Science of Imaging

The trend towards larger screens has led to the making of images on horizontal 35 mm film, and even 70 mm film for the giant Imax screens. The soundtrack has expanded, too, with seven-track optical sound on many modern films.

Slow motion and time lapse

There is so much slow motion on television wildlife programmes that some folks must think large birds and animals really do move like that. And, at the opposite extreme, they may perhaps imagine that toadstools grow and tropical flowers open in a few tenths of a second. It would be a good idea if TV film companies were to put a little 'S' in one corner of slow motion sequences, and a 'T' for time-lapse.

To make a slow motion sequence you simply shoot at a higher speed, usually at twice or four times the normal framing rate, and project at the normal rate. In TV and video recording you can, of course, replay a normally recorded sequence at any speed you like, but the result may be jerky. Time-lapse photography needs more careful preparation, and some calculation. For example, if a flower takes six hours to open fully, to show this in twelve seconds at a framing rate of 25 pps, you need a picture frequency of $(60 \times 60 \times 6) \div (25 \times 12) = \frac{1}{72}$ pps or 1 picture per 72 seconds. With a time as long as this one usually needs to rely on artificial lighting, preferably flash. Most high-quality cine cameras, and some camcorders, have facilities for making single exposures at timed intervals.

High-speed cine

Up to about 500 pps it is possible to use a conventional cine camera system, usually with a strengthened shutter and a modified claw movement with two claws, and possibly a beater in the bottom loop to aid pull-down of the film. However, there is a physical limit to a stop–start action, usually dictated by the stress on the film, and above about 500 pps the film needs to be moved continuously. This means that the image must also be moved in synchronism with the film.

The usual way to achieve this is to replace the normal shutter by a rotating parallel-sided glass block (a square prism). The rotation of this prism (Figure 8.4) moves the image down at the same speed as the film, and at the same time acts as a shutter. This system also allows the maximum exposure duration. Some models use a many-sided prism with a synchronised sector shutter, and this permits higher framing rates, up to 10 000 pps.

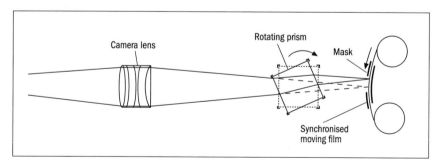

Figure 8.4 High-speed framing camera with continuous film movement.

These cameras provide sufficient speed to capture most natural phenomena such as insect flight, a chameleon's tongue or an exploding seedcase, and show them in slow motion with excellent resolution. The highest framing rates demand many-sided prisms, and the frame is usually halved in height. At these speeds it becomes difficult to obtain adequate exposure using a steady light source, as sufficiently powerful lamps would quickly fry the subject matter, so it is usual to employ a synchronised stroboscopic flash. This may be a xenon discharge lamp, or, increasingly commonly, a solid-state laser. Such a laser can produce pulses lasting less than a nanosecond, and can 'freeze' the fastest motion.

The highest framing rates put a severe strain on the capabilities of film. At over 10 000 pps the responsibility for imaging is usually handed over to electronic devices. Here, image intensifiers can increase the effective exposure by several magnitudes, though with some loss of resolution. I shall deal with these devices in Chapter 13. However, there are important uses for film still available, provided the film can be held stationary with respect to its support.

Mirror and drum photography

As far as resolution is concerned, film still has the edge over electronic recording media (at the time of writing!), but there is a limit to the speed at which you can run a film through a gate even at a uniform speed. One solution is to fix the film to its support, i.e., inside a drum, and to rotate either the image or the drum. In both cases, of course, we still need to stabilise the image with respect to the drum for each exposure interval.

Mirror cameras In a mirror camera the drum is stationary and the beam of image-forming light rotates (Figure 8.5). The separate images are formed by a ring of lenses cut to a rectangular shape to save space. The optical principle underlying the technique is called the Miller principle, after its inventor David Miller. The image is formed by the main camera lens on to a field lens that is itself focused on an entrance stop. The light passing through this stop is focused by a relay lens on a further field lens just in front of a rapidly rotating mirror. The image of the entrance stop fills each of a ring of lenses in turn, and these in turn focus the final image on the stationary film. By careful adjustment of the distance between the second image and the rotating mirror, the final image can be made to remain stationary as the image of the stop sweeps across each lens.

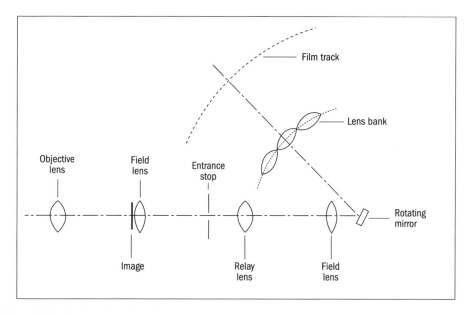

Figure 8.5 Optical configuration of a rotating mirror camera using the Miller principle.

The Science of Imaging

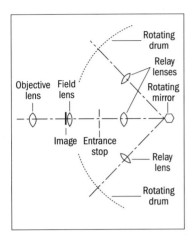

Figure 8.6 Optical configuration of a rotating drum camera.

Drum cameras In a drum camera the film is fixed to the drum as before, but the drum itself rotates. The system again uses the Miller principle, but omits the ring of final lenses. Instead, the image is focused finally by a single lens. As the drum is moving, the image has to move too. If the second image is formed ahead of the rotating mirror it will move in one direction, and if it is formed behind the mirror it will move in the other direction. By adjusting this distance the image movement can be synchronised to match the film movement. The second field lens is omitted. The camera uses a multi-faceted mirror that takes in only the width of the final relay lens in its scan. Thus as one exposed frame moves out of the exposure path the next mirror facet begins to record the next frame. If there are two relay lenses, as in Figure 8.6, one above and one below the mirror and offset horizontally, two sets of frames can be recorded side by side, and the framing speed effectively doubled. Up to four relay lenses have been used, giving up to 200 000 pps.

Smear and streak photography

These two methods of imaging are somewhat similar, but differ radically in principle from those of conventional photography. The photo finish camera is the nearest in principle, and that, as we have seen, uses an unusual photographic technique, recording motion in one direction using film movement that matches the image movement. Thus there is resolution in one spatial dimension and in time, instead of the usual two dimensions. Smear and streak photography employ a similar principle, but with a subtler use of optics.

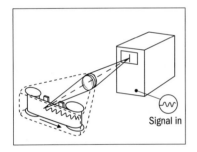

Figure 8.7 Principle of the oscillograph camera.

Smear photography This is the more basic of the two methods. In its fundamental form it is a camera used with an open shutter and film moving through the gate at a fixed speed. Such a system has been in use for many years for making oscillograph records, where the circumstances rule out pen recorders. The camera records the movement of an oscilloscope spot as a graph. The movement of the spot along the x-axis is switched off, so that the spot only moves up and down, and the movement of the film contributes the time component. There is usually a device to place timing pulses on the bottom of the film (Figure 8.7).

The second mode of use for the smear camera is analogous to image motion compensation in aerial photography. The camera system produces an image in the normal way but, since the object is moving, the film has to move synchronously with the image. This is fine when you can predict the speed of the object (as, for example, a bullet in flight), but of little use when you can't. The speed criterion is less critical when you use a very brief flash for the exposure instead of a continuous light. In this case you can also obtain an acceptably sharp image of objects that are not travelling in quite the same direction as the test object (e.g. fragments of smashed glass). You can obtain a succession of images like a cine record if you move the film somewhat faster than the image, and use ultrashort strobe pulses from a pulse laser as your light source.

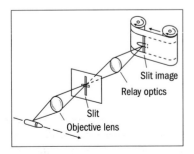

Figure 8.8 Cross slit streak camera for ballistics photography.

Streak photography This is an adaptation of smear photography that resembles photo finish photography rather more closely. It can operate in two modes, *cross slit* and *parallel slit*. Instead of being in the film plane, the slit is in the focal plane of the main objective lens, the image of the slit being projected on the moving film by a relay lens (Figure 8.8). This slit image is as narrow as is feasible in order to give optimum resolution: ideally, the width of the projected slit should match the resolution of the optical system. In cross slit photography the slit is at right angles

Motion and high-speed photography

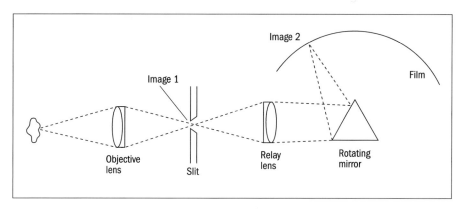

Figure 8.9 Schematic of a streak camera with rotating mirror. Alternatively, the mirror may be fixed and the drum rotate.

to the motion of the subject and the film, and the image, as it were, paints itself on the film during the time of transit, just as the competitors do in a photo finish camera. The result is a sharp image of everything that is moving in roughly the right direction.

For very high time resolution the streak camera has a rotating mirror system, with the mirror sweeping the slit image round the film mounted inside a stationary drum. Alternatively, the image may be projected on the film mounted in a revolving drum (Figure 8.9). Such systems can achieve a time resolution of just a few nanoseconds.

In the parallel slit configuration the subject moves parallel to the slit, whereas the film still moves perpendicular to the slit image. The result is in principle similar to the oscilloscope camera above. The movement of the various parts of the subject is recorded in the manner of a time–distance graph.

You will seldom see any recognisable image on the negative; but if you wind the developed film past the slit you can replay the movement. Figure 8.10b is an impression of the sort of image you would get on the film from a bullet striking an armour plate.

Of course, with such very high-speed events the film is used up in a fraction of a second, and the moving parts of the camera have to be brought up to speed, so some kind of capping shutter is essential unless the exposure is made by a very brief flash or the event is self-luminous (as with an electrical discharge). Most events need very precise synchronisation, and there is a whole range of techniques for achieving the start of the exposure. To avoid over-writing the film the exposure has to be terminated after one revolution of the mirror or drum, and in some cases this is done by a so-called blast shutter, which fires a small explosive charge to shatter a mirror in the optical path, or to render a plastic window opaque.

Lighting for high-speed photography

For medium- to high-speed cine photography, continuous light sources such as metal halide or xenon arc lamps are the rule; but for very high speeds it is more usual to employ a flash source. For running times longer than about 40 milliseconds (the effective duration of a flashbulb) a set of flashbulbs can be arranged to fire successively ('ripple firing'). For much shorter exposures the

You can see this effect occasionally in photo finishes of horse races where a horse's hoof comes to the ground exactly in line with the camera slit. There are diagonal smears at the beginning and end of the movement, as the hoof goes down and up, and a longer horizontal smear indicating the time it is on the ground (Figure 8.10a).

Henry Talbot made a spark photograph in 1851 of a rapidly rotating disc bearing part of a newspaper. The exposure duration has been estimated at less than 1 ms.

99

The Science of Imaging

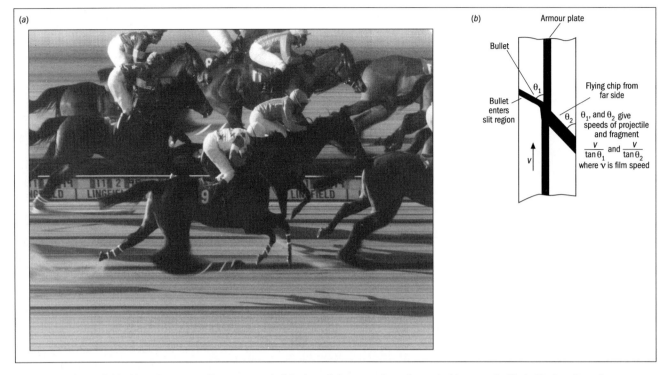

Figure 8.10 (a) Photo finish. Note the smear effect on several of the horse's hooves, where the optical image coincided with the slit as they were on the ground and momentarily stationary. (Photograph by Racetech Ltd.) (b) Parallel streak record of a bullet striking armour plate, and a fragment flying off the opposite side of the plate.

earliest, and one of the simplest, is an electric spark from a capacitor charged to a high voltage, which ionises the air and gives a very brief, intense white flash (I expect you have noticed the sparks from the contact shoes of London Tube trains going over points).

The most commonly used source today for exposures of around $10\,\mu s$–$1\,ms$ is electronic flash, which can be very accurately timed. You can also use electronic flashtubes in stroboscopic mode, which means that you can employ them for framed cine photography, with appropriate synchronisation.

The most dramatic flash source is the *argon bomb*. It is used mainly in ballistics photography, as it is so powerful that it can swamp even the light from an explosion. It consists of a cardboard or plastics tube with a clear window at one end, filled with argon gas. At the other end is a small explosive charge. Detonating this causes a shock wave to travel through the gas, ionising it and producing a brilliant flash of white light with a duration of about $100\,\mu s$. Needless to say, if you are going to use such a device you need a little more protection than just goggles and a white coat.

Stroboscopic effects sometimes appear gratuitously in movies: for instance, when you see an aircraft propeller starting up, and it seems to turn first one way, then the other. An unfortunate example occurs regularly in Westerns, where wagon wheels always seem to be turning backwards while the wagon goes forwards.

Stroboscopy

As I said earlier, electronic flashtubes can be programmed to produce short repeated flashes. If you want a sharp photograph of an object that is in a steady state of vibration or rotation, you can make it appear stationary by synchronising the flash rate with the cycles of its motion. You can also make it appear to move slowly forward by setting the rate of flash repetition slightly slower than the cycling time, or slightly backward if you set it faster.

This technique is handy for examining the behaviour of loudspeaker diaphragms, valves in internal combustion engines, even human larynxes, by synchronised cine photography or television. Strobe light sources are usually either a continuous source interrupted by a chopper (many-bladed fan shutter) or a repeating electronic flash. A comparative newcomer to this field is the laser, which can be designed to operate up to very high speeds (up to 100 kHz) and very short flash durations (down to 100 femtoseconds, or 10^{-13} s).

Another type of stroboscopic photography puts all the images on to one composite photograph. You may have seen multiple images of falling objects in physics textbooks, or, more interestingly, of golf swings or tennis serves (Figure 8.11). Again, you usually need some kind of trigger system to start the strobe action at the beginning of the operation and switch it off at the end.

Digging deeper

There don't seem to be many books around dealing with the principles of cine photography *per se*. There are dozens of books on the practicalities of film and video: Focal Press and its associate label George Newnes have a large section of their title list devoted to all aspects of vision and sound recording. Most of today's manuals are concerned with the practicalities of movies and video, and go into the underlying principles only sketchily, if at all.

There are two notable exceptions, however, and between them they cover the field completely. *High Speed Photography and Photonics* (ed. Sidney Ray, Focal Press, 1997) is a multi-author work and, as might be expected, there is some overlap of material. But it does no harm to see the same concept through different eyes, and the sections relevant to this chapter are wide ranging and insightful. Sidney Ray's own book *Scientific Photography and Digital Imaging* (3rd edition, Focal Press, 2001) is also invaluable. It contains a long chapter on cine and high-speed photography, with some 140 references at the end of the chapter. If you want to keep right up to date, the SPIE holds at least one conference each year on high-speed photography and photonics, usually publishing the proceedings within six months. The society's address is The International Society for Optical Engineering, PO Box 10, Bellingham, WA 98227, USA. You can obtain all the SPIE's publications on the Internet from www.amazon.com. So if you are enthusiastic about the subject, there is plenty of ground to dig.

The term 'strobe' is often used loosely in the USA to refer to an ordinary (single) electronic flash. If you use an American textbook, don't let this mislead you.

Figure 8.11 Stroboscopic photograph of a martial arts exponent in action. (Photograph by Sidney Ray.)

Chapter 9 The photographic process

First, a brief overview of the whole photographic process, so that you have some idea of where we are going as this chapter progresses.

To begin with, you focus the camera lens so that the optical image is sharp at the film plane. Next, you set the shutter and iris diaphragm to control the amount of light energy that reaches the light-sensitive material, and make the exposure. Finally, you make the image on the film visible by the process called development, and if the result is a negative, you produce a positive print by a method that is similar in principle. The sensitive material consists of a suspension of silver compounds in gelatin, coated on a polyester film base. It is called an *emulsion*.

> In this chapter I will be considering only traditional materials; digital cameras are considered in Chapter 13.

> Incorrectly, as it happens. An emulsion is a stable mixture of two or more liquids (such as oil and water) that do not normally mix: mayonnaise is a familiar example. A photographic 'emulsion' is actually a solid suspension.

The emulsion records the light intensity at each point on the optical image, in the form of a chemical change that produces a *latent image*. ('Latent' means 'unrevealed', as you cannot see any change at this stage, except perhaps with an electron microscope.) The next stage is to render the latent image visible as a *photographic image*, by a process called *development*, which produces a negative image in metallic silver – negative, because the most light energy (the highlights of the subject) results in the most silver, and the least energy (the shadows) results in the least silver. After development, the remaining silver compounds are still light sensitive, so a further process called *fixation* removes them. It only remains to wash out the leftover chemicals and dry the negative.

You make the print by exactly the same route. This time the emulsion is coated on white paper or some similar base, and you project an image of the negative on to it in a photographic enlarger, an optical projector using the same principles as a camera. The print is a negative of the negative, i.e., a positive, and you process it in the same way as a negative, chemically speaking. That is the photographic process, in a nutshell. Now we need to go into the underlying principles in more detail.

The uniqueness of silver halides

Silver occupies a unique position among the metallic elements. It forms compounds with other elements more readily than the other 'noble' metals such as gold, platinum and iridium. On the other hand, its compounds are much less stable than those of the 'base' metals such as sodium, magnesium and aluminium. But its real uniqueness lies in its being the only element whose compounds are readily broken down by light energy. Silver, symbol Ag, forms compounds with the halogens chlorine, Cl, bromine, Br, and iodine, I. In practice we usually form these compounds by adding a solution of a halide such as sodium chloride, NaCl, to a solution of silver nitrate, $AgNO_3$. (N is the symbol for nitrogen, Na for sodium and O for oxygen.) Silver chloride, AgCl, forms immediately, and, being heavy, sinks to the bottom of the vessel as a curdy white or yellowish solid. This quickly goes a dark purple if you put it in the sun, and gives off the characteristic pungent smell of chlorine (or bromine or iodine if you use the other halides).

> There are two other halogens, fluorine, F, and astatine, At. These are not used in photography as astatine is unstable, with a half-life of barely 8 hours (!), and silver fluoride is soluble in water.

What is happening is that the light energy absorbed by the compound is splitting it up into its component elements. The dark coloration is finely-divided silver. This phenomenon fascinated the eighteenth-century chemists, and both Joseph Priestley in England and Karl Wilhelm Scheele in Sweden actually produced shadowgrams of leaves and paper cutouts on silver chloride precipitated in a beaker. They were unable to make the image permanent.

It is easy to duplicate this experiment: all you need is some silver nitrate and a little common salt, and some distilled water; but if you do your experiment in a beaker your image will disappear as soon as you disturb the surface. Early workers such as Thomas Wedgwood (son of the great Josiah) impregnated white leather and other surfaces with silver chloride and made shadowgrams on them, but again were unable to fix them. Eventually, Henry Talbot managed to do so with a strong solution of common salt. Soon afterwards, John Herschel discovered that sodium thiosulphate, $Na_2S_2O_3$, then called hyposulphite or 'hypo', would dissolve silver halides (or rather, turn them into a soluble compound), so that the remaining light-sensitive material could be washed away, leaving a 'fixed' photographic image.

There was still a need for a good binding medium to hold the silver halide. Talbot used sized paper; Daguerre formed silver iodide directly on silver plates by the action of iodine vapour. By 1848, Nièpce de Saint-Victor, a relative of Nicéphore Nièpce (the first successful photographer) had discovered the merits of albumen (egg white). In 1851 Scott Archer introduced collodion as a substrate, and this proved an excellent medium, provided it was not allowed to dry out before processing.

The real breakthrough came in 1871 when Richard Maddox introduced the gelatin dry plate. Gelatin is a form of collagen, a substance found in skin, bone and sinew, and is manufactured by boiling these inedible parts of cattle and pigs and refining the resulting goo (sorry, but I had to tell you). You can appreciate the miraculous behaviour of this substance if you just add a little gelatin dissolved in hot water to your silver nitrate solution. When you pour in the sodium chloride solution, instead of a curdy precipitate you get a pearly translucent liquid. The gelatin prevents the precipitation and causes the silver halide to form as microscopic crystals. No other substance that can do this so efficiently has ever been found.

The manufacture of an emulsion is fairly complicated, and usually involves several 'ripening' stages, during which the crystal size and sensitivity increase. Most black-and-white films are coated with two layers of emulsion, one 'slow' (low sensitivity) and one 'fast' (high sensitivity), to increase the exposure latitude.

The latent image

In order to understand what goes on when we expose an emulsion to light, we need to look more closely at those microscopic crystals. (To save having to keep listing the halogens, I shall simply say 'silver bromide' as this is the main halide used in negative emulsions.) The tiny crystals of silver bromide that form the light-sensitive part of the emulsion are made up of alternate atoms of silver and bromine in a structure called a *face-centred cube* (Figure 9.1).

The atoms are held in position by interatomic forces. To see how these operate we need to look at a model for the structure of atoms themselves. Nearly all the mass of an atom is in the *nucleus*, which is made up of particles called *protons*, which carry a unit positive electric charge, and particles called *neutrons*, which have no

Joseph Priestley (1733–1804) was one of the earliest chemists (as distinct from alchemists). He is best known for his discovery of oxygen. He was an outspoken religious dissenter, and his opinions were so unpopular that he was forced to emigrate to Pennsylvania. **Karl Wilhelm Scheele (1742–1786)** was an apothecary; he discovered many previously unknown elements (including oxygen, independently), and showed how to synthesise a large number of compounds.

Sir John Herschel (1792–1871) was the equal of his distinguished father Sir William Herschel, famed organist, astronomer and discoverer of the planet Uranus. He spent four years in South Africa, where he mapped the whole of the southern sky. He was well read, and coined much of the nomenclature of photography such as positive, negative and snapshot, as well as many of the '-scopes', '-graphs' and '-ologies' of contemporary science.

Collodion is a solution of nitrocellulose in a mixture of ether and ethanol. It is porous as long as it remains tacky. It is dangerously flammable (nitrocellulose is also known as 'guncotton').

Richard Leach Maddox (1816–1902) was an inventor who, like Du Hauron, published his ideas for the public good instead of reaping a fortune from patents. Like Du Hauron he died a pauper.

Latitude means, roughly, what you can get away with in terms of over- or underexposure and still get a good-quality print. It is discussed in detail in Chapter 10.

The Science of Imaging

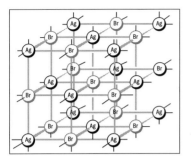

Figure 9.1 Silver (Ag) and bromine (Br) in a face-centred cubic crystal. This is a 'lattice' diagram; in a space-filling model the atoms are much closer together.

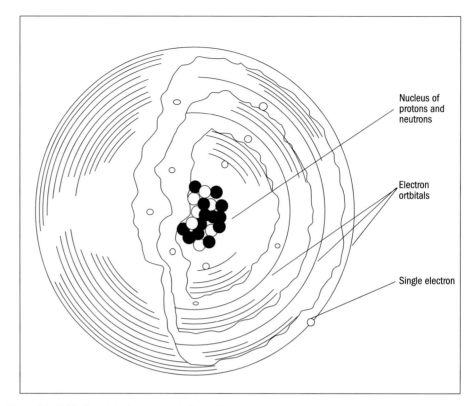

Figure 9.2 'Orbital' model of an atom of silver, with one electron in outer orbital.

An atom that has lost an electron is called a *positive ion* or *cation* (symbol Ag^+ for silver) and an atom that has gained an electron is called a *negative ion* or *anion* (symbol Br^- for bromine).

charge. Around the nucleus are roughly spherical volumes called *orbitals*, containing particles called *electrons*, which carry a charge equal and opposite to that of the protons. The mass of an electron is very small compared with that of a proton, but electron orbitals take up nearly all the volume of an atom (Figure 9.2). There is the same number of electrons as protons, so the net electric charge on an atom is zero.

Each orbital is associated with a specific energy level for the electrons in it, and the laws of quantum physics permit only a certain number of electrons in each orbital. For the outermost orbitals of both silver and bromine the permitted number is eight. The halogens all have seven outer electrons, and silver has just one. There is a strong tendency for this odd electron to be 'lent' to the bromine atom, so that both the silver and the bromine now have complete outer shells. But the silver atoms now have a positive charge (as they have lost a negative charge with the lost electron); and the bromine atoms have acquired a negative charge from the extra electron. As positively and negatively charged bodies attract one another, a bond is developed between adjacent pairs. This is called an *ionic* bond.

In the face-centred cube of a silver bromide crystal each silver ion is surrounded by six bromine ions, and each bromine ion is surrounded by six silver ions. You may find it helpful to think of the crystal as being buttoned together by electrostatic press-studs, with the male studs on the silver. However, no crystal is perfect, and there are places in the structure where there are dislocations (Figure 9.3). Here, electrons may be unpaired and, as they are held only loosely by the silver atoms, they can wander about within the crystal. In addition there are specks of impurity present, and these contribute both dislocations and electrons. It takes only a small

The photographic process

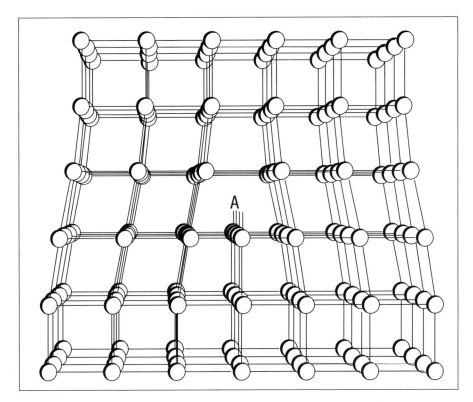

Figure 9.3 Example of a lattice distortion. The atoms at point A have an unattached electron.

amount of energy to detach an unpaired electron and allow it to find a silver ion and complete a silver atom.

This energy can be provided by a photon, a unit of light energy. One photon can disrupt a bond, leaving a silver ion free to pick up a stray electron and become a free atom. This can move to a position of low energy such as a crystal dislocation or an impurity speck. Once three or four atoms have gathered in such an 'energy sink' they will form a stable speck of silver, and will attract further atoms. This is the *latent image*. In theory you need only three or four photons to create a stable latent image provided they arrive simultaneously, but in practice you need several times this amount arriving within a millisecond or so, as broken bonds quickly re-form spontaneously, emitting the absorbed energy as either a further photon, or as heat. Plainly, the larger the crystal, the greater the chance of its acquiring a stable latent image for a given exposure. So high-speed emulsions have large crystals, and low-speed emulsions have small crystals ('fine grain').

Speed and inherent contrast

In a general-purpose emulsion the silver halide crystals vary in mean diameter from about 50 µm to 5 µm. During the ripening process the mean crystal size increases as some crystals grow at the expense of smaller crystals. As the largest crystals are the most sensitive to light, only these bear a latent image in the areas of lowest exposure. In the areas of greater exposure the smaller crystals will also be affected, and in the highlight areas even the smallest crystals will bear a latent image.

Photographers refer to both undeveloped and developed crystals as 'grains'. This can be confusing, and I am using the term 'grain' to refer only to the developed image.

The Science of Imaging

In fine-grain emulsions the ripening process is curtailed, so that the crystals are both smaller and more nearly uniform in size. This means that the inherent contrast of the emulsion is higher, because the threshold between 'latent image' and 'no latent image' is at a similar level for all crystals of a single size.

Manufacturers of gelatin emulsions discovered quite early in the history of the process that the traces of impurities in the gelatin, particularly sulphur and phosphorus, had a pronounced effect on the sensitivity of the emulsion. At one time there were attempts to control these impurities by supervising the diet of the animals whose spare parts were to be used subsequently for making photographic gelatin (Eastman Kodak had its own herds). Today the raw gelatin is purified, and the resulting inert material is doped with precisely controlled 'impurities'. Although the original silver halide crystals are microscopic, the final developed grains are much larger, and with a high-speed emulsion they are visible under quite low magnification. There is therefore a limit on the acceptable size (and sensitivity) of the crystals. Furthermore, most of the impurity specks are on the surface of the crystal, and the surface area does not grow as fast as the volume (unlike the grain size). Recent advances in emulsion technology have countered this problem by forming the crystals thin and flat, in effect two dimensional. These tabular crystals are called 'T-grains' (Figure 9.4).

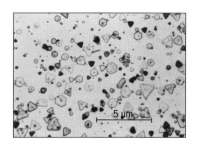

Figure 9.4 Photomicrograph of a general-purpose emulsion. The large triangular and hexagonal crystals are tabular. In a modern T-grain emulsion all the crystals are tabular.

Colour sensitivity of emulsions

As I explained in Chapter 1, the energy of a photon is proportional to its frequency. The lattice strength of a silver chloride crystal is fairly high, and needs an energetic photon to break a bond. A plain silver chloride emulsion is sensitive to only blue-violet and UV radiation. Silver bromide and iodide have weaker bonds that can also be broken by blue and blue-green light energy. In order to sensitise an emulsion to longer wavelengths the crystal surfaces are coated with dyes that absorb low-frequency photons and pass the energy into the crystal. The first of these dyes to be discovered, around 1860, was erythrosine, a red dye that sensitises an emulsion to green light. Such an emulsion is termed *orthochromatic*, from Greek words meaning (somewhat optimistically) 'correct colour'. As orthochromatic emulsions are insensitive to red they can be handled in red light (called a 'safelight') without becoming fogged. Around 1890 it was found that blue and green dyes of the cyanine family would sensitise emulsions to red (and, later, to infrared), and such emulsions became called *panchromatic*, Greek for 'all colours'. Figure 9.5 shows the spectral sensitivities of the various types compared.

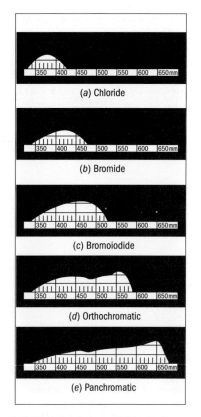

Figure 9.5 Spectral sensitivities for the various types of emulsion.

Gallic acid is a close relative of pyrogallol, a classic developing agent.

Development

When Henry Talbot made his first negatives, they were printed out, that is to say, the lens caps on his 'mousetraps' were left off until the light had actually decomposed his silver halide into a visible image, a process that could take several hours. He discovered that he could strengthen his weakest images with a solution of gallic acid. This was the first historic example of the use of a chemical developer.

Some textbooks treat development of a negative as analogous to amplification of an electronic signal, which turns inaudible millivolts into tens of volts and powers them out as ear-splitting kilowatts. On the surface this seems a reasonable simile, but it is not an ideal one, as an amplifier operates on an analogue basis, whereas a developer, at the silver grain level, is nearer to an all-or-nothing process: if a crystal bears a latent image the developing solution turns it into an opaque silver grain. The process is very slow as chemical reactions go. Just as well, perhaps, as if it were to be allowed to go to completion almost the entire emulsion would turn black. In practice what happens is that the largest crystals, those with the most latent image specks, begin to turn into silver grains first, then the medium-sized ones, and finally the smallest ones. First the contrast is built up, then the whole image gradually becomes darker, and eventually even the crystals with no latent image begin to be attacked. Chapter 10 analyses this process in terms of the image itself.

Constituents of a developer

The purpose of a developer is to turn the silver halide crystals into grains of metallic silver if, and only if, they bear a latent image. There are some restrictions on the type of substance we can use for this. First, it has to work sufficiently slowly for us to be able to stop the process before it goes too far. Secondly, we have to choose a substance that will attack the crystals that already have a latent image, but not the ones that do not have one. The process of removing the halide ions and leaving silver atoms behind is called *reduction*, and the active constituent of the developer is called the *reducing agent*. In conventional chemistry, a reducing agent is any substance that can give away electrons. i.e., is an *electron donor*. Its opposite number is an *oxidising agent* or *electron acceptor*. We shall meet some of those later. For our purpose, a reducing agent is something that will separate silver ions from their halogen partners and make good their missing electrons.

> In dietetics, reducing agents are called 'anti-oxidants', and are alleged to be good for your long-term health. However, this does not apply to all reducing agents. You won't extend your lifespan by taking a large swig of photographic developer, even if some developers do contain Vitamin C.

Oxidation and reduction

When an atom of silver loses an electron to an atom of (say) bromine, it has lost a unit of negative charge and become a positive ion, Ag^+. The atom of bromine has become a negative ion, Br^-. In an ionic crystal such as silver bromide the bonds are the electrical attraction between the oppositely charged ions of silver and bromine. A photographic developer can break such a bond and supply electrons to the silver ions, but only in the presence of a speck of metallic silver (i.e., a latent image). The silver is in effect a catalyst. By donating electrons it turns the Ag^+ ions into Ag atoms, which extend the latent image. Eventually the whole crystal is reduced to a tangle of filaments of silver metal.

> A catalyst is a substance that promotes a chemical reaction without itself being changed.

In order to function efficiently, a developing solution needs more than just a reducing agent. Apart from the solvent (water) there are usually four main constituents in a developer: the *developing agent*, *alkali*, *preservative* and *restrainer*.

Developing agent   This is the reducing agent that turns the exposed silver halide crystals into opaque grains of silver. The number of compounds that will do this job is limited.

> Not all that limited, though. Henney and Dudley list 211 of them – and that was in 1939.

A developing agent is a rather special kind of reducing agent. Some reducing agents cannot discriminate between exposed and unexposed crystals, and blacken the entire emulsion; others do not attack the crystals at all. Nearly all practical

The Science of Imaging

developing agents are closely related to benzene. The benzene molecule is a bit of an oddity: it has six carbon atoms in a ring, each attached to a single hydrogen atom. Some of these hydrogen atoms can be replaced by other groups that are electron donors, and it is molecules of this particular type that form the great majority of developing agents. (See Box.)

> ## Developing agents and molecular structure
>
> You have already come across ionic bonds, where an atom lends an electron to another atom, and the resultant electrostatic forces hold a crystal together. Some elements, however, instead of lending electrons, share them. Carbon, C, is the prime example. It has four electrons in its outer orbital, and it readily shares these with hydrogen, oxygen, nitrogen and the halogens. These shared bonds are called *covalent bonds*, and they may be single or double, or occasionally triple. We indicate them conventionally by straight lines between the atomic symbols. Thus ethane, C_2H_6, is shown as in Figure 9.6a, ethene, C_2H_4, as in Figure 9.6b, and ethyne, C_2H_2, as in Figure 9.6c. (Just because they are depicted on flat paper, don't assume that these molecules are necessarily flat. Methane, CH_4 (Figure 9.6d), for example, is a tetrahedron.) In each case, as you see, the carbon atoms have four bonds and hydrogen one. Oxygen has two (the number of electrons in its outer orbital) and nitrogen three.
>
> An unusual property of the carbon molecule is its propensity for joining up in rings. The most stable of these is a ring of six. The most fundamental of the molecules with this structure is benzene, C_6H_6, which appears as in Figure 9.7a (and this one *is* flat). Molecular structures based on benzene are known as *aromatics* (many of its members have pleasant smells); they turn up so frequently in organic chemistry that the benzene molecule is usually depicted in shorthand form (Figure 9.7b).
>
> A large number of compounds based on this ring are simple substitutions of groups of atoms called *radicals* for one or more hydrogen atoms, groups such as $-OH$ (hydroxyl), $-NH_2$ (amine) and $-CH_3$ (methyl). Many of these radicals dissociate (split up) in solution, and some of these are developing agents. A simple example is hydroquinone (1,4-dihydroxybenzene), Figure 9.8a. In solution this

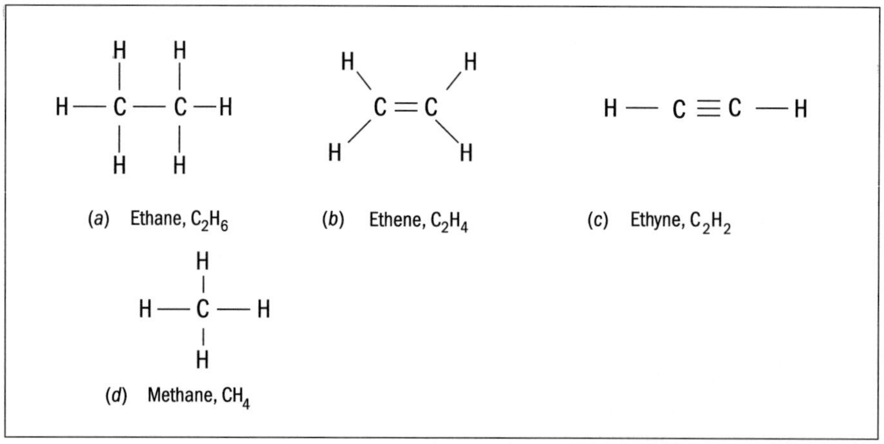

The carbon atoms in a benzene ring are identified by the numbers 1 to 6, going clockwise from the top.

Figure 9.6 Structure of some simple hydrocarbon molecules.

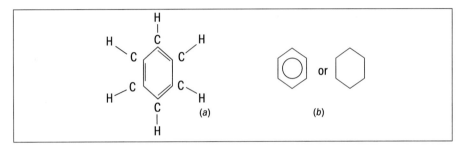

Figure 9.7 (a) Structure of the benzene molecule, C_6H_6; (b) the shorthand depiction.

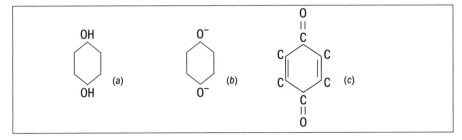

Figure 9.8 (a) Hydroquinone (1,4-dihydroxybenzene) molecule; (b) molecule dissociated in aqueous solution; (c) bond rearrangement into quinone.

dissociates as in Figure 9.8b, which is an electron donor and a developing agent. When it has given up its electrons it becomes quinone, Figure 9.7c. As you see, the bonds have been rearranged so that the carbons still have four bonds and the oxygens two. The electrons have gone to neutralise the silver ions.

As I said earlier, almost all developing agents are based on substituted benzene molecules. Notable exceptions are phenidone and ascorbic acid.

Quinone (also known as parabenzoquinone or PBQ), is not an aromatic, and has in fact a foul and choking smell.

Alkali   The loss of electrons from the developing agent releases hydrogen ions (H^+), which combine with the freed halogen ions (Br^-) to form hydrobromic acid, HBr. Few developing agents will work in acid solutions, so we need to add an alkali to neutralise this acid. This is usually sodium carbonate, Na_2CO_3, though stronger or weaker alkalis may be suitable for special developers.

Preservative   Left on their own, most developing agents in solution absorb oxygen from the air and quickly become useless. The preservative is a further reducing agent that protects the developer from oxidising but has little effect on the development process. The most commonly used preservative is sodium sulphite, Na_2SO_3, which eagerly mops up oxygen to become sodium sulphate, Na_2SO_4. As a solution of sodium sulphite is mildly alkaline, it sometimes plays the role of alkali in low-energy developers.

Restrainer   The development process produces free halides, which tend to hinder the action of the developing agent, and may produce image artefacts such as streaks or haloes. The addition of a halide (usually potassium bromide) to the original solution evens out the effect and slows the development process, lowering the risk of fog. Developers containing phenidone are unaffected by the presence of halide ions, and for these the restrainer is an organic anti-foggant such as benzotriazole.

The Science of Imaging

Fixing, washing and drying

After development the emulsion contains unchanged silver halide, which is still sensitive to light, so it needs to be removed. Sodium and ammonium thiosulphates ($Na_2S_2O_3$ and $(NH_4)_2S_2O_3$) react with silver halides to form soluble compounds that can be washed out of the emulsion with water. You can see the process of fixation in the clearing of the milky appearance of the negative, the received wisdom being that you need to give twice the time taken for complete clearance, as the first reaction products are not very soluble. Ammonium thiosulphate acts much more rapidly than sodium thiosulphate. Washing is a diffusion process that takes upwards of ten minutes. Washing thoroughly is important, as the compounds of silver with thiosulphate break down after a few weeks to form brown silver sulphide.

You need to make sure the emulsion dries uniformly, without any water drops on the emulsion, as the gelatin shrinks on drying, and spots that remain damp will be stretched out as the surrounding gelatin dries, and will show as pale spots on the negative. It is common practice to add a few drops of wetting agent to the final wash water.

Printing

Print paper works in the same way as black-and-white negative film, so when you project the image of the negative on it with an enlarger, it produces a negative of the negative, i.e., a positive. Although often called 'bromide' prints, the emulsion is usually mainly silver chloride. As it is insensitive to red and green light, you can handle it under a yellow safelight. Printing emulsions come in up to seven grades of contrast to match the contrast of the print to that of the negative (this aspect is dealt with in detail in Chapter 10); but many (probably most) photographers prefer to use variable-contrast (VC) paper, as this saves having to keep several boxes of different grades. Variable-contrast papers have a second, green-sensitive layer of emulsion (demanding an orange safelight) of low contrast. Consequently, you can control the contrast by filtering the enlarger light with magenta for higher contrast and yellow for lower contrast. (Most enlarger heads are nowadays fitted with filter drawers; the more sophisticated ones have graduated filters in cyan, magenta and yellow for colour printing, the cyan being set to zero for black-and-white variable-contrast paper.)

Most print papers for general use are resin coated (RC). They have a base of paper impregnated with a polymer, and need only a few minutes' wash. Some papers, though, more especially types of paper intended for exhibition prints, are fibre based, and prints made on this base need up to an hour's washing in running water, as the fixation products adhere strongly to the paper fibres, and any residual sulphur will eventually cause the image to fade and discolour.

Colour emulsions

Colour film is much more complicated to manufacture than black-and-white film. In principle, a colour emulsion contains three layers of emulsion, an inner layer sensitive to red light, a middle layer sensitive to green light and an outer layer sensitive to blue light. In practice all three layers are sensitive to blue, so blue light

The photographic process

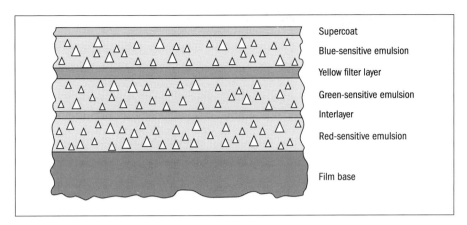

Figure 9.9 Construction of a colour emulsion (simplified).

has to be blocked off from the inner two layers, by including a layer of yellow dye beneath the blue-recording layer (Figure 9.9).

To extend the response of the film, as in black-and-white emulsions two emulsions of different speeds are coated for each colour, and with further layers to inhibit migration of dyes between emulsions, an opaque-dyed undercoat to suppress internal reflections and an abrasion resistant supercoat, a colour emulsion may have as many as fourteen layers, still with a total dry thickness of less than 50 micrometres.

Processing of colour emulsions

The processing of a colour negative film is broadly similar to that of a black-and-white film, except that the final stage is not simply fixing but bleach-fixing, that is, the removal of all silver and silver halide, using a solution that oxidises the negative image to a substance soluble in thiosulphate, which may be either in the same bath or in a subsequent one. The developer is a special colour-forming developer giving reaction products that couple to latent dyes in the emulsion layers wherever an image is formed, so that the red recording layer produces a negative image in cyan, the green an image in magenta and the blue an image in yellow. When the silver image is removed along with the unchanged silver halide the dyes remain, and in the final negative the hues of the original are reversed as well as its tones.

The print material is broadly similar, though the layers are in a different order, and there is a gap in the yellow sensitivity to allow the use of a sodium safelight.

Transparency material is processed in a different manner. The initial development is with a more or less traditional black-and-white developer, after which the remaining undeveloped silver halide is fogged and developed in a colour-forming developer. Bleach-fixing then leaves a positive image in dyes.

Printing from a transparency is by a similar method. However, in one process (the *dye-bleach* process) all the dyes are present to begin with, and are selectively bleached by exposure and processing.

Kodachrome processing is different. The dye-couplers are in the developers, and there are separate developments for cyan, magenta and yellow (Chapter 6).

Digging deeper

For a more detailed description of emulsion making you need to consult a work such as *Basic Photographic Materials and Processes*, by Ira Current, John Compton, Leslie Stroebel and Richard Zakia (2nd edition, Focal Press, 2000), which lays out the whole process in detail. Latent image theory is trickier. People are still arguing about it in the pages of *The Imaging Science Journal* (Royal Photographic Society, bi-monthly). You can find a well-balanced view in *Science and Technology of Photography* (ed. Karlheinz Keller, VCH, 1993), but it is not a very easy read. As usual, for a general exposition, as well as a more detailed description of development chemistry, go to *The Manual of Photography* (9th edition, ed. R E Jacobson, Focal Press, 2000). *The Theory of the Photographic Process* (4th edition, ed. T James, Macmillan, 1977) has it all, but it is overdue for a new edition, and is strictly for academic study. You should regard any literature with a title such as 'Emulsion chemistry made simple' with caution. At best it will be unsatisfying; at worst, misleading.

(a)

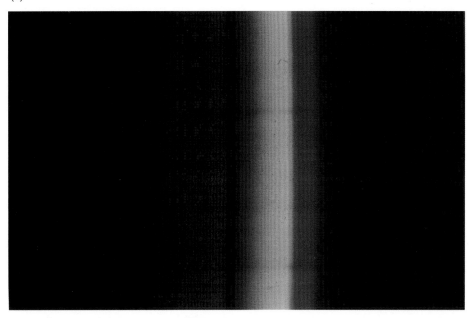

(b)

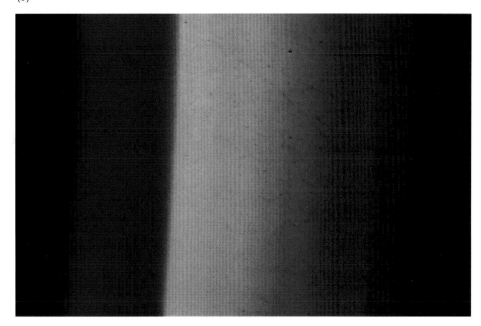

Figure 1.9 Spectra formed by (a) a prism; (b) a diffraction grating.

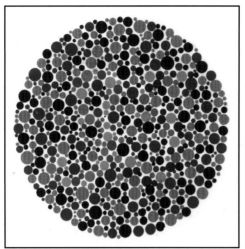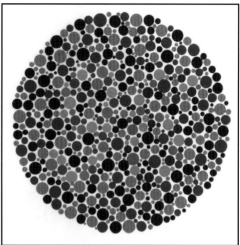

Figure 3.7 A page from the Ishihara colour vision test book. Normal subjects will see the figures 35 and 96. Subjects with protanopia (red cone deficiency) will see only the figures 5 and 6, and subjects with deuteranopia (green cone deficiency) will see only the figures 3 and 9. Note that this reproduced plate is not a good test for colour deficiency. (© Isshinkai Foundation, published by Kanehara Trading Co.)

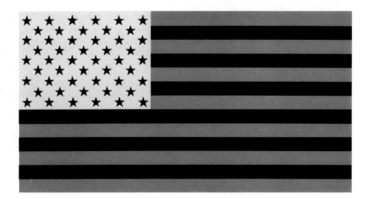

Figure 3.8 If you stare at the centre of this flag in a bright light for about half a minute and then look away at a white surface, you will see the US flag as an after-image, in its correct colours.

Figure 4.17 The best of lenses will produce ghost images if you shoot straight into the sun. Here the edge of the iris diaphragm appears as a rainbow arc centred on a pale area.

Figure 6.5 Self-portrait of Dr Hans Bjelkhagen made using the Lippmann technique. Unfortunately it is not possible to do justice to the subtle colours of the original in a printed reproduction. (Photograph courtesy of Dr Bjelkhagen.)

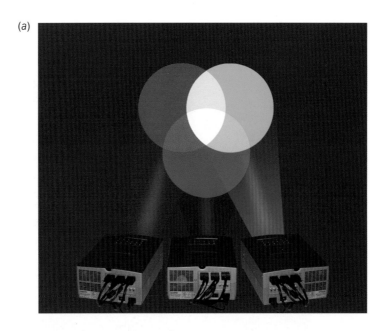

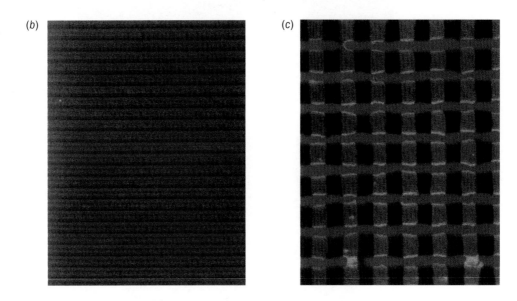

Figure 6.6 (a) Additive colour patches, using projected light; (b) enlargement of a section of Polachrome transparency film, showing colour raster; (c) enlargement of a Dufaycolor reseau on the same scale.

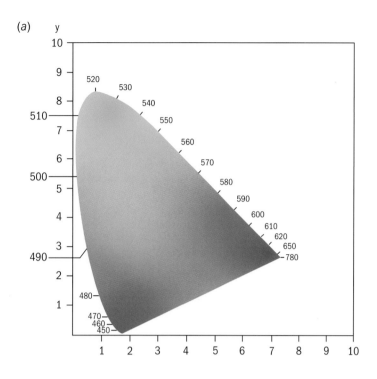

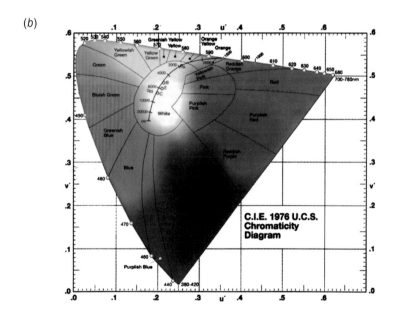

Figure 6.7 (a) 1947 CIE chromaticity diagram, showing approximate wavelengths and corresponding colours; (b) 1978 modification, representing equal perceived colour shifts in hue and saturation by equal distances, and showing the position of various colour temperatures.

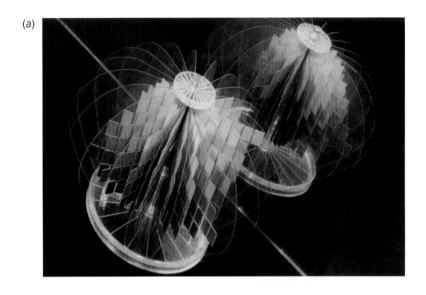

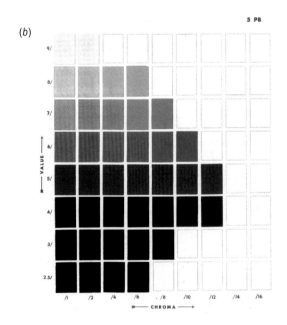

Figure 6.8 (a) The full Munsell colour globe; (b) a page from the Munsell colour atlas (from Measuring Colour, 3rd Edition by R W G Hunt, courtesy Fountain Press).

(a)

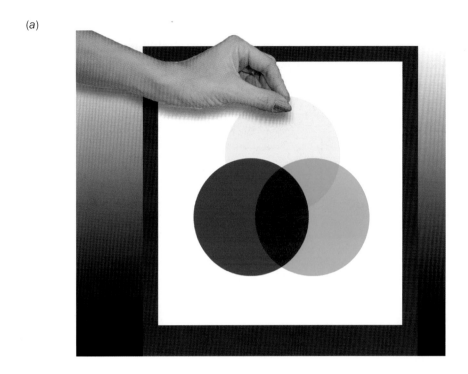

(b)

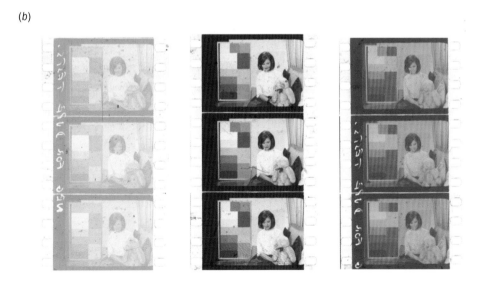

Figure 6.9 (a) Subtractive colour filters, using transmitted light; (b) Technicolor separations.

Figure 6.10 The Kodak colour patches and grey scale.

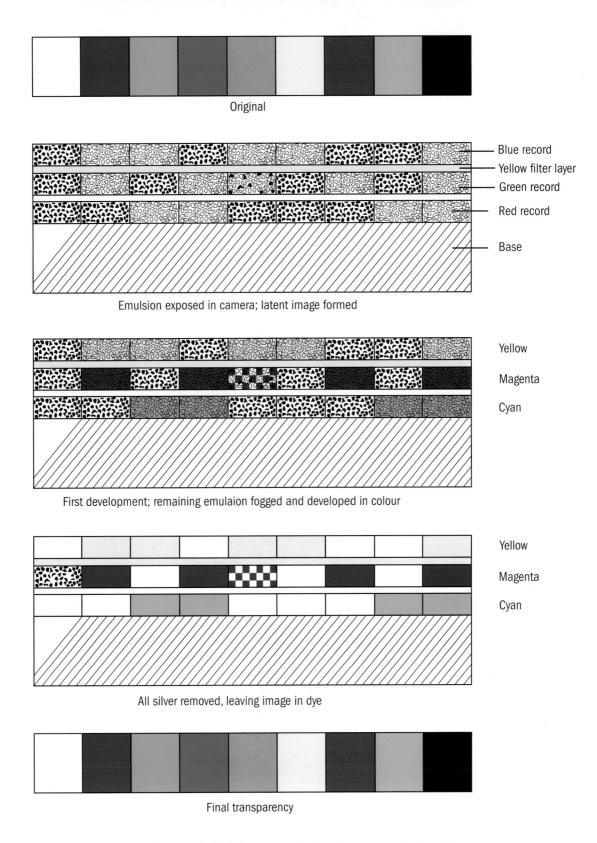

Figure 6.13 Colour reproduction in a reversal colour film.

Figure 6.16 Infrared false-colour aerial photograph.

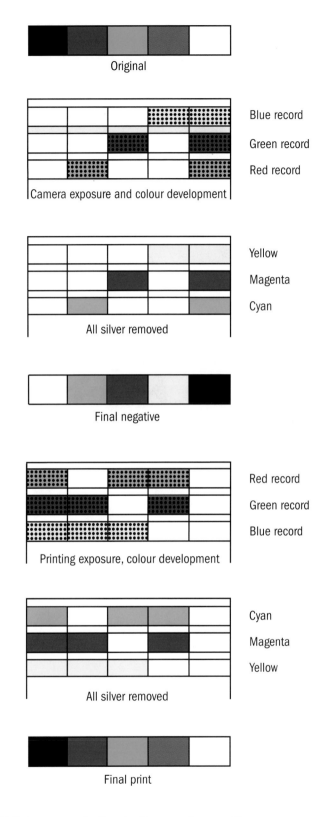

Figure 6.17 Colour reproduction by the negative–positive process.

Figure 8.2 This is a simple form of zoetrope. To operate it, make a photocopy, mount it on card and cut around the outside, including the slots. Push a pencil through the centre. Now hold the disc in front of a mirror so that you can see an image through one of the slots, and rotate the disc clockwise.

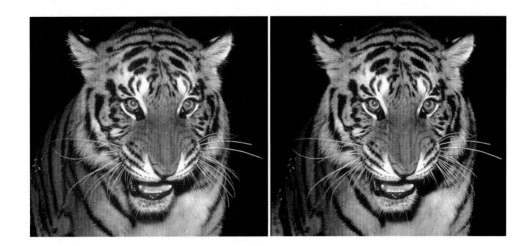

Figure 17.2 To view these images stereoscopically without an optical aid, hold the book about 40 cm away, just below eye level, while you look past it at a distant object. Then bring the book up into your line of vision, keeping your eyes relaxed. You will see one (blurred) image in the middle, flanked by two other (also blurred) images. Now slowly bring the middle image into focus without losing fusion. This may take some practice. You may find it easier to deliberately cross your eyes, in which case the stereoscopic depth will be reversed.

Figure 17.1 Notice (1) the relative size of the two pillars; (2) the overlap of objects; (3) the slight haze reducing the contrast of the distant fort; and (4) the modelling and texture resulting from the side light on the nearer pillar.

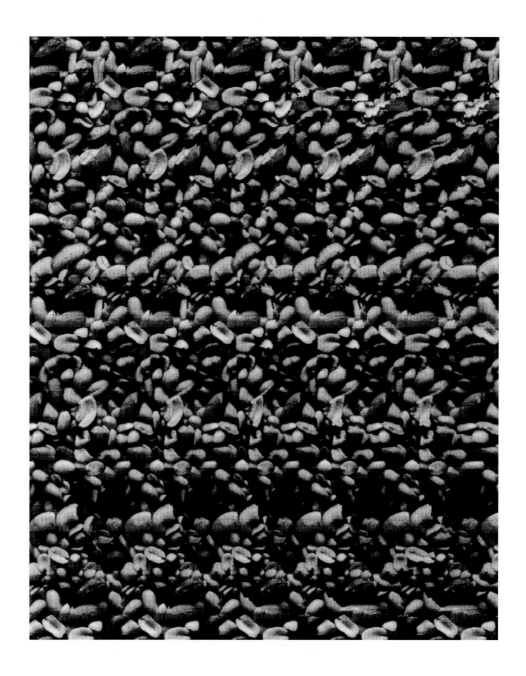

Figure 17.12 An interlaced autostereogram. Hold the book in front of you about 40 cm away, and stare steadily at the figure. The stereoscopic image will jump out after a short while. (Image courtesy of David Burder, 3-D Images Ltd.)

(b)

(c)

Figure 17.13 If you examine the (unscreened) images with a magnifying glass you can identify the individual frames. To view these two figures in their final form, place the lenticular screen (in the back pocket of the book) over the figures and align it with the horizontal lines (b) and vertical lines (c). Adjust until the moiré pattern disappears from the upper and lower border lines. (Images courtesy of David Burder, 3-D images Ltd.)

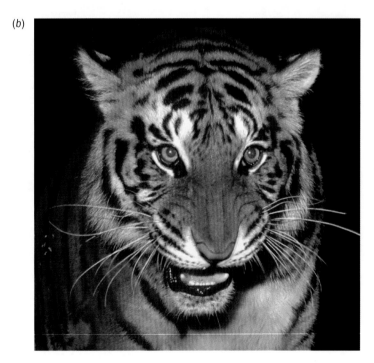

Figure 17.14 (a) Monochromatic and (b) colour anaglyphs. To view these, use the coloured spectacles from the back colour pocket of the book. (Images courtesy of David Burder, 3-D Images Ltd.)

Figure 18.13 (b) Natural colour hologram made using three laser beams. (Photograph courtesy of Hans Bjelkhagen.)

Figure 19.7 Pinhole photograph of the Royal Crescent in Bath, with fingers, by Justin Quinnell. The depth of field here extends from near infinity to about 3 cm.

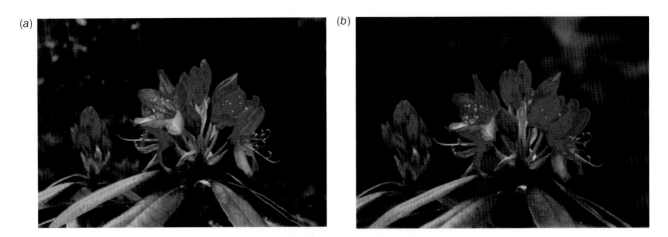

Figure 19.11 Comparison of (a) short-focus and (b) long-focus macro lenses, taken at the same f/no. Note the difference in the backgrounds.

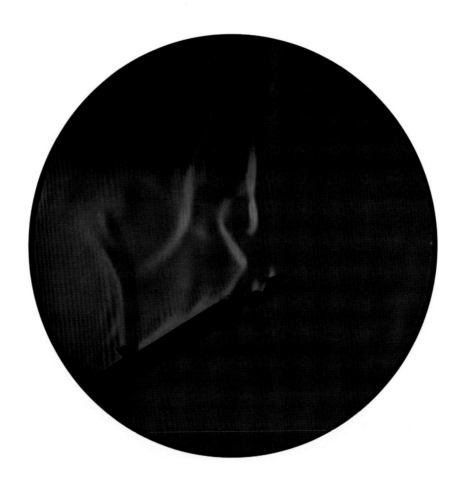

Figure 20.17 (b) Schlieren image of the airflow round a hot soldering iron, using colour filters to show up density gradients. (Photograph by Jon Tarrant.)

Chapter 10 How photographic films behave

When you buy a film for a specific purpose you want to know whether it is the best possible choice. If your purpose is social photography, you need a film that can cope with a range of exposures that may not always be correct; if it is sports photography, it needs to give results with very short exposures, possibly in bad light; if it is for making copies of, say, paintings or photographs, it needs a highly accurate tone reproduction, or, for line diagrams, a very high contrast. In short, you want to know the *characteristics* of the film; and if you are to expect the best possible results you need to know them with some precision.

The measurement of these characteristics bears the rather grandiose name of *sensitometry*. Basically all it involves is measurement of the luminances of the details in the subject matter, the corresponding transmittances of the resulting negative and, finally, the corresponding reflectances of the print. The scientific foundations of sensitometry were laid by two brilliant scientists, F Hurter and V C Driffield. Their research resulted in the establishment of a characteristic curve for an emulsion, still known as an *H & D curve*, and a speed rating system, the *H & D System*. Their results were published in 1880.

Ferdinand Hurter (1844–1898) was a Swiss-born chemist who worked in a British chemical factory with Vero Charles Driffield (1848–1915). They were both keen amateur photographers and wanted to market an exposure meter, but in order to do so they had to establish a rigorous system for establishing the speed and contrast of a photographic emulsion. For their pioneering work in sensitometry they were awarded the Royal Photographic Society's Progress Medal, its highest honour, in 1898.

Hurter and Driffield's task

Hurter and Driffield needed to establish meaningful figures for the speed and inherent contrast of an emulsion. As has since become standard practice, they decided to do this on a basis of graphically plotted data. However, plotting the transmittance of the negative against the luminance of the subject gave a very nonlinear result. After much experimentation they settled on a plot of the logarithm of the reciprocal of the transmittance (a quantity we may call *opacitance*, by analogy) against the logarithm of the exposure (i.e., duration × illuminance). With contemporary emulsions and processing techniques, this gave a straight line with only a short curved 'toe' (Figure 10.1).

When the straight line was continued downwards, the point where it crossed the horizontal axis they called the *inertia point*. This defined the emulsion speed, and the slope of the straight line defined the inherent contrast of the negative, that is, the contrast of the negative as compared with that of the subject. They called this quantity *gamma* (the Greek symbol γ). There was (at least to begin with) no specified development time, as the position of the inertia point did not alter with changing development times as long as the standard pyrogallol developer was used (Figure 10.2).

Hurter and Driffield obtained their numerical value for emulsion speed by dividing the exposure (in foot-candle-seconds) corresponding to the inertia point into 34. This gave a figure of about 200 for a general-purpose emulsion of their time (equivalent to ISO 8/10). The figure became known as the *H & D speed*. The system continued in use for photographic plates until the 1950s, by which time the manufacture of plates was being phased out except for special purposes; pyrogallol, too, had fallen out of general use, and the increasing use of restrainers in the developer had made the concept of an inertia point meaningless.

It seems odd to me that the two scientists should not have tried this immediately. Every budding junior scientist knows that if you have got a regular curve when what you wanted was a straight line, you just plot the logs of one or both of the variables instead. (The reason this works is revealed in Appendix 1.) Furthermore, it seems to have been simply trial and error that brought the two scientists to the use of logarithmic values. Had they been familiar with the Weber–Fechner Law (see Chapter 3) they would have seen this strategy at once.

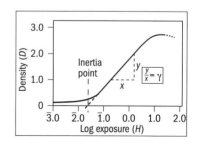

Figure 10.1 Hurter and Driffield's emulsions had a straight-line characteristic, with only a short toe.

The Science of Imaging

Pyrogallol is 1,2,3-trihydroxybenzene. It hardens gelatin, and was at one time very popular as a developing agent. It is now little used except in holography (see Chapter 18).

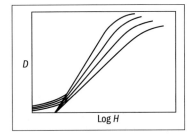

Figure 10.2 Gamma increased with increasing development time, but with the standard pyrogallol developer the inertia point remained stationary.

Strictly, *photographic density*, to distinguish it from density of a substance (kg/m^3), and from optical density, which is an alternative term for refractive index.

Practical units of measurement

I discussed the Weber–Fechner Law in Chapter 3. It relates the intensity of a stimulus to the intensity of perception, and one way of stating it is that when a stimulus increases in equal multiples (e.g. 1, 2, 4, 8, 16...) what we perceive is equal increments (e.g. 1, 2, 3, 4, 5...). This is important both for hearing and vision. Now this relationship, between a multiplying progression and a linear progression, is the same as the relationship between a number and its logarithm. (See Appendix 1.) As the log of 1 is 0 and the log of 2 is 0.3, the series

	1	2	4	8	16	32	64	128
has log values	0	0.3	0.6	0.9	1.2	1.5	1.8	2.1

and so on. The log of 10 is 1 and the log of 100 is 2. So to plot the H & D curve for a film, we expose strips of it for a standard time (in practice around 50 ms) to a range of illuminances, and measure the corresponding transmittances of the processed negative. The horizontal axis represents the logs of the illuminances. The vertical axis is also logarithmic values, the logs of the opacitances (the reciprocals of the transmittances). This gives a curve with a positive slope that fits our ideas of perception (we tend to think of negatives in shades of darkness rather than lightness). The log of the opacitance is called *density*, and can be measured directly using an instrument called a *densitometer*.

How the H & D curve is produced

To construct an H & D curve for a monochrome negative material you need to give the material a series of exposures in logarithmic steps (typically steps of ×1.4, so that every two steps represents a doubling up, or one stop increase. You then process the film, measure the densities of the steps in exposure, and plot them against the log of the exposures in lux seconds. You have to expose and process the film under precise conditions laid down by the International Standards Organisation (ISO).

These standards have been agreed by the former organisations in the USA (ANSI, formerly ASA), Britain (BSI), Continental Europe (DIN) and the former Soviet Union (GOST). By the way, Hurter and Driffield's 'foot-candle' was approximately 10 luxes.

The ISO speed index

The international standard specifies the processing formulae to be used, and requires that the development be such that a log-exposure increment of 1.3 units from the speed point shall have produced an increment in density of 0.8 units. The emulsion speed is then computed from the formulae

$$\text{Arithmetic speed} = 0.8/H_M$$

$$\text{Logarithmic speed} = 10 \log 1/H_M$$

where H_M is the exposure at the speed point in lux seconds. Those are the ISO figures you see on the film box, e.g. ISO 200/24. The first figure is the arithmetic speed and the second is the logarithmic speed. A doubling of film speed is indicated by a doubling of the arithmetic index and an addition of 3 to the logarithmic index, so that a film twice the speed of the example above would be ISO 400/27. You may care to verify that the arithmetic and logarithmic figures are the same at two points (answer on next page).

Figure 10.3 shows the way these figures are determined. The *speed point* is defined as the log-exposure value corresponding to the point on the H & D curve where the density is 0.1 above 'base plus fog' density (i.e., above the density of a piece of film that has been processed but received no exposure).

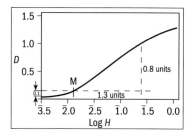

Figure 10.3 Determination of the speed point M (here approximately ISO 64/19.

Average gradient (Ḡ)

The concept of 'gamma' became largely meaningless when multi-coated emulsions were introduced, as the H & D curve no longer had a straight-line portion. The modern method of assessing the inherent contrast of an emulsion is to measure the gradient of a line joining two specific points on the H & D curve. The two points are the speed point and a point on the curve 1.5 log-exposure units farther along the horizontal axis (Figure 10.4). This slope is called Ḡ (pronounced gee-bar) (the bar is a symbol indicating an average value). Typical values of Ḡ for general photography are 0.6 to 0.8, depending on the requirements of the subject matter.

Gamma is still used for some special-purpose emulsions such as X-ray and lithographic materials.

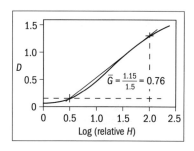

Figure 10.4 Determination of Ḡ (here approximately 0.76).

What the H & D curve can tell us

The rigorous method described above is needed only if you need an exact figure for emulsion speed in ISO units. For most purposes we can be content with figures for the log of the *relative* exposure. We can learn a great deal from the H & D curve, especially when we compare curves made using different films or different processing techniques:

- Effect on speed and contrast of variations in development times and temperatures.
- Effect of varying the composition of the developer.
- Tonal reproduction characteristics of a film.
- Exposure latitude of a film.
- Variation of speed and contrast with wavelength.
- Effect of after-treatment of a negative.
- Reciprocity Law failure.

Robert Wilhelm Bunsen (1811–99) is best known for the Bunsen burner, which he didn't in fact invent but only refined. He did, though, invent the Bunsen cell, the Bunsen ice calorimeter and the Bunsen grease-spot photometer. He isolated magnesium using electrolysis, and pioneered the use of spectroscopy in analysis, discovering rubidium and caesium. He never married; contemporary comment attributed this to an alleged aversion to soap and water. It seems more likely that his unattractive aura stemmed from lengthy research into certain malodorous organic compounds of arsenic (which nearly killed him). Henry Enfield Roscoe is a much more shadowy figure in the annals of photochemical history.

The formal version of the Law states that a photochemical effect is proportional to the total incident light energy.

A do-it-yourself H & D curve

The usual way to obtain the data for plotting an H & D curve is to expose the film to a controlled series of exposures. Now, you can control exposure by varying either the light intensity or the duration. Bunsen and Roscoe's Law of Reciprocity states that exposure, H, is directly proportional to both illuminance E and duration t, i.e.,

$$H = Et.$$

The answer is at ISO 12/12 and ISO1/1. Did you get them?

The Science of Imaging

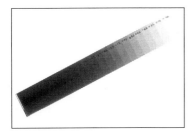

Figure 10.5 A Kodak density step tablet.

Figure 10.6 A simple densitometer (X-ograph).

Finding a best curve can be a tedious process involving error estimates and statistical calculations. In this case we know that the curve will be a smooth one, so steer an average path, and ignore any obvious rogue points.

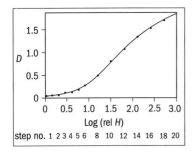

Figure 10.7 A typical plot of an H & D curve of a modern general-purpose emulsion.

This is the principle that says that if you open the aperture one stop and at the same time halve the exposure duration, the exposure (i.e., the effect on the emulsion) will be the same.

The easiest way to obtain a suitable series of exposures is to make a contact negative of a Kodak density step tablet (Figure 10.5). This is a strip of developed black-and-white film with a set of steps of densities that increase in steps of 0.15, i.e., 0.15, 0.3, 0.45, 0.6 and so on, up to 3.0. This represents an arithmetic contrast of 1000:1 and covers the whole usable range of a general-purpose negative emulsion. If you have an enlarger, make the exposure on the easel, with a piece of black paper underneath and a piece of glass on top to weigh the sandwich down (or you can use a contact-sheet print holder). You will need to find an exposure that will give you some density on about the third or fourth step. (Try 1 s at $f/16$ to start with, and open up the lens aperture if you need to, but do not go much above 1 s.) Alternatively you can expose by means of an electronic flash. (Start at about 1 m distance.) Process the film, and then read off the densities with a densitometer. Densitometers vary from very simple gadgets resembling an exposure meter to elaborate equipment that measures monochrome and colour densities and the reflection density of prints and even plots H & D curves via its own computer. A simple and reliable version is shown in Figure 10.6.

You can use an ordinary exposure meter as a densitometer if you mask its aperture down to the tablet step size. You can calibrate it from the Kodak step tablet. Photographic exposure meters also use a logarithmic scale, where each successive light value (LV) represents an increase in log-luminance of 0.3. Simply place the developed film on a light table and take a reading for each step. (Make sure the light is bright enough to give a reading on the darkest step.)

Now plot your curve. Begin by marking out the tablet steps on the horizontal axis, counting the densest one as step 0, and marking each step as 0.15 log-exposure units. Mark out the vertical axis, also in steps of 0.15 units, and plot your graph. The result should look something like Figure 10.7. Not all your points will lie exactly on the 'best curve', but from its nature the curve has to be smooth, so draw a smooth curve and don't just 'join the dots'.

Effect of varying the development time

Manufacturers of developers usually give lists of recommended development times for various types of film. Sometimes these may be in the form of a graph. So far we have discussed only a single H & D curve derived from a single development time and at a standard temperature. If we vary the development time and produce an H & D curve for each time, we shall finish up with a family of curves similar to Figure 10.8. A short time results in a low value for $\bar{G}$, and longer development times give progressively higher values.

At the same time the speed point ($D = 0.1$ above fog) moves to the left, indicating an effective increase in emulsion speed, until finally the value of $\bar{G}$ becomes stationary and the fog level begins to rise (Figure 10.9). Raising the temperature of the developer has a similar effect, and you can construct time/temperature curves for various values of $\bar{G}$ (Figure 10.9). Most manufacturers of developers supply these data.

How photographic films behave

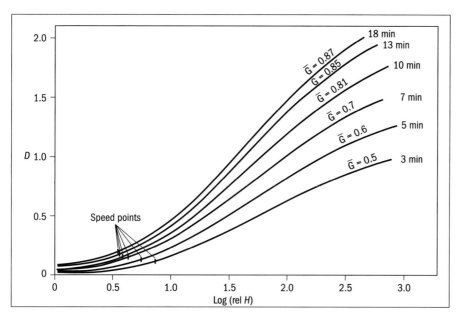

Figure 10.8 Family of H & D curves for varying development times, showing increase in effective emulsion speed and $\bar{G}$ with increasing development time.

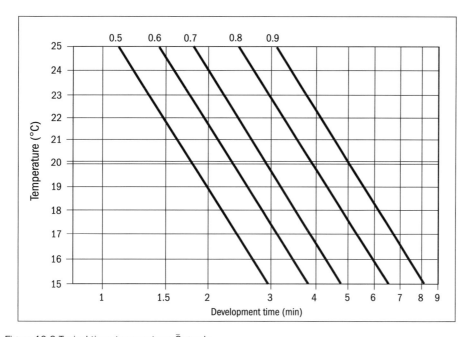

Figure 10.9 Typical time–temperature–$\bar{G}$ graph.

Effect of varying the developer composition

If you process your own negatives you probably have a favourite developer: most photographers have one. Actually, a great deal of nonsense is talked about developers, especially about developers that are alleged to increase film speed. One or two (e.g. pyro-metol) undoubtedly do so, usually at the cost of distorted tone values in the denser parts of the negative coupled with a large increase in graininess.

The Science of Imaging

If you believe your own magic brew really does increase your film speed, do a fully controlled comparative test with the standard developer (see BS 1380, Part 1 1973). You will quickly learn the truth of the matter.

Most developers give a somewhat lower speed than the standard ISO test. This is particularly true of 'fine-grain' developers of low alkalinity, or containing silver halide solvents such as potassium thiocyanate or thiosulphate, or large amounts of sulphite; it is also true of developers containing comparatively feeble developing agents such as the diaminobenzenes (also called phenylenediamines). The errors arise because (a) the standard procedure allows that an increase in development time will increase effective emulsion speed, and (b) the speed index contains a built-in safety factor of more than half a stop. Furthermore, if your subject matter has a contrast at the film plane of less than 32:1 your exposure meter will be indicating more exposure than you need (see Latitude, below).

Tonal reproduction of a film

In the earlier days of photography the official line was that in order to obtain perfect tone reproduction you had to get the exposure range wholly on the straight-line portion of the H & D curve. This belief not only resulted in some very dense and grainy negatives (and long exposures), but was also quite erroneous. In the 1930s the American Standards Association (ASA) was looking for a way to establish a valid speed point. The H & D criterion no longer worked, and the European Scheiner system used an unrealistic development procedure; and in any case its speed point was difficult to determine. So the Association had a large number of prints made, from negatives given a range of exposures, and asked thousands of members of the public, as well as professionals, to assess the prints for quality. From the statistical results they picked the 'first excellent print', i.e., the picture possessing excellent tone quality that had the least exposure, and measured the density of the deepest shadows recorded in the negative to find the 'speed point'. As we have seen earlier, this proved to be well within the toe region. This appears to mean that people will tolerate a loss of contrast in the shadows; but this doesn't take into account the characteristics of the print material, as we shall see later.

The resulting ASA criterion for finding the speed point was based on the slope of the H & D curve at that point, and needed a complicated geometrical construction to find it. The European Standards Institute DIN had decided to fix its speed point at a density of 0.1 above fog, and as the ASA criterion in practice came to the same result in nearly all cases it was agreed that this should be the new criterion. With the formation of the ISO, it was formally adopted universally.

What the H & D curve helps us to see is the way in which a particular emulsion handles tones and contrast for highlights, mid-tones and shadows; it allows us in addition to compare the tone reproduction of different films. Figure 10.10 gives some idea of this.

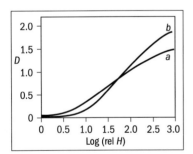

Figure 10.10 Comparison of two emulsions: (a) fast general-purpose emulsion; (b) fine-grain emulsion with higher inherent contrast.

Comparison of two films

To compare two films directly, you need to ensure they both get precisely the same exposure under the step tablet. The best way to do this is to cut both films together down the centre line with a guillotine, and expose them butted together under the step tablet. Process them to the same value of $\bar{G}$ (the developer manufacturer will give this information), plot the H & D curves on the same sheet, find the speed points and read off the difference in log units. To get the logarithmic speed difference, multiply this figure by 10; to get the arithmetic speed difference, take the antilog.

To compare the tone reproduction you will have to shift one of the curves until the speed points coincide – or you can simply do the comparison by eye. You will be able to judge, for example, which film has a better tone separation in the shadows, or a more uniform tone range throughout, than the other.

Exposure latitude

An ideal exposure should fit on the H & D curve so that the deepest shadow where you want to record detail receives an exposure that takes it precisely to the speed point (M on Figure 10.11); anything less will result in the shadow exposure being too far into the toe region, where it will be recorded at a very low contrast, if at all. However, with an average subject you can give a good deal more exposure than the minimum and still be sure of a result having acceptable tone quality (though with some increase in graininess).

Exposure meters (including through-the-lens meters) are calibrated to give a correct exposure for a subject that has a contrast at the film of 32:1, or 1.5 in logarithmic terms. Halfway up this scale is a log-illuminance of 0.75. This corresponds to a subject luminance that is roughly 18 per cent of that of a full white. This is why you can take an exposure reading on an 18 per cent grey card.

The upper limit of usefulness for a given H & D curve is somewhat arbitrary. It depends on how much graininess and flare you can tolerate. Plainly, the higher the subject contrast, the less latitude in exposure you have. Equally plainly, if your subject matter has a very low contrast you can move down below the meter indication and get away with it, because the criterion is still that the deepest shadow should correspond to the speed point (Figure 10.12).

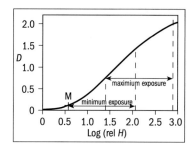

Figure 10.11 With general-purpose emulsions there is an exposure latitude of about three stops (0.9 on the log-exposure scale) before the quality deteriorates unacceptably.

The grey card actually *looks* more like 50 per cent – the reason for which you should now appreciate!

This is the origin of one of the fallacies concerning the uprating of film speeds. The film speed does not change, of course. It is the meter that is giving a false reading, because of the nature of its calibration.

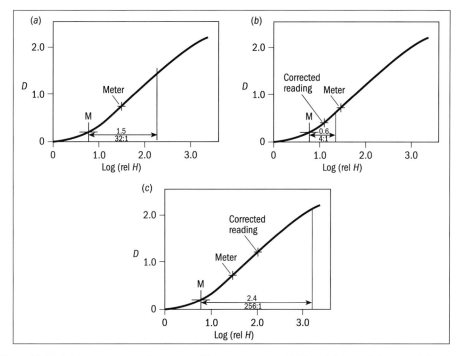

Figure 10.12 Subject contrast and exposure: (a) normal contrast (321 or 1.5 log units); meter indication correct; (b) low contrast (4:1 or 0.6 log units); meter indication high; indicated exposure can be reduced by 50 per cent; (c) high contrast (256:1 or 2.4 log units); meter indication low; indicated exposure must be increased by 100 per cent.

The Science of Imaging

Variation of speed and contrast with wavelength

This may be important when you are using strong filters with black-and-white film. If you have constructed a set-up for producing your own H & D curves (see earlier Box), all you need do is fit your filters in turn to your light source and expose and process your film strips identically. Comparing the speed points and $\bar{G}$ values (including an unfiltered film strip) will give you the relative contrasts and filter factors for that light source. This will usually be 3400 K (tungsten) for an enlarger lamp. For a 'daylight' result you will need to fit an 80B (−130 mirek) filter to your light source, or use electronic flash.

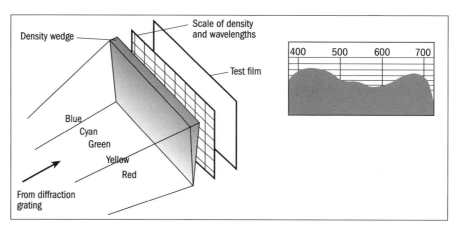

Figure 10.13 Schematic of wedge spectrograph, and a typical result.

Film manufacturers use a broad density wedge, and project a spectrum on to it using a diffraction grating (Figure 10.13). The result can be read off for any wavelength, and you can get an idea of the spectral sensitivity of the emulsion simply by looking at the result (it is called a *wedge spectrogram*). These are often included in manufacturers' data.

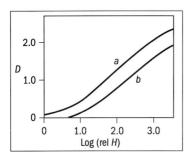

Figure 10.14 Effect on the H & D curve of use of Farmer's reducer (a) before treatment, (b) after treatment.

Farmer's original formula used the two solutions mixed together. This gives an unpredictable result with present-day emulsions, and the bath becomes useless after a few minutes.

Effects of after-treatment

At one time it was common practice to make any negatives that had been incorrectly exposed or processed undergo 'after-treatment'. This was useful when you could only get the finest quality print papers in a single grade, or when you wanted to make a print by a process such as platinum or carbon, which required a precise density range in the negative. The two types of treatment were known as *intensification* and (density) *reduction*. We scarcely ever resort to intensification today, but reduction can be useful on occasion, and is often used by experienced printers to clean up highlights in prints. The most frequently used reducer is Farmer's reducer, called after its inventor, Howard Farmer. Its most controllable version uses two baths, the first being a 1 per cent solution of potassium hexacyanoferrate(III) (potassium ferricyanide), $K_3Fe(CN)_6$, followed by a bath containing a 10 per cent solution of sodium thiosulphate, $Na_2S_2O_3$, repeated as necessary. This gives uniform density reduction. Figure 10.14 shows the effect.

How photographic films behave

Reciprocity Law failure

Bunsen and Roscoe's Reciprocity Law is only approximate. It breaks down for very short and very long exposure durations. The former circumstance is not usually significant, as the only result is a slight reduction in contrast, and for general-purpose emulsions the effect is negligible for exposure durations down to around 0.1 ms. Long exposure durations are another matter. At very low light levels so much recombination takes place within the emulsion that latent image formation is inefficient. In the shadow regions of the optical image the light level is even lower, and the effect is aggravated. Thus not only is emulsion speed lowered, but contrast is increased (Figure 10.15).

H & D curves for colour negatives

To plot H & D curves for colour films you have to make separate measurements through red, green and blue filters. An 'ideal' colour film would show three identical curves. In practice the characteristics match well over the main part of the curve, though less well in the regions of highest and lowest density, so that over- and underexposed negatives may show colour casts in the highlight or the shadow regions respectively. All colour negative films have a low inherent contrast; this helps to extend the latitude, and allows for lighting that does not have the correct colour temperature.

H & D curves for transparency materials

Black-and-white reversal materials for home processing have not been around for a long time, but you can get good results from fine-grain films if you use the manufacturers' formulae for reversal development. The method is to develop the film in the normal way, but to remove the metallic silver instead of the unchanged silver halide by turning it into a soluble compound such as silver sulphate. You wash this out, then expose the film to light and develop the remaining silver halide, which is, of course, a positive image. The resulting H & D curve will look like Figure 10.17. As you can see, with this method, more exposure results in less density. There is little or no latitude, as a transparency needs to have close to 100 per cent transmittance in the brightest highlight. (This is one case where Farmer's reducer can put a little sparkle into a dull result.)

Colour transparencies are processed in a broadly similar way, except that all the silver is removed in the last processing step. As with colour negatives, you need to measure red, green and blue densities, and a correct exposure places the brightest highlight close to zero density. This means that on exposure all (or nearly all) of the silver halide in the brightest highlight has to bear a latent image, so that there is hardly any left for the second development. So instead of exposing for the shadows as you do for negatives, you expose for the highlights. This means that through-the-lens metering is less reliable, particularly with subjects that are not average in luminance or contrast. You have therefore to make allowances for dark subjects, which would otherwise come out overexposed and too pale, and light subjects, which would otherwise come out underexposed and too dark. Low-contrast subjects may need extra exposure and high-contrast subjects less, depending on what you require of the final transparency (Figure 10.18).

There is more on reciprocity failure in Chapter 12.

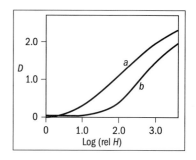

Figure 10.15 H & D curves showing reciprocity failure: (a) Bright image (1/125 s); (b) dim image (5 s).

When taking density readings you need to have the filter between the negative and the densitometer head or over the densitometer light source. The filters are the narrow-cut variety, and you need to re-zero the densitometer for each filter. Manufacturers' data often show the three curves zeroed to neutral grey, which means that the green and blue densities start much higher because of the masking dyes (Figure 10.16).

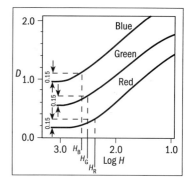

Figure 10.16 H & D curves for the three emulsions of a colour negative showing speed points. The considerable differences in basic density are due to the orange masking dyes. The speed point for colour negative materials is 0.15 above fog level.

121

The Science of Imaging

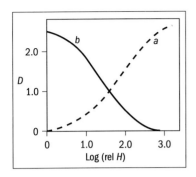

Figure 10.17 H & D curve for a reversal-processed black-and-white film: (a) initial development; (b) final result.

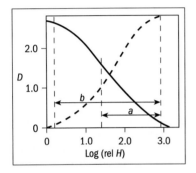

Figure 10.18 Correct exposure in a colour transparency: (a) low-contrast subject; (b) high-contrast subject. The broken line shows the initial negative development.

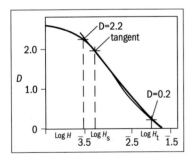

Figure 10.19 Highlight and shadow speed points of a colour reversal emulsion.

Note: to simplify the above figures I have ignored the insensitive remnants that escape the first development and would cause a degradation of the final image. In practice these are removed by solvents in the first developer.

Push processing

In the early days of Ektachrome, the first successful subtractive colour reversal material intended for home processing, users were given stark warnings by the manufacturer (Eastman Kodak) that any variation from the set times and temperatures would result in tone and colour distortion. This, of course, was regarded as a direct challenge to some of us, and we soon found that it wasn't true. We discovered that we could compensate for exposure errors of up to one stop either way by varying the first development time. This technique is now routine, and the makers condone 'push processing' and even publish data on techniques for increasing effective film speed. It works because the first development is a conventional black-and-white type negative development. Increasing the first development time increases the overall density of the negative image, leaving less silver halide to form the positive colour image and reducing its highlight density. It also reduces the maximum black to some extent, but it still allows you to uprate the film speed and to get a usable result under adverse lighting conditions. Most modern emulsions permit push processing up to two stops, by adding up to 40 per cent to the first development time. At the other end of the scale, you can get away with up to one stop overexposure by shortening the first development time by up to 20 per cent, with no loss of quality.

Speed criteria for colour materials

Colour negatives In general the speeds of the red, green and blue recording emulsions are not exactly the same, and the speed point is taken as a density of 0.15 above base plus fog, taking the average of the green and the slower of the other two curves, usually the red (Figure 10.18). Processing is according to the manufacturer's instructions.

The speed point H_M is given by

$$\log H_M = (\log H_G + \log H_R)/2$$

and the arithmetic speed is

$$\sqrt{2} \div H_M$$

where H is the exposure in lux seconds.

Colour transparencies Films are again to be processed according to the manufacturer's instructions. This time, however, two speed points are established. The *highlight point*, $\log H_t$, is at a density of 0.2 above base plus fog, and the *shadow point*, $\log H_s$, is at the point where a straight line drawn from $\log H_t$ is tangent to the shoulder of the curve (or 2.0 above the minimum density on the vertical axis if this is less). This is shown in Figure 10.19.

Sensitometry is a big subject, and a full account of it would fill a medium-sized book. So that you can have a breather, I have divided it into two separate chapters. We have now seen how a photographic emulsion behaves in the camera, and the way you can quantify its properties so as to obtain useful information. The next step will be to discuss the properties of print materials, and the way we match their characteristics to the result of the first stage.

Chapter 11 How photographic print materials behave

A cautionary note: can sensitometry harm your image?

Because this book is about the science that underlies the making of an image, it is concerned with producing an image that is as faithful as possible to the tones and colours of the original subject matter. This chapter lays a great deal of stress on this. Faithful recording is certainly the goal in most scientific, technical and medical applications of imaging. If, though, you are a pictorialist (a term sometimes used deprecatingly, though not by me), your aim will be to produce a beautiful or striking image, and you don't always get such images by slavishly following the 'rules'. But when you actually know the 'rules', you will be in a far better position to bend them to the advantage of your image. In this chapter I assume, just as other manuals of photographic theory have always done, that the ideal print has both a full black and a full white somewhere in it, but that no detail is lost in either shadow or highlight areas; and, moreover, that all the intermediate tones are spaced out in exactly the same way as they are in the original subject. Quite a tough order, as we shall see.

The H & D curve for print materials

There is an important difference between a negative and a print: the negative is only an intermediate stage, so it doesn't matter how it actually looks; but a print is made to be looked at. This is particularly true of colour negatives which, as I showed in Chapter 6, are made with deliberate colour distortions in order to counterbalance those of the print emulsions. The other, less obvious, difference is that a negative can have a very high contrast, up to 400:1, or a density range of 2.6, whereas a print is limited to not much more than 32:1, or a density range of 1.5. This is because although a print is exposed and developed in much the same way as a negative, you view it by reflected light, not transmitted light. We define *reflection density* as the log of the reciprocal of the reflectance; and although in theory this can go as high as the transmission density of a negative, in practice there is scarcely any visible difference between reflection densities of 1.6 and 2.6: they both look black.

In order to measure reflection density you need a modification to the densitometer. The illuminating light needs to be on the same side of the print as the densitometer head, and it must be at such an angle that none of it is reflected directly off the print surface into the head. However, if you are going to make your own measurements and don't possess a reflection densitometer (there is no cheap version), you can do equally well by buying a Kodak calibrated grey scale, and matching it against your test strip by eye. It sounds crude, but the human eye is remarkably good at matching grey tones, and you should be able to estimate reflection densities to two decimal places after some practice.

Another difficulty is that whereas it is easy to define 100 per cent transmittance, which equates simply with total transparency, it is less easy to define 100 per cent reflectance, as the directionality of the reflectance comes into consideration. In practice the zero is usually established by taking the best practical totally diffuse reflector, freshly scraped magnesium carbonate, as the standard.

The Science of Imaging

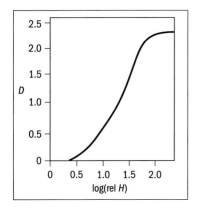

Figure 11.1 H & D curve for a typical print paper.

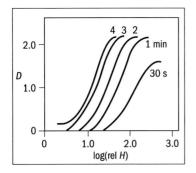

Figure 11.2 Effect of development time on H & D curve for a print.

As with negative material, print emulsions have an H & D curve, with reflection density plotted against log of relative exposure, but there are important differences in the shape of the curve. For one thing, it reaches its maximum value quickly, and with longer development time its slope and density range increase little if at all. The curve has a sharply angled shoulder, and a maximum density only a little higher (Figure 11.1).

You will notice that the curve becomes steeper as it approaches maximum black. When you remember that this part of the curve is the part that is going to reproduce the lowest densities of the negative (which are on the toe region of the negative H & D curve) you will see that this is going to help to compensate for the low shadow contrast of the negative. (I mentioned this in the previous chapter.)

If you increase the development time, the H & D curve behaves somewhat differently from that of the negative. Instead of increasing in slope, the curve stays the same shape, but moves towards the left. This indicates that the effective speed of the material is increasing with increasing development time (Figure 11.2). There is thus a latitude in exposure controlled by varying the development time. You can get a good-quality print even though your exposure in the enlarger was somewhat less or more than the 'correct' one, by giving a somewhat longer or shorter development time ('stewing' or 'snatching' in darkroom parlance).

The density range of a paper

When talking about negatives I used the term 'contrast' to mean the density range of a negative; and it is a valid term for a print, too. However, it is more often used in this case to mean much the same as 'gamma', i.e., the slope of the H & D curve. In the days of Hurter and Driffield both negative and print emulsions had long straight portions, and it was possible to say with some truth that to get an accurate tone reproduction, the product of the negative and print gamma values should be 1. This is still approximately true when you are making a duplicate negative via an intermediate positive (or vice versa); but the procedure involves discarding the toe regions of both negative and positive, not an appropriate procedure for a negative–print combination. What we do in practice is to try to get a full tone range print without blocked up shadows or black highlights, and this means matching the density range of the negative to the log-exposure range of the print material. This latter is the log-exposure range that will give precisely the gamut of available print tones. Figure 11.3 illustrates this. A negative with too large a

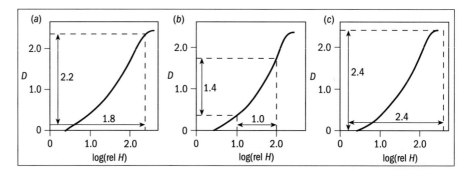

Figure 11.3 Matching the negative density range to the print log-exposure range: (a) correct; (b) too small; (c) too large.

How photographic print materials behave

density range will give a print with burnt-out highlights or blocked-up shadows or even both; a negative with a small density range will not give a full black and a full white at the same time: its best compromise will be a narrow range of greys.

Some print processes, for example bromoil, platinum and some special types of display print material, have log-exposure ranges that cannot be extended or compressed by variations in processing, so the negative density range has to be carefully controlled to match the print characteristics. This is also broadly true of colour print processes.

Present-day print materials are made with a large range of log-exposure ranges, and by tradition these are called 'soft' (log-exposure range 1.3 to 1.6), 'normal' (0.9 to 1.2) and 'hard' or 'contrasty' (0.6 to 0.8). Grades go from 00 (very soft) to 7 (very hard), though the significance of these numbers varies from one manufacturer to another. In theory, provided the log-exposure range of your paper is equal to the density range of your negative, you should be able to produce an 'excellent' print, at least by the criteria set out above.

Of course, any good printer is capable of controlling print densities locally by dodging and burning in tricky areas, but this involves manipulation of the image for aesthetic reasons, and is not really relevant to the points I am making here.

Contrast control of prints

Photographers are sometimes frustrated by a negative that appears to require a print paper with a contrast grade exactly halfway between two available grades. There are two ways of dealing with this problem. The first lies in the nature of the enlarger light source, and what is known as the *Callier effect*. Most enlargers have condenser lenses that avoid wasting light by directing it towards the enlarging lens. As this light passes through the negative, some of it is scattered by the silver grains that make up the photographic image, and consequently doesn't reach the enlarging lens at all. This effect is greater the more grains are present, and this results in an increase in apparent contrast of the negative by a factor called the *Callier coefficient*. This is around 1.2 for most condenser enlargers, and is slightly higher for the shadow tones. You can eliminate the Callier effect by putting a diffuser such as a ground glass or a piece of tracing paper under the condenser. This is equivalent to going down half a grade of contrast.

Some enlargers don't have a condenser, but employ a diffuser head. So you can't use this method. Instead, you can change the developer formula. By using a plain metol developer you can lose half a grade of contrast, and by using a plain hydroquinone developer you can gain half a grade. This doesn't work with all papers, though, as some print emulsions already contain a small quantity of a developing agent.

Small-source projection printers for special purposes may have a Callier coefficient of 1.5. However, these show up grain and surface defects very badly, and most amateur and professional enlargers use a fairly large source, so that the Callier effect is somewhat muted.

Variable contrast papers

At one time a photographer tackling a variety of assignments needed to keep at least one box of each grade of paper in order to produce a satisfactory print from every negative shot. Today this need has all but vanished. All the most popular surfaces of paper are now available with variable contrast (VC). As I explained in Chapter 9, a variable-contrast print paper has two layers of emulsion: one is blue sensitive and has a high contrast; the other is green sensitive, and has a low contrast. Varying the proportion of blue to green light in the enlarger by means of yellow or magenta filtration varies the effective contrast of the paper. Most papers

The Science of Imaging

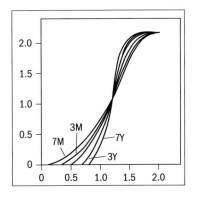

Figure 11.4 Some of the grades obtainable on a variable-contrast paper.

are balanced to give a result that is slightly softer than normal when exposed without a filter (Figure 11.4).

Cascading H & D curves

If you are involved in scientific, technical or medical photography you will no doubt be concerned with more than just shadow and highlight densities. You will also want to know how the rest of the tones are depicted; that is, how truly the tones, as recorded, relate to those of the original subject matter. This is important, too, when you are making a duplicate via an interpositive or internegative. To obtain the information you need, you have to 'cascade' the characteristics of negative and print, or of negative, interpositive and duplicate negative, and compare the result with the original.

The usual way to do this is with a *quadrant diagram*. Figures 11.8 and 11.9, on later pages, show two types of diagram cascading characteristics. Figure 11.6 shows the response of the negative material matched on to the enlarger characteristics, and this response is in turn matched on to the print characteristics; these are then compared with the original in Figure 11.8. Figure 11.9 shows the duplication of a negative via an intermediate positive, transferred the same way.

These diagrams may seem a bit daunting, swallowed whole. So let us take them a bit at a time. Here is the response of the negative material. You have seen this before, of course. But notice the way the log-exposure values for the subject are recorded as densities: the tones are squashed up at the lower end because we have used the toe region of the curve. This is evident on the vertical scale (Figure 11.5).

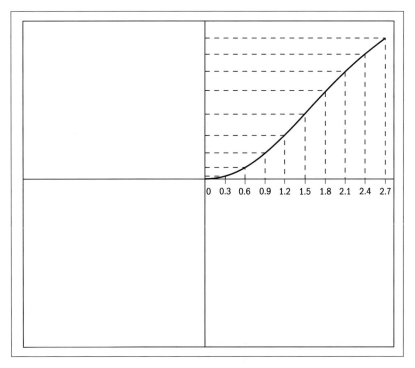

Figure 11.5

How photographic print materials behave

These values are projected as log-illuminances by the enlarger. Because of the Callier effect (see above), the contrast is increased a little in the toe region, and consequently the enlarger adds its own characteristic (Figure 11.6).

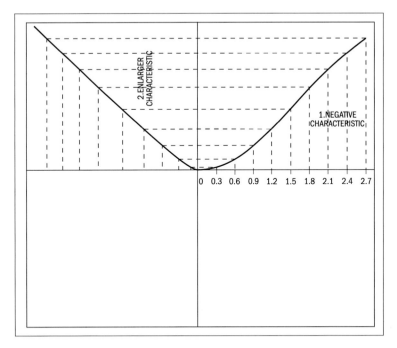

Figure 11.6

This new log-exposure series is now projected on the paper, and modified by the paper's own characteristic (Figure 11.7).

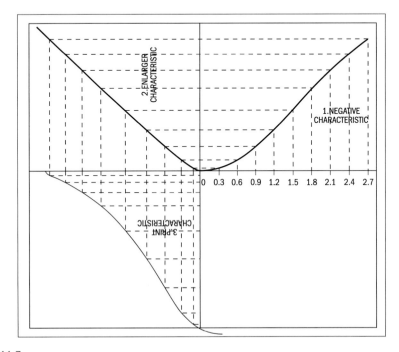

Figure 11.7

Finally, these are compared with the uniform spacing of the original subject matter. We now have all four quadrants (Figure 11.8).

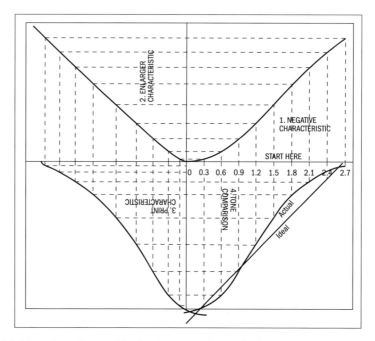

Figure 11.8 A full quadrant diagram. The fourth quadrant (lower right) matches the final print, tone by tone, to the original subject matter. Note: Intervals of 0.3 log units have been chosen to match steps on the Kodak step tablet. Each step represents a doubling of the illuminance.

This is, of course, something a skilled printer would be able to correct by shading and burning in during the printing exposure.

As you can see, the final print (in this instance) is a fair match to the original scene, but has lost some tone separation in the highlights and shadows, and has excessive tone separation in the mid-tones.

To satisfy yourself, see if you can follow through a similar diagram for making a duplicate transparency via an internegative (Figure 11.9).

Figure 11.9 is simplified. The H & D plot of duplicating film is not a dead straight line, its curve being designed to compensate for deficiencies in the printing system. But for the purpose of this exercise I have approximated it to a straight line.

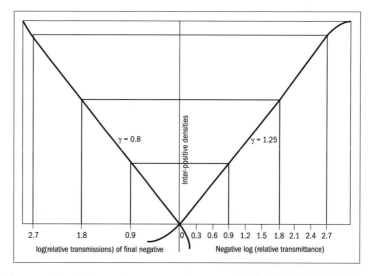

Figure 11.9 Quadrant diagram for duplicating a negative via an interpositive. As duplicating emulsions have a nearly linear response it is only necessary to ensure the product of the two gammas is unity.

So it is not all that difficult, really. But if you have found this all somewhat heavy going the first time round, let me give you a simpler example: the making of a positive print directly from a positive transparency. This time the H & D curve of the transparency is as in Figure 11.10.

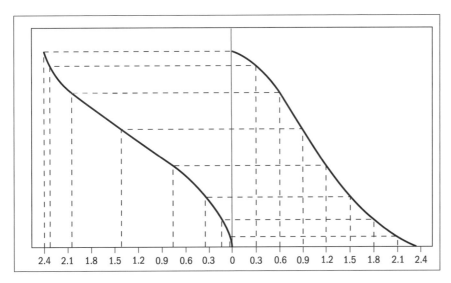

Figure 11.10 Quadrant diagram for a one-stage transparency copy.

When this is transferred to the print H & D curve (which is much the same shape) the result is as in Figure 11.10.

Even without the transfer lines you can see the way the shadow and highlight tones are bunched up and the middle tones spaced out. This is a permanent hazard in positive–positive printing, and in the direct one-step duplication of transparencies.

Colour print papers

Papers for direct positive printing As I discussed in Chapter 6, there are two types of print material for producing direct positives from transparencies, the reversal-process type and the dye-bleach type. The former employs a process similar in principle to that of transparency material, with a first developer, a colour developer and a bleach-fix stage. The latter employs a developer followed by a dye-bleach bath that destroys the dyes already present in the emulsion in proportion to the amount of silver present. The final stage is a bleach-fix that removes the silver. All positive papers have a long exposure scale (i.e., a low inherent contrast) with a response that is as linear as possible.

However, as we have seen, there will always be small tonal distortions at the ends of the scale, and any colour errors in the original transparency will be aggravated, too. Because of the yellow cast of the magenta dye, the contrast of the yellow image has to be lowered a little to balance it, so that yellows are too light and greens bluish. For the cyan with its spurious red content, the red contrast (i.e., yellow and magenta) has to be lowered too, lightening the yellow further and lowering the saturation of blues and greens. This sounds terrible, but in practice

Unfortunately, these deficiencies are all too obvious in prints made from transparencies of flowers, when you compare them against the original. Seedsmen's catalogues are notorious for their inaccurate colour renderings. That is why botanical textbooks still prefer watercolour illustrations to photographs.

the effects are slight, and you have to go through several generations of copies before the effect becomes obvious.

Papers for negative–positive printing As colour negative material contains integral masks to correct for errors in the magenta and cyan dyes, you are not likely to encounter colour errors of the type described above. The print material has a high inherent contrast to match the low value of $\bar{G}$ in the negative material. This means that when you use CP (colour print) filters to balance the final colour, it takes only a very small change in filtration to produce a large swing in the colour balance of the print.

Both negative and positive print materials have a spectral sensitivity different from that of camera film. Instead of the red, green and blue sensitivities adding up to a continuous, roughly constant sensitivity throughout the spectrum, the emulsions respond only to comparatively narrow bands of radiation, much like the narrow-band filters of Figure 2.14b, but farther apart. This prevents the loss of saturation that would otherwise occur through overlap. It also provides a complete gap in sensitivity around 590 nm, so that it is possible to use a low-pressure sodium lamp as a safelight, with some caution. (The sodium spectrum does contain lines other than the yellow-orange pair at 589.0 and 589.6 nm, though they are not nearly so bright.)

The emulsions of colour print papers are matched approximately to tungsten light at 3200–3400 K; for negative print papers this has to take into account the overall orange hue of the dye mask. In addition, there are usually small errors that require initial corrections. These are usually on the yellowish side, suggesting a small amount of yellow filtration. This is deliberate, as it mostly avoids the need for cyan filtration (cyan CP filters carry filter factors that increase the required exposure by amounts that need to be calculated).

In general, you would not be advised to vary processing times arbitrarily with colour print material, as this leads to changes in colour balance. However, provided you allow for these changes, you can increase the contrast of a negative/positive print by extending the development time by up to about 40 per cent.

Digging deeper

In this area of imaging science you can dig very deeply indeed. The study of the characteristic curves associated with electronic recording and reproduction is somewhat simpler, and I shall be dealing with this later, in the appropriate chapters. There is a good section on photographic sensitometry in *The Manual of Photography* (9th edition, ed. R E Jacobson, Focal Press, 2000), which deals more thoroughly with the practice of image measurement than I am able to do here. A crude but highly practical way of looking at sensitometry for those who are diffident about a graphical approach is afforded by the Zone System, a method of approaching exposure and print quality in terms of numbered subject brightnesses or 'zones'. The Zone System was adopted wholeheartedly by the great photographer and teacher Ansell Adams, and you could do worse than read his two books *The Negative* and *The Print* (Little, Brown, republished 1995); the first of the trilogy, *The Camera*, is also worth reading. However, the Zone System was designed for black-and-white photography, and does not fit very well into colour transparency work. If you are already familiar with the Zone System and have discovered its limitations, *Beyond the Zone System*, by Phil Davis (4th edition,

Focal Press, 1998) is an excellent sensitometry textbook. There is also a fairly comprehensive discussion of sensitometry in Stroebel, Compton and Current, *Basic Photographic Materials and Processes* (Focal Press, 1996). Perhaps the best of all is Jack Eggleston's *Sensitometry for Photographers* (Focal Press, 1984). Now out of print, but still easy to find through your local library.

Chapter 12 Image modification

It used to be said that the camera never lies. Your camera can certainly furnish better evidence of some event than your memory can. But at best a camera can still tell only part of the truth. We have already seen in Chapter 5 that the detail in a photograph is limited by the resolving power of the camera's optical system and the recording medium. We have seen, too, how the optical transfer characteristics of the system can generate spurious image artefacts. We need to look at the ways we can tackle these errors by modifying the image.

Of course, by 'modifying' I don't mean putting anything into the image that wasn't there in the first place. There are plenty of techniques around that *can* make images lie. Indeed, modern digital methods can produce extravagant visual fabrications that would have gladdened the hearts of the artists who modified the news photographs during Stalin's wicked reign (see Digging Deeper).

So let's begin again. Even though a photograph can't show the whole truth, almost all of it may be buried there in the image, if only we could get at it. Image modification helps us to make it visible. We can go a long way towards finding what *is* there from the optical transfer function (OTF) of the system, which enables us to obtain an objective assessment of the way the system depicts an image.

This chapter looks at the deviations from a 'true' representation by a camera, and shows how these effects can be controlled and minimised.

We can do this for the human eye too, though, unfortunately, not very well for the brain.

Contrast control

A photographic subject may have a contrast that is too high for the recording material, so that its range of luminances cannot be recorded in a linear manner; in a photographic print, the shadows and/or the highlights are blocked up. Again, if the contrast of the subject is low, the recording material may only be able to show a muddy image with inadequate tone separation for the distinguishing of detail. Before the Second World War, many print papers were available in only one grade of contrast, and if the negative didn't have the right density range, there was little chance of obtaining a good-quality print. The only way to salvage your image was to change the gamma of the negative by intensification or reduction. Intensification involved depositing an opaque substance such as metallic chromium on the silver grains; reduction involved the removal of silver by the combination of an oxidising agent with a solvent (e.g., Farmer's reducer). Nowadays, a more reliable method is to make an intermediate contact positive on film, then a duplicate negative, developing these to give the right final contrast. You can obtain special duplicating film with very high resolution for this purpose. You can also duplicate colour negatives and transparencies using internegative film, but this technique allows less control over the contrast.

Shading, dodging and burning in

These terms are photographers' jargon for the selective control of exposure when making a print with an enlarger. Shading means holding back the exposure over a large area in order to print up a region that would otherwise be too light, say an

Image modification

overexposed sky. Dodging is holding back the exposure in a small area in order to prevent its printing too dark and losing detail; and burning in is its opposite, giving extra exposure to a small area; shading, on a smaller scale. Some photographers keep a stock of little pieces of card mounted on wire handles for dodging, and various shaped holes for burning in.

Shading, dodging and burning in are ways of modifying the negative's H & D curve locally, so as to bring all the negative densities within the print paper log-exposure range. By using these devices the printer is enabled to use a more contrasty grade of paper and thus to get a good tone separation without sacrificing tone quality in the highlights and shadows. These 'tricks of the trade' are in fact a simple form of unsharp masking, as we shall see.

Shading techniques with VC and colour print materials If you have exposed your original material incorrectly, you may find you have muddy shadows (underexposure) or blocked-up highlights (overexposure). In both cases you will need normal contrast print paper for the main part of the tone scale, and higher contrast for the shadows or highlights. In the old days there were after-treatments that purported to balance up these errors in the negative, but the results were not very reliable.

With today's variable-contrast papers it is possible to vary the contrast as well as the density of the print by shading with colour filters: magenta to shade up with increased contrast, and yellow to shade up with reduced contrast. In colour printing, you can correct small local errors in hue by dodging with pieces of CP filter of the appropriate colour.

Unsharp masking

Unsharp masking is a subtle and ingenious method of increasing the sharpness of an image. It may at first seem illogical that you can make an image sharper by coupling it with an unsharp one, but it is indeed logical enough. What you do is to make a slightly unsharp positive transparency from your negative, with a low contrast (say a $\bar{G}$ of about 0.3). You then fix it in register with your negative and print the combination on paper with a contrast a grade higher than you would have used for the original negative. As the mask is unsharp, it will have lowered the contrast in the larger areas only, leaving the fine detail unaffected. Hence this fine detail will be printed in increased contrast, while the larger features will be at the original contrast. The sharpness will be noticeably improved (Figure 12.1).

The principle is easily appreciated in terms of the modulation transfer function of the combined system. Figure 12.2 shows diagrammatically the way it works.

You can see now why I suggested that dodging and burning in amounted to a crude form of unsharp masking. By varying the exposure time locally you are in effect lowering the contrast of large areas of the negative.

The traditional method of unsharp masking is to make a contact print from the negative on duplicating film using a diffuse light source, with the negative and print film base to base so that the emulsions are separated by twice the base thickness. You then develop the positive in a low-contrast developer such as glycin or phenidone (POTA), and tape the finished mask to the negative in register, emulsion

Professional printers usually shade with their hands. It helps if you are good at making rabbit shadows at children's parties.

Some years ago I took part in a project to test all the published intensifier and reducer formulae sensitometrically. Only one of them worked as it was supposed to do, namely the quinone–thiosulphate formula for proportional intensification. All the others gave results that differed (in some cases wildly) from those described in contemporary textbooks.

Sharpness is a subjective concept. It is correlated to some extent with acutance and resolving power, and has been quantified in these terms, though without much justification (see Chapter 5). Unsharp masking has no effect on resolving power.

The Science of Imaging

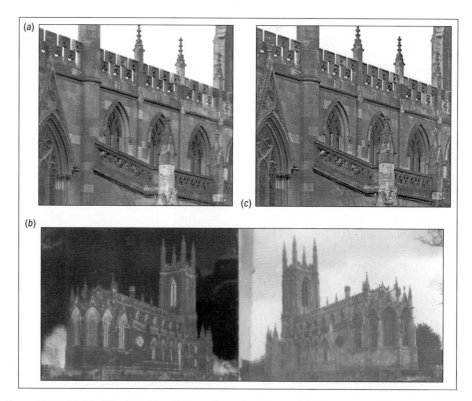

Figure 12.1 (a) Print without masking; (b) negative and unsharp mask; (c) masked print.

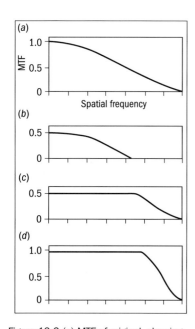

Figure 12.2 (a) MTF of original, showing severe fall-off in contrast; (b) unsharp mask of low contrast and resolution; (c) combined MTF of original and mask; (d) MTF normalised by printing on contrasty paper: microcontrast boosted.

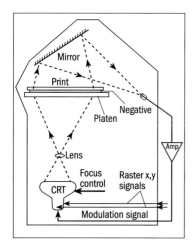

Figure 12.3 Schematic layout of an electronic (flying spot) printer.

to emulsion, for printing on paper one grade more contrasty than would be appropriate for the negative alone.

There are other ways of achieving the same result. One way is to scan the negative in the contact printer or enlarger stage with a flying spot scanner instead of a full-area illuminant. A photocell monitors the transmittance of the negative, and the intensity of the spot (or its speed of scanning) is controlled via a feedback path (Figure 12.3). The spot size determines the degree of unsharpness, and the feedback gain level controls the contrast. In order to avoid visible raster lines the scans overlap considerably.

Adjacency effects as unsharp masking

Electronic printers are expensive, and masking by unsharp positives is tedious. We can obtain a similar (though less controllable) effect by a chemical ruse. Some developing agents, in particular metol, are inhibited by the presence of bromide ions. So if you make up a metol developer without any bromide, the bromide generated during development will hold back the development wherever it appears in the emulsion. The densest areas of the developing negative are held back in comparison with the lightest areas, so the overall contrast will be held down. But – and this is the subtle bit – bromide-rich solution will diffuse from the edges of the dense areas into the light edges, and bromide-weak solution will diffuse from the edges of the light areas into the dense areas. The result is that at the edges of details (i.e., high spatial frequencies) the contrast will be boosted. You can see how this happens in Figure 12.4.

Image modification

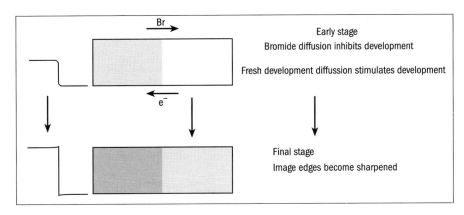

Figure 12.4 Operation of an adjacency developer.

This is one of the reasons for adding potassium bromide to developers containing metol. If bromide ions are already present in the developer, the small amount generated during development will have little additional effect.

This is called an *adjacency effect*. There are a number of adjacency effects, some named after the persons who first described them in print, and these are shown in the Box. This type of development is called high-acutance development, and commercial developers of this type are given names like Acutol and Definol. They work best with fine-grain emulsions of inherently high contrast.

Adjacency artefacts

Adjacency effects are caused by the diffusion of bromide ions from dense to less dense regions of a negative, holding back development locally. Although they may be useful, as in the acutance enhancement described above, they can also lead to unwanted image artefacts:

- *Mackie lines* Where there is a high-contrast edge in the image, the adjacency effect results in a dark line at the edge of the dense area and a light line at the edge of the light area.

- *Streamers* These were a common occurrence in old newsreel footage passed through a continuous processing unit. Bromide produced in heavily developed areas left light streamers behind, and sometimes there were dark streamers behind light areas.

- *Eberhard effect* The density of small dark areas in the negative is greater than that of large dark areas owing to the inward diffusion of fresh developer in the small areas. This can be troublesome in astronomical photography, where the density of images is of crucial importance.

- *Kostinsky effect* The accumulation of bromide between two adjacent point images that are close together causes the distance between the centres of the images to increase. This is also important in astronomy, particularly in the measurement of double stars (Kostinsky was an astronomer).

Modern developers based on phenidone are unaffected by the presence of bromide, and are now invariably used where these effects are important; these artefacts are today seen only on old negatives. However, the beneficial effect of acutance enhancement developers ensures their continued popularity among amateur photographers.

The Science of Imaging

Adjacency effects in colour emulsions

Adjacency effects in the processing of colour emulsions can be arranged to be beneficial, too. They affect not only the adjacent areas of a single emulsion layer but also the emulsions above and below it. In modern emulsions they have an important effect on both sharpness and colour balance, especially in transparencies, where it is desirable to provide some compensation for the inherent errors in the cyan and magenta dyes. A full discussion of this aspect of colour reproduction is outside the remit of this book (but see Digging Deeper).

Colour masking

In Chapter 6 there was a Box explaining the way that intrinsic colour masking is used in colour negatives to compensate for inherent errors in hue and saturation of the cyan and magenta dyes. It would be a good idea to go back and look at this Box again, as we are going to analyse this problem more closely. A colour negative consists of three colour-separation negatives superposed, and it is easier to see exactly what is needed if we think of three actual colour-separation negatives made separately on black-and-white film and then printed in printers' pigments (cyan, magenta and yellow).

In a three-colour separation printing process the red-filter negative (the red record) is printed in cyan ink, the green record is printed in magenta, and the blue record in yellow. But printers' inks suffer from the same errors in hue and saturation as the dyes in a colour film, and to a greater extent.

In particular, the magenta ink absorbs blue light that it should reflect, and the cyan dye absorbs both blue and green. The error with standard printing inks is more than that of colour film: up to 30 per cent for each ink. Considering first the magenta ink: this absorbs around 30 per cent of blue light, and thus behaves as if it was mixed with 30 per cent yellow. So whenever magenta and yellow occur together, we can compensate for this error by removing 30 per cent of the magenta value *from the yellow printer*. The way to do this is to make a positive of 30 per cent contrast from the green record (the magenta printer) and bind it in register with the blue record (the yellow printer). That takes care of every situation where both green and blue records bear an image.

The cyan ink is both desaturated and bluish. It behaves as if it contained some 30 per cent magenta and 15 per cent yellow. So, in a similar manner to the magenta correction, we make a 30 per cent positive mask from the cyan printer (red record) and bind it in register with the green record, to reduce the magenta content where cyan is present, and a 15 per cent mask from the same cyan printer, to bind in register with the blue record, to reduce the yellow content by this smaller amount where cyan is present. This corrects the errors in hue and saturation sufficiently for most purposes.

However, if you are to obtain full correction for every colour, you need to take into account the small absorption of red by the magenta dye, and the small absorptions of red and green by the yellow dye; so in theory you need six masks, two for each separation negative. As if this were not enough, none of the three inks can produce a fully saturated colour, so the three dyes printed 100 per cent still don't produce a true black, only a dark muddy brown. So a grey printer has to be

It does not take care of a pure magenta where no yellow is present. Photographs of magenta flowers with highly saturated colours *do* look reddish, and there is nothing you can do about this.

added, just to give a good black, and this is traditionally made by a fourth negative taken using a palish yellow filter.

If you think that this sounds like an immense amount of work you are not far wrong. In practice, masking was never taken to these lengths; most of the final correction was left to the fine-etcher, who worked from the proofs and made small corrections to the printing plates with a fine paintbrush and nitric acid. Today nearly all printing is carried out either by offset lithography or by electronic or electrostatic printing methods, where such fine corrections are not possible. Fortunately they can all now be made electronically.

Undercolour removal

Three-colour inks are expensive, and, as we have seen, not 100 per cent efficient. Wherever cyan, magenta and yellow appear together, we could substitute grey. So why not do so? With electronic analysis and control it is easy. All the computer has to do is to take away equal quantities of cyan, magenta and yellow until there are only two colours present, and then print those two colours in appropriate proportions, plus the required amount of grey. In most cases this can do away with the necessity for the fourth exposure too.

Unbalanced lighting

In colour negative films the sensitivities of the three layers are more or less equally balanced. Almost all amateur films are balanced for daylight and short exposure durations. You can, however, use them in tungsten illumination provided you are careful to give a full exposure. Tungsten light is rich in red wavelengths and poor in blue, and this effectively shifts the H & D curves relative to one another, as you can see from Figure 12.5.

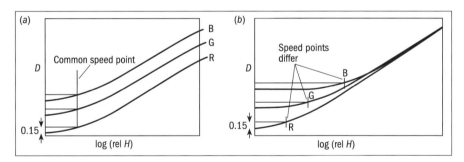

Figure 12.5 H & D curves for a colour negative emulsion (a) for daylight; (b) for tungsten light.

Provided you expose to the level where all three curves become parallel, you will be able to obtain a satisfactory print by using yellow and magenta CP filters. But if the blue record is still on the toe of its curve the print will have orange shadows that you can't compensate for without turning the highlights blue. This fault is almost impossible to correct photographically (though you may be able to botch it by digitising the image and playing among the pixels – see later). The safest way to avoid this problem is to use the proper correction filter, an 80A or 80B blue filter on the camera. You will still have to give about two stops extra exposure, but you

The Science of Imaging

would have to do that anyway, and by using the filter you retain the full latitude of the film. The 80A is a little stronger than the 80B, and is a better choice for general-purpose filament lamps, whereas the 80B is matched to photographic lamps of 3400 K colour temperature.

Correction filters

Tungsten filament lamps for displays and in photographic studios operate at a colour temperature of either 3200 K (312.5 mireks) for large bulbs, or 3400 K (294 mireks) for the small halogen type. The 80A (−130 mireks) is matched to the former, the 80B (−112 mireks) to the latter, in each case converting the colour temperature of the light to approximately 5500 K (182 mireks), to which 'daylight' film is matched. (Household lamps operate at around 2800–3000 K.)

Because of their high contrast, reversal (transparency) films are much more sensitive to shifts in lighting balance than negative films, and it is common practice among professional photographers to use pale filters of around ±10 mireks to warm up or cool down the overall tone of a scene. Tungsten-balanced reversal film is available. Many professional photographers use nothing else for studio work, and add an 85B (+115 mireks) amber filter for the odd daylight shot. However, fluorescent light, as you will remember from Chapter 2, does not have a uniform spectrum. Although it may look the same as daylight to the eye, it has a greenish cast on colour film, and needs something like a CC30M magenta plus a CC10Y yellow filter.

It should go without saying that using a mixture of tungsten and daylight or fluorescent light will result in an odd colour balance, probably varying over the subject area. Nevertheless, many photographers in editorial or advertising work do use such mixtures for special effects.

Skewed curves

Skewed curves are H & D curves for colour material where the individual colour curves have different slopes. There can be several reasons for this. Most commonly, the film is out of date or has been badly stored. It can also be a consequence of reciprocity failure with very long exposure times, or faulty processing. Old images sometimes fade in a similar way. Figure 12.6 shows examples of such curves for negative and reversal films.

This fault is difficult to correct photographically. There have been formulae published for lowering the contrast of individual colour layers, but they are not very reliable. The only possible treatment for a skewed colour negative by photographic means is to make a separation (colour) positive using the appropriate narrow-band tricolour filter, then make a *sharp* (colour) negative mask using the same filter, both developed to a reduced contrast. Thus if the green curve was of too low a slope, a green filter would produce a low-contrast duplicate of the green record which would boost the slope of the curve. You can use a similar technique for a transparency, using a complementary filter. You can then make a print or a copy with the correct colour balance. If this sounds tedious and difficult, it is. You need sensitometric control throughout, too, or you are going to have a good many misses before you have a hit. Before there was digital manipulation this was the

Filter numbers of this type do not have any arithmetical significance. They were originally numbers in the old Wratten catalogue, and were retained when Wratten was taken over by Kodak. Other manufacturers may label their filters by mirek (or mired) shift, and you may find an 80A labelled 13B (for 130 blue) or some similar code number. But photographers are a conservative lot, lexically speaking, and you are much more likely to hear '80A' quoted.

The CC scale *is* a logical one. CC means 'colour compensation', M stands for magenta, and the 30 means a density (to green light) of 0.30. Y stands for yellow and C for cyan.

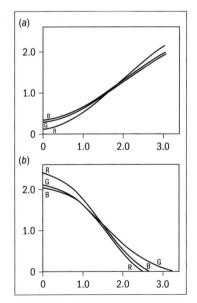

Figure 12.6 Skewed H & D curves: (a) colour negative. The red record has too high a $\bar{G}$. This will result in reddish highlights and cyan shadows in the print. (b) Colour transparency. The red record has too high a $\bar{G}$, and the green record too low. This will result in magenta highlights and cyan shadows. In both cases the exposure range will need to be confined to 1–2 log H units, producing a density range of ~0.6–1.3 for the negative and 0.6–1.6 for the transparency.

only way to salvage valuable historic material that had faded, without going right back to the beginning and making a fresh set of separations in black and white. But read on, as they say.

Digital manipulation of images

In the mid-90s things suddenly became much simpler. Computers equipped with scanners and printers moved into the home, and soon afterwards programs for the digital modification of pictures became available. The Windows facilities include a program called Paintbox that can carry out a good deal of modification, but it is slow and somewhat limited in scope. If you are going to do serious afterwork on your images you need a program such as Adobe Photoshop or the equivalent. As this book is about science, not technique, it would be inappropriate to give instructions on how the various tools work and how to use them. In any case, more than five updated versions of Photoshop have appeared in as many years, and there is a plethora of manuals around. (See Digging Deeper.) However, I have summarised some of the more important capabilities of programs of this type below.

- *Contrast*   You can change the image contrast overall, or you can change the shape of the characteristic curve, and alter the contrast in the shadows, the highlights or in any part of the tone scale.

- *Colour balance*   You can alter the colour balance by as little or as much as you need, and even deal with skewed curves by altering the slopes of the individual curves to match one another.

- *Mixed lighting*   You can shift the colour balance locally, where part of the subject was illuminated by a different type of lighting from the rest.

- *Colour masking*   You can produce colour masks, sharp or unsharp, and add them to correct inherent colour errors.

- *Unsharp masking*   You can do this directly, and if you wish you can carry it to the extreme, finishing with a bas-relief or a line drawing.

- *Tone separation*   You can 'posterise' an image, separating it into just three or four tones.

- *Shading, dodging and burning in*   Not as quick as doing it in the darkroom, but you need a lot less skill, and you can keep changing things until you get it right.

- *Removal of tiny blemishes*   You simply target the blemished pixels and replace them with tones that match the surrounding pixels. You can take out red-eye effects, put new reflections into eyes, and even change their colour. You can remove telephone poles, TV aerials and other unsightly objects, too.

- *Combination prints and montages*   In a photographic darkroom, unless you are very skilled, your combination printing will probably be limited to adding clouds to landscapes. Digital manipulation allows you to combine images in all kinds of ways far beyond the dreams of the Soviet backroom boys with their scissors and airbrushes.

This is only a small part of what digital manipulation can do.

Rectification

At some time or other, you will no doubt have photographed a building with the camera tilted upwards (see Chapter 4), with the result that in the image the verticals, which you know to have been parallel, converge. If the amount is very small, it may not be important (after all, they *do* converge). But our brains have a built-in template that tells our perceptual mechanism that the apparent convergence is an optical artefact, so we perceive the actual building as having parallel sides. Unfortunately, the camera sees it as it is.

The usual darkroom strategy, when one is faced with this problem, is to tilt the enlarging easel until the verticals in the projected image become parallel, or nearly so (and what a surprising amount of tilt it needs!), and close down the lens aperture until the image is sharp overall. However, this results in a convincing image only when the convergence in the negative is very slight. If it exceeds a few degrees the building appears noticeably elongated.

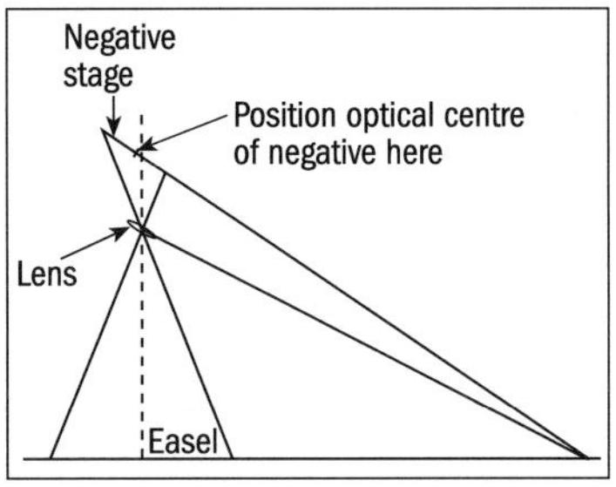

Converging verticals are a form of geometrical distortion known as *keystoning* (see Chapter 4). This is a perennial problem in aerial survey photography. It led to the development of the rectifying enlarger, in which a system of levers fixes the negative and easel such that the planes they lie in satisfy the Scheimpflug Rule (Figure 12.7).

Figure 12.7 Principle of rectifying enlarger for correction of keystoning (angles exaggerated for clarity).

But that is not sufficient. The centre of the negative needs to be in the correct position, which is displaced from the optical axis of the enlarger lens. There is a complicated formula for working this out, and the top-quality rectifying enlargers that are used in survey photography have a built-in computer that works out the correct position and moves the negative stage accordingly. If you are simply rectifying a keystoned image of a building, move the negative so that the top of the building moves up by about 10–20 per cent. This should correct the vertical scale. If you digitise your picture, however, there is a program for correcting converging verticals that needs no guessing and no tilting of easels.

Other image effects

There are many other ways in which a photographic image can be affected. Some, such as exposure to X-radiation, are now, thank goodness, rare; but there are plenty of others. Among the most important are the following:

Reciprocity failure   I mentioned this in Chapter 10. The Law of Reciprocity is a very general law concerning light energy, and was first framed by Bunsen and Roscoe in 1880.

It states that photochemical effects are directly proportional to the absorbed light energy. In photographic terms this boils down to the equation

$$H = Et$$

where H is the exposure, E is the illuminance at the plane of the sensitive material and t is the duration of the exposure. All camera shutter calibrations make this assumption, and for modern films this holds well for exposure durations between a few microseconds and several seconds. However, as we saw in Chapter 9, the formation of a latent image depends on a certain critical number of photons being absorbed by a silver halide crystal within a very short time, as the broken bond caused by a single photon is not stable and quickly re-forms, emitting another photon of lower energy, unless several similar events occur simultaneously in the neighbourhood, in which case the disconnected silver ions trap free electrons, and the resulting atoms coalesce to form a stable speck. When the illuminance is very low, the rate of photon impacts during the calculated exposure time may be so low that the smaller crystals do not catch enough photons within the required time to form a latent image. The result is underexposure. Increasing the exposure time compensates for this, of course, but the effect also produces increased contrast (see Figure 10.15). Photographers call the effect *reciprocity failure*, and with a calculated exposure of 60 s or so, you may have to increase the exposure by three or four times. Photographs of star fields may require several hours. The astronomer Karl Schwarzschild investigated low intensity reciprocity failure in 1900 and modified the Bunsen–Roscoe Law to

$$H = Et^p$$

where p, the *Schwarzschild exponent*, may vary from 1.1 to 2, depending on emulsion characteristics.

Reciprocity failure is most serious with colour reversal films, where long exposures can result in colour balance errors. Some manufacturers offer tables of correction filtration and exposure increase factors for exposure times in excess of 1 s. Colour negative films are affected less, and you can obtain professional colour negative film specially designed for long exposure times.

Karl Schwarzschild (1873–1916) was the first person to use density measurements in astronomical photography. He anticipated Einstein in suggesting that the geometry of space could be non-Euclidean, and just before his early death he published the first exact solution of Einstein's field equations. He was also a pioneer of stellar spectrography.

Intermittency effects When you make an exposure that is discontinuous, the total effect may be different from what it would be for a single exposure of the same total duration. In general, the effective exposure lies between the actual exposure and the exposure that would have been received had the shutter been open for the full time. The effect is small, but you may need to allow for it, for example when you are photographing the inside of a cathedral with a long exposure time and can open the shutter only when visitors have moved out of frame. In such a case your exposure gain might be around half a stop. When the intermittency rate is high, as with stroboscopic illumination, the effective exposure is determined by the time-averaged intensity level.

Pre-flashing If you expose a film to a small amount of light before shooting, your effective emulsion speed can be approximately doubled. This technique is known as *pre-flashing*. It was a popular trick used by press photographers in the days when the highest speed obtainable was around ISO 100/21. The pre-exposure takes the emulsion up to the point where the toe of the H & D curve begins, the idea being to overcome the 'inertia', i.e., the exposure you need to give the emulsion before anything happens. (As you might guess, the reason this treatment works is more complicated than this.) There is some loss of contrast, and, as I mentioned earlier, pre-flashing is one method of avoiding an over-contrasty result when copying a colour transparency using standard colour film.

Figure 12.8 The building at the epicentre of the Hiroshima atomic bomb explosion (now a national memorial). I took this photograph using a locally-bought and outdated film, on the fifth anniversary of the dropping of the bomb. At that time the ionising radiation was still sufficient to cause a partial reversal of the image. The whole film was affected.

Post-flashing This mostly *doesn't* work. In fact, in some circumstances it may even lead to a partial reversal of the image. Other fogging agents such as ionising radiation can have a similar effect (Figure 12.8).

Solarisation Gross local overexposure, such as might be produced by a direct image of the sun, a lightning strike or a bright street lamp, can actually destroy the latent image, resulting in a 'black highlight'. This is uncommon with modern emulsions, and usually occurs only in connection with long exposures to otherwise dim light. When a simple emulsion is fogged to the solarisation point, a further exposure to long-wave radiation will lower the density, resulting in a direct positive. So-called direct positive papers operate on this principle (the exposure is to amber light).

Sabattier effect This is a partial or total reversal of an image caused by exposure to light when the emulsion is already partly developed. It is largely a print-through effect, the image on the surface acting as a contact-printing negative for the lower emulsion layers. The full explanation is certainly more complicated, as outline artefacts are also generated. The effect has been exploited by generations of creative photographers. It is particularly striking when used with colour materials. Unfortunately, the outcome of any single experiment is not entirely predictable, and (once again) you can obtain the result you want with more confidence by digital manipulation.

The photographer and model Lee Miller claimed to have introduced the effect to the surrealist artist Man Ray by accidentally switching on the light in his darkroom while he was developing negatives.

Artefacts in digital images

These are almost non-existent, apart from the effects of underexposure, which are similar to those of traditional photography, and of *aliasing*, which usually takes the form of 'staircase' edges to sloping boundaries in the image. The effects can be minimised by adding *dither*, a process that blurs the edges sufficiently to eliminate the staircase effect without visibly affecting the sharpness. It is difficult – even impossible – to produce some of the other photographic effects directly. However, you can feed a digital image directly into your computer, where you can apply all the tricks you like, including a believable Sabattier effect.

Digging deeper

To find out more about contrast control you should try the sensitometry section of *The Manual of Photography* (9th edition, ed. R E Jacobson, Focal Press, 2000). Almost any darkroom manual contains the basic techniques of shading, dodging and burning in, and often multiple printing too; but there is nothing like practice, especially if you have an expert printer standing by your elbow. It is not easy today to find books with formulae for making up developers for special purposes. Here, the older the book the better: for example the *Manual* when it was the *Ilford Manual*, or the older copies of the *British Journal of Photography Annual*. These contained formulae for high and low contrast developers, intensifiers and reducers, and reversal processes. Kodak publishes data of all types, from leaflets to slim one-topic books for the aspiring professional.

The principles, equipment and practice of unsharp masking, together with other masking techniques, are set out in some depth in Sidney Ray's book *Scientific Photography and Applied Imaging* (Focal Press, 1999). The same chapter contains the fundamentals of image manipulation by digital methods, and gives a very comprehensive reading list. R W G Hunt's *The Reproduction of Colour* (5th edition, Fountain Press, 1995) discusses colour correction by adjacency effects in colour reversal emulsions. For a fuller account of rectification in aerial photography, complete with all the formulae, and of electronic printers, find a copy of Ron Graham and Roger Read's *Manual of Aerial Photography* (Focal Press, 1986, now, unfairly, out of print). If you want to read more about the machinations of the Soviet Press under the Stalinist regime, *The Commissar Vanishes*, by David King (Canongate, 1997) is a revelation.

Image modification

If you aspire to ownership of Adobe Photoshop, the most recent manual is *Adobe Photoshop 6.0 for Photographers*, by Martin Evening (Focal Press, 2001). If your Photoshop is the previous version, a very good manual for both Windows and Mac is *Photoshop 5.5 in Easy Steps*, by Robert Shufflebotham (Computer Step, 2000). But the best way to learn digital manipulation is to take part in a hands-on workshop. The Royal Photographic Society and the British Institute of Professional Photography run one-day workshops several times a year in various parts of the UK.

Chapter 13 Digital recording of images

The digital principle

Picture an ornamental gate topped with spikes, something like Figure 13.1, with their upper outline tracing an elegant curve. But they don't, do they? They are only a row of spikes. But the designer must have given a template to the craftsman, who would need to know the exact length to make each spike. Just as the craftsman would have fitted the spikes to the template, so now, looking at the spikes, you can visualise its shape as a continuous curve.

This is the fundamental principle of the analogue–digital relationship. If you have a quantity that varies smoothly with distance or time, such as light intensity, voltage or any similarly measurable varying quantity, you can represent its behaviour by a series of sample values, and still recover the original smooth curve, provided the samples are sufficiently close together. The smooth curve is called the *analogue signal*, and the row of samples is called the *digital signal*.

Of course, the sampling principle is much older than its modern applications. You use it every time you plot a graph of a continuous function. You use it, for example, when you plot an H & D curve as described in Chapter 10. In this case you have a set of discrete values (density readings) corresponding to another set of discrete values at equal intervals (the log-exposure values). If these latter values are close together you can get a reasonable result just by joining the dots, like a puzzle in a child's picture book. But logic tells us that the result should be a smooth curve, so we get a better picture by drawing a smooth curve through the dots instead. We have now constructed a continuous function from a set of discrete values.

If you know that the result has to be a smooth curve, you can probably manage with fewer points: in the case of an H & D curve, there are usually 20 readings available from the standard step tablet, but because of the regularity of the curve you can easily manage with only 10. Generally speaking, the more regular the curve the fewer readings you need. As we saw in earlier chapters, the most regular curve of all (apart from a straight line) is a sinusoid. If we know that our analogue signal is a sinusoid, we should be able to reconstruct it by taking only one sample per cycle, at the peak value. But of course this only works if you already know the frequency, in which case the exercise is trivial. In any case, a wave of exactly twice the frequency would fit the data equally well. It turns out that we have to take samples at not less than twice the maximum frequency at which the wave is operating in order to be able to build a reliable replica of the original sinusoidal waveform from the sample data. This is known as the *Nyquist criterion* (see Box).

Digital recording of luminance

Digital photography involves electronic measurement of the luminance of the optical image, at a closely set mosaic of points all over the image. But how closely set must they be? This is where the Nyquist criterion comes in. The designer has to ask 'What is the highest spatial frequency the device needs to record?' and the

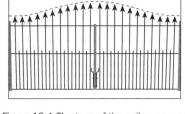

Figure 13.1 The tops of the spikes appear to trace out a curve.

This is an idealised example, as you will know if you have actually plotted H & D curves from measured data. What you have to do in fact is to draw a *best curve*, i.e., the curve that is the best match to all the points. To satisfy yourself that this is indeed the correct curve, there are various devices: invoking the limits of experimental error; deleting the two values that look the dodgiest; finding a polynomial expression that fits both the curve and the theory; and other dubious stratagems that you can find exemplified in almost any research paper.

Remember, *any* signal, however irregular, can be analysed into a combination of sinusoids. It is the highest frequency present that determines the sampling frequency needed.

Digital recording of images

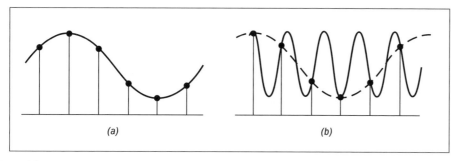

Figure 13.2 Aliasing: (a) adequate sampling rate; (b) inadequate rate: reconstruction produces the wrong waveform (broken lines).

> **The Nyquist criterion**
>
> If the sampling rate is slightly higher or slightly lower than the highest frequency present in a waveform, the samples themselves will fluctuate sinusoidally at the difference frequency. In radio or sound signals this will result in beats, and in spatial frequencies it will result in moiré fringes. This effect is known as *aliasing*. Nyquist showed that aliasing would not occur as long as the sampling frequency was at least twice the highest frequency present in the signal (Figure 13.2).

Nyquist limit is twice that frequency. So if the requirement is for a resolution of 50 cycles per mm there must be at least 100 discrete readings per mm, or 2400×3600 on a 35 mm standard frame. Each point is called a picture element (*pixel* for short); and this spacing represents a total of 8.6 megapixels. This is the kind of performance one would expect from an ISO 800/30 emulsion used with a high-quality lens. At the time of writing (2001) this has already been achieved. In digital cameras the sensitive elements are arranged in rectangular or trilinear (hexagonal) arrays (Figure 13.3). Each pixel sees the luminance level of its own tiny area as part of a continuous function; that is to say, it is in itself an analogue recording device. But it makes everything simpler and less prone to glitches if its signal can be coded digitally. So the light levels recorded are digitised at the storage stage: 4096 levels (around the limit of human visual discrimination), represented by 12 bits of information.

This is obviously not the same as the 'staircasing' of diagonal edges referred to earlier; but it is all part of the same general phenomenon.

You should remember that in small digital cameras equivalent to 35 mm conventional film cameras the picture area is much smaller, and the resolution requirement correspondingly higher. However, for most purposes the performance of a digital camera can be compared with its nearest equivalent in the 35 mm field.

Exactly the same situation arises in digital sound recording. Here it is usual to store 20 bit levels, though reproduction may be only at 16 bits.

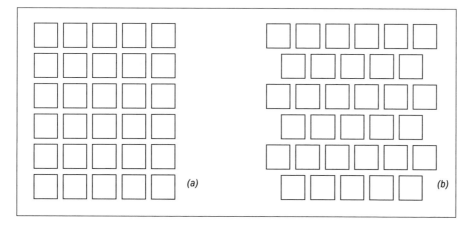

Figure 13.3 Arrays of pixels: (a) rectangular; (b) trilinear.

The Science of Imaging

Read aloud as 'ten base ten'.

Bits, bytes and binary arithmetic

For people who count on their fingers, the decimal or denary (base 10) system is a convenient form of arithmetic, so much so that it is the universal counting system, at least on this planet. But for electronic circuitry it is unnecessarily complicated. There is no reason, apart from the obvious physiological one, for counting in tens: it would be equally easy to use a system of twelves or eights. But the simplest system of all is the binary (base 2) system. There are only two digits, 0 and 1. The 'place' convention is exactly the same as in denary notation, but the number 10 comes round very quickly, as you can see from Table 13.1, which shows the equivalent binary and denary numbers up to 10_{10}.

Table 13.1 Binary and denary equivalents

Denary	0	1	2	3	4	5	6	7	8	9	10
Binary	0	1	10	11	100	101	110	111	1000	1001	1010

The thing about binary numbers that makes them so useful in digital calculations and measurement is that because there are only two digits they can be represented by the 'on' and 'off' positions of a switch. One switch codes for 0 and 1; two switches code for up to 3 (00 to 11), three switches for 0 to 7 (000 to 111) and so on. In order to count up to 10 you need four switches (four binary digits or *bits* give you up to 16 choices). To code all 26 letters of the alphabet you need five switches (up to 32 choices).

In binary arithmetic you can carry out all the normal operations of arithmetic such as addition, subtraction, multiplication, division, powers, fractions (vulgar and 'bicimal') and so on. All these operations are greatly simplified in binary notation as compared with denary.

You will probably have realised that each successive place in a binary number represents a step in powers of 2, just as each successive place in a denary number represents a step in powers of 10. To see how the correspondence works, look at the number 269.325 expressed in full, first in denary, then in binary notation:

$$2 \times 10^2 + 6 \times 10^1 + 9 \times 10^0 + 3 \times 10^{-1} + 2 \times 10^{-2} + 5 \times 10^{-3}$$
$$= 269.325 \text{ (in denary)}$$
$$= 1 \times 2^8 + 0 \times 2^7 + 0 \times 2^6 + 1 \times 2^5 + 0 \times 2^4 + 1 \times 2^3 + 1 \times 2^2$$
$$+ 0 \times 2^1 + 1 \times 2^0 + 0 \times 2^{-1} + 1 \times 2^{-2} + 1 \times 2^{-3}$$
$$= 10010110101.011 \text{ (in binary)}$$

So, as you can see, the rules for notation are identical. But as you can also see, binary notation is comparatively long-winded, so programs often count in eights (octal notation) instead, leaving the machinery to convert this to binary – something it can do much more easily than converting from denary.

The principle of octal notation gave rise to the concept of 'octal bits', which became known as *bytes*. Originally a byte was a three-bit number; but gradually the meaning of the term has expanded to take in the idea of a single piece of information or 'syllable' without a rigid length specification. As it makes more sense to specify the memory of a computer in terms of the information it can hold rather than the amount of binary data it can store, computer memories are usually quoted in megabytes (or giga- or even terabytes!). On the other hand, when discussing the number of distinguishable levels in a digital sound reproduction system, or illuminance levels in a digital photodetector system, it makes sense to talk in terms of, for example, 12-bit (4096) levels or 16-bit (65 536) levels.

Principles of electronic information storage

Just as you can understand the physical principles of what goes on in a car engine without necessarily knowing all the details of its electronic ignition system, so you do not need to know every last theoretical detail about the electronic circuitry of a digital camera in order to understand its principles. Most active electronic components these days are microchips of one sort or another, often containing thousands of elements, and even at quite an advanced level of understanding one is really only concerned with the relationship between what goes in and what comes out. The great-grandfather of all microchips, the 741 operational amplifier, contains hundreds of transistor, resistor and capacitor elements; but, seen from the outside, it merely has a current supply, a signal input, a gain control and an output. In effect, it is just a transistor with style.

However, in order to grasp what goes on in electronic image recording, you do need to have some inkling of the principles underlying modern electronic components. They are all based on *semiconductors*. Now, although a semiconductor has an electrical resistance somewhere between an insulator such as porcelain or PVC and a conductor such as copper or silver, it is not just a partial conductor like wood or damp earth. Conducting materials have many free electrons in their structure, and can thus carry an electric current freely. Insulators have all their electrons bound up with their structure, and there are none free to carry a current.

Semiconductors also have their electrons bound up with their structure, but they can be shaken out of their bonds by the application of sufficient electrical energy (voltage), so that the material becomes able to carry a current. The most commonly used chemical element is silicon, in the form of a single crystal. This has a structure of atoms with four outer electrons in orbitals locked with those of adjacent atoms, so that each atom 'sees' a full orbital of eight electrons.

The semiconductor diode   A pure silicon crystal is called an *intrinsic* semiconductor. It will conduct electricity only if the voltage and the temperature are high enough. However, if we add a very small controlled amount of the right kind of impurity, its electrical properties change considerably. If we add an element with five outer electrons instead of four, such as phosphorus, there will be extra unbound electrons in the crystal, and it will now conduct electricity when only a low voltage is applied, as the electrons can be readily freed. If instead we add an element with only three outer electrons, such as aluminium, there will be spaces in the orbitals that can be filled by any stray electron, leaving a 'hole' where it has come from – effectively a *positive hole*, as the donor atom has lost an electron. This crystal will also conduct electricity, by the rather laborious-seeming method of positive holes moving in the opposite direction to the electrons.

The two types of 'doped' semiconductor crystal are called respectively *n-type* and *p-type semiconductors*. On their own they are, of course, electrically neutral, as there is no external source of either electrons or holes. The exciting thing is when a *p*-type and an *n*-type semiconductor are in contact (Figure 13.4).

An electric current is just a movement of electrons in a conductor from the negative to the positive terminal of a voltage source. A current of 1 ampere (A) represents 6×10^{18} electrons entering and leaving the conductor per second (not the same ones, of course).

Although a 'hole' is just the absence of an electron, it represents a positive charge equal and opposite to that of an electron, and can be thought of in the same terms. A 'hole' current is represented by holes moving from the positive to the negative terminal.

The crystal structure has to be continuous across the junction, so the *p*-doping and *n*-doping have to be carried out on a single crystal of silicon.

The Science of Imaging

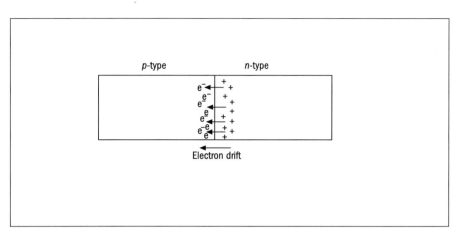

Figure 13.4 A *p-n* junction. Electrons drift from the *n*-type zone to fill the lattice gaps in the *p*-type zone, resulting in the accumulation of a negative charge in the latter and a positive charge in the former. Applying a positive/negative voltage to the *n*-side and a voltage to the *p*-side increases the field opposing it, and no current flows. Reversing the voltage overcomes the field and a current flows.

Electrons from the *n*-type material are now available to the *p*-type material, and as the attraction to complete the orbitals is high, many of the electrons migrate from the *n*-type to the *p*-type, giving the former a positive charge and the latter a negative charge. An electric field is thus set up opposing the drift of any further electrons towards the *p*-type material. If an external positive voltage is now applied to the *p*-side and a negative voltage to the *n*-side, the field will be overcome and a current will flow.

Remember that the actual flow of electrons is from negative to positive.

On the other hand, if the voltage is applied in the opposite direction, the field will increase and oppose any initiation of a current. We have a one-way device, a rectifier that will allow electrons to flow only in the direction *n*-type to *p*-type. It is called a *semiconductor diode*.

The junction transistor This is a semiconductor device that is capable of amplification. It consists basically of two *p–n* junctions back to back, in a single crystal. There are two types, *p–n–p* and *n–p–n*. The very thin central region is called the *base*, and the outer regions are called the *emitter* and the *collector*. In an *n–p–n* transistor the emitter is connected to a negative voltage and the collector to a positive voltage. When the base (the *p*-region) is given a small positive voltage ('bias') a large number of electrons will pass into the base from the emitter, and as the base is very thin many of these will be carried over into the collector region, so that a current of electrons flows from the emitter to the collector. The base current has a large influence on the collector current, and a small change in base current can result in a 20- to 100-fold increase in collector current, thus achieving amplification. A *p–n–p* transistor operates in the same way, but with holes instead of electrons. There are several different types of transistor operating on somewhat different principles; this type is known as the *junction transistor*. Figure 13.5 illustrates the principle.

This is, of course, a much-simplified picture, but this book is about imaging, not electronics technology. If you are anxious for a more rigorous treatment, you will need to go to one of the sources listed in Digging Deeper.

The field-effect transistor (FET) A transistor, by its nature, is an analogue device. But if the bias is high enough it is effectively a full conductor, and it can be operated as a switch; and this is its main function in digital devices. However, it is sensitive to changes in temperature, and there is always a leakage current, so for many purposes it has been supplanted by one or other of the forms of *field-effect transistor* (FET).

Digital recording of images

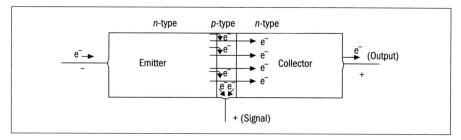

Figure 13.5 Junction transistor (*n-p-n*). When the number of holes in the base is increased by applying a small positive voltage, a large number of electrons pass from the emitter into the base, and as these are now in a majority over those on the collector they flow into it, completing the circuit. The base current is very small and the collector current large, providing amplification.

The FET fulfils much the same functions as the junction transistor, but is much more efficient. Its construction is different, however. In its basic form, instead of a *p–n–p* or *n–p–n* configuration it has a narrow channel of intrinsic semiconductor material joining two heavily doped layers of *p*-type material and *n*-type material respectively. These are known as the *source* and the *drain*; the channel is called the *gate*. In one common type, the *MOSFET* (*metal oxide semiconductor field-effect transistor*), the gate has a thin layer of an insulator with a conductor attached to its upper surface. When a positive voltage is applied to the gate, an electron current flows along its surface. There is no 'base current' as such. The schematic construction of a MOSFET is shown in Figure 13.6.

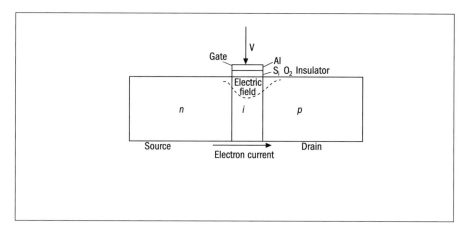

Figure 13.6 Principle of the field-effect transistor (FET). Instead of a base current there is an electric field resulting from a voltage on the gate. The strength of the field controls the current flowing through the substrate.

The charge-coupled device (CCD)   A capacitor is a device for storing an electric charge. In its most basic form it consists of two metal plates or *electrodes*, separated by an insulator (the *dielectric*). The very earliest capacitor was called a Leyden jar, and was simply a glass jar lined inside and out with metal foil (Figure 13.7).

Application of a voltage between the outside and inside foils caused electrons to flow into one side and out of the other, creating a large potential difference between the two foils. When the applied voltage was removed the charge would remain for a considerable time.

In the mid-nineteenth century, before Faraday's revelations, the only way to produce high voltages was with electrostatic generators such as Wimshurst's machine. This produced little current but, with a Leyden jar, you could store an accumulated charge, and obtain quite large currents from it for a brief time. Crude though it was, you could get a very nasty shock off a charged Leyden jar.

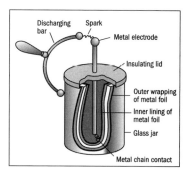

Figure 13.7 A Leyden jar, the earliest capacitor.

149

The Science of Imaging

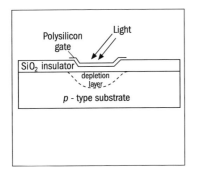

Figure 13.8 Principle of the metal oxide semiconductor (MOS) capacitor. The gate is made of a material that releases electrons into the substrate when stimulated by light. The electron charge remains until it is read out. This device forms one pixel of a CCD array.

Modern capacitors have a very thin dielectric layer and a large area of foil, folded or rolled up; they can store large numbers of electrons in a small space. The thinner the layer of dielectric, the greater is the capacitance (charge storing ability). A variant of the MOSFET, known as a MOS capacitor, is the basic component in the *charge-coupled device* (CCD) that forms the heart of the digital camera. The MOS capacitor consists basically of a layer of dielectric material (silicon dioxide, SiO_2) only a few molecules thick, deposited on a substrate of *p*-type silicon. The uppermost electrode, or gate, is of transparent polycrystalline silicon. When light falls on the device the photon energy transfers electrons into the substrate, where they remain as a surface charge (Figure 13.8).

This type of MOS capacitor represents a single picture element (pixel) of the photosensitive array in the focal plane of a digital camera. Each pixel records the light intensity of the optical image (to be exact, the exposure) at that point of the focal plane, coded in the form of an electric charge. The next stage, of course, will be to read and store those values in such a way that they can be re-created on a screen or printed on paper.

Getting the image out of the camera

Linear CCD arrays The simplest kind of CCD array for forming an image is the single-line linear array, recording in one dimension. This type of array is appropriate for scanning and photocopying, where the array moves across the subject matter, or for the equivalent of a slit camera. There are several systems for retrieving the information, all of a similar nature.

The term 'charge coupling' refers to the way signal charges are transferred from one electrode to the next. In the three-phase system, at any one instant only one out of every three electrodes carries a high potential, and any charge will be localised under that electrode. If an adjacent electrode has a potential applied and the potential on the original is cancelled, the charge moves to the new position. At each full cycle of clock voltages each charge moves one pixel forward. Charges can thus be passed along the line in what has been called 'bucket brigade' action, to be

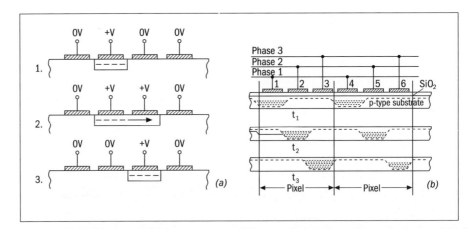

Figure 13.9 Principle of signal transfer. (a) The voltage pulses are on a three-phase system. The charge is always under the voltage, so when the voltage is transferred to the next cell, (2), the charge moves with it and (3) becomes established in the next cell, and so on to the end of the line, where it is recorded. (b) 'Bucket brigade' movement of charges.

Digital recording of images

recorded sequentially at the exit end. Three separate but interconnected clock generators control the potentials. (Three are necessary to ensure the charges move in the right direction.) Figure 13.9 illustrates the principle.

More recent developments include 'buried-channel' CCDs, which carry the charge at a deeper level. They have a higher transfer efficiency and permit higher clocking frequencies. By employing asymmetrical doping it is possible to operate a two-phase system that can shift in only one direction, improving resolution by 50 per cent (Figure 13.10).

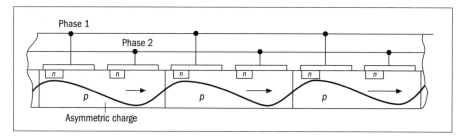

Figure 13.10 Two-phase CCD principle. The asymmetrical doping ensures charges move one way only.

Two-dimensional arrays Linear CCD arrays are employed in flatbed scanners, and for recording moving images as in satellite imagery. But for straightforward photography we need a two-dimensional array. There are several methods for transferring charge in two dimensions, the full-frame transfer array being preferred for still cameras. In this system all the horizontal line readouts are fed to a shift register, where a separate clock performs a vertical shift. The output from the line readout is amplified as it emerges, pixel by pixel, then fed to a 12-bit (or higher) analogue-to-digital converter, and its value and location coordinates stored on the camera's software. This was formerly a hard disk, but now is more often a card. From here the reconstructed image can be displayed on a small screen at the back of the camera if required, and downloaded directly into a computer, where it can be enhanced if necessary, and printed out.

Colour in a digital camera

Monochrome digital cameras are used only for special applications. General-purpose cameras are always colour systems. There are two ways of acquiring colour images in a digital camera. One is to separate the red, green and blue components of the optical image by dichroic filters as described in Chapter 6 under 'separation negatives'; this demands the use of three CCD arrays, and gives the highest resolution, but is expensive and comparatively bulky. A typical arrangement using totally reflecting prisms to orientate the images correctly is shown in Figure 13.11.

A more common method, and one used in all amateur digital cameras, is to use a three-colour filter arrangement over the pixel array. A simple system used in rostrum cameras and scanning devices (including satellite survey cameras) is the triple linear array (Figure 13.12a). General-purpose cameras have a colour mosaic filter with one filter element to each pixel. A common arrangement is as in Figure 13.12b. The filter densities need to be selected so that the colour sensitivity of the array matches that of the eye. In Figure 13.12b you will notice that there are twice

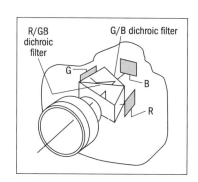

Figure 13.11 Colour digital image capture using three CCD arrays.

The Science of Imaging

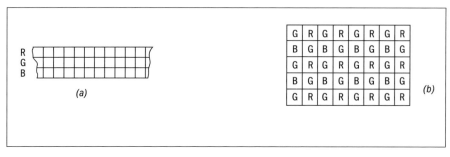

Figure 13.12 (a) Triple linear colour CCD array for scanning devices. (b) Mosaic filtered colour CCD array for general-purpose digital cameras.

as many green-filtered pixels as red or blue. This helps to make a match to the spectral sensitivity of the eye.

You will also notice that in this latter case, when a saturated red or blue is imaged, only one-quarter of the pixels are operational. In practice the spectral transmissions of the filters overlap, so that all the pixels are active to some extent even for a saturated colour. In addition, an interpolation algorithm is built into the camera software which, when it detects a uniform area of a saturated colour, fills in the lowest pixel outputs so that there are no actual gaps. This does degrade the resolution by about 25 per cent, but the result is preferable to a pattern of tiny holes in a uniform colour area.

Compression

If you associate with hi-fi enthusiasts, you will be aware that the data compression that enables a minidisc to play for as long as an audio CD is a contentious matter. Fastidious photographers tend to have a similar attitude towards data compression in visual image recording. Camera software can hold only a limited amount of data, and – particularly as digital cameras become progressively smaller and lighter – the images that can be held in the camera's memory are limited in number. Each manufacturer uses a different compression technique, and supplies appropriate decompression software. The simplest types of compression store only the data that are changing from one pixel to the next. Thus a scene containing large areas of uniform tone and/or colour takes up less memory than a complex scene containing a wealth of colour and detail.

The compression system for sound recording involves deleting data for any sound that is masked by louder sounds. As this treatment is likely to wipe out, for example, the harpsichord continuo from loud passages in any early Haydn symphony, perhaps the hi-fi buffs have a point!

The future for digital cameras

At the time of writing, the resolving power of a CCD array is still below the capabilities of a typical fine-grain 35 mm silver halide colour film. But there is no theoretical reason why the resolution should not approach the best that conventional films can achieve. The technological problems are being tackled, and there are regular announcements of yet another megapixel being added to the latest CCD array. Transfer efficiency is now at 99.9 per cent, and the number of recording levels has in the past few years increased from 8 bits (256 levels) to 12 bits (4096 levels). Card memories are replacing disks, so that cameras can be made smaller and lighter – and cheaper. The earliest digital cameras were little more than (and often literally) digital backs grafted on to conventional camera bodies.

Digital recording of images

Indeed, this is still the norm for studio cameras. After only a few years the very first purpose-built digital cameras have become dinosaurs. The convenience and cheapness of a camera that uses no film, offers instant viewing and re-takes, downloading directly on to a computer, and – for news photographers – affording direct transmission to a central office, seems a sure indication that digital photography will take over both the amateur and photojournalist fields before long. And digitally stored images do not fade. Mind you, I don't think that the next edition of this book will have silver halide photography relegated to an appendix. There are many things traditional methods can do that digital technology is unlikely to do better or even equal. But the future for digital cameras does look very bright indeed.

Scanners and scanning methods

The scanning principle has a long pedigree. The earliest scanners appeared in the 1930s, and were used for sending black-and-white pictures by wire (later by wireless telegraphy). A collimated light beam was focused on the surface of a rotating drum bearing the material to be transmitted, and an associated photocell picked up the reflected beam. A lead screw caused the lamp and photocell to scan the material helically at about 80 turns per inch, and the signal went down the line in binary form: 'on' for black and 'off' for white. Such a system couldn't cope with continuous tone material, but it could manage fairly well with halftones, provided the spacing and orientation of the dot structure didn't result in moiré patterns. In the 1950s the principle was adapted to produce copies of drawings or typescript as stencils for ink duplicating machines, the stencils being cut by an electric spark.

Today's scanners have come a long way from these primitive beginnings. Modern scanning equipment belongs in a chapter on digital imaging for two reasons, namely that it employs a linear CCD array, and that scanners are used routinely to digitise silver halide photographic prints. I have already discussed the principles of linear CCD arrays. Apart from their use in moving-film cameras ('slit cameras') such as photo finish cameras, aerial and satellite imaging, rotating panoramic cameras and peripheral cameras, they are also useful where the subject is stationary and the detector does the moving. In one type of scanner, the drum scanner, it is the subject that moves under the CCD array; but most scanners are now flatbed.

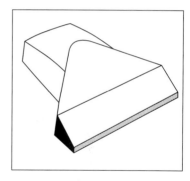

Figure 13.13 A hand held scanner.

The hand held scanner resembles a miniature vacuum cleaner head (Figure 13.13). The scanner head contains a row of coloured light-emitting diodes (LEDs) to illuminate the original, and a linear colour CCD array to record the subject. To operate the scanner you simply draw it over your document. By sweeping overlapping sections you can scan documents that are wider than the scanner. The accompanying software stitches the swathes together to give a single image. The ability to scan areas of almost any size is one of the main advantages of the hand held scanner.

The more familiar flatbed scanner operates more like a photocopier. Typically, the original, positioned face down on a glass platen, is illuminated from beneath by a tubular light source, and a system of mirrors moves down the original, focusing an image on the CCD array via a lens system (Figure 13.14). Colour scanners may operate either by making three separate passes under different filters, or by using coloured tubes that flash in synchronism with the CCD clock; but the preferred system now seems to be a triple-CCD array with red, green and blue filters.

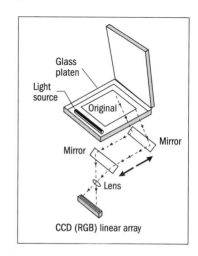

Figure 13.14 Typical flatbed scanner (schematic).

The Science of Imaging

The retention of these obsolete measurements is a result of the American origin of much of the equipment, and of a stubborn resistance to change in the printing trade.

Optical resolution is usually quoted in dots per inch (dpi) horizontally, the figure depending on the number of CCD elements per inch, and in lines per inch (lpi) vertically, this figure depending on the pitch of the stepping mechanism. Most flatbed scanners have an adaptor for scanning 35 mm transparencies and negatives, but this represents only a compromise, and a dedicated 35 mm scanner is essential for acceptable results.

Frame grabbers

The frame grabber is the successor to the still picture extracted from a movie sequence, or, more closely, to a photograph of a TV screen. Physically, it is a printed circuit board (PCB) which takes the output from an analogue TV picture, digitises it, then feeds the signal into a computer in a form similar to that given by a digital camera. Presumably, once TV becomes wholly digital, the task of the frame grabber will be greatly simplified.

Digging deeper

If you know little or nothing about electronics, or have had difficulty in following the explanation of CCD mechanisms given in this chapter, try *Starting Electronics*, by Keith Brindley (Newnes, 1999), which begins from scratch. It even teaches you how to solder connecting wires and terminals. A somewhat more advanced book, leaning to the applied side, is *Electronic Products*, by Barry Payne and Daniel Rampley (Collins Educational, 1997). *High Speed Digital Design*, by Howard Johnson and Graham Martin (Prentice-Hall, 1993) is an undergraduate-level text. A more controversial choice is *The Art of Electronics*, by P Horowitz and W Hall (2nd edition, Cambridge University Press, 1989). You will either love its friendly, helpful approach or hate the way important concepts seem to be left only half explained. Some electronics engineers swear by it. Borrow it from a library before you make up your mind.

In a subject that is moving so fast, it is difficult to provide a booklist that is fully up to date. The best I have come across is Ron Graham's *Digital Imaging* (Whittles Publishing, 1997). The explanations of the workings of CCD arrays are so clear and well expressed that I have found it difficult to avoid quoting directly from them in this chapter. Some of the equipment described is already going out of circulation, but a book as good as this deserves frequent updates in this respect, and no doubt will get them. For a more recent (though also more superficial) treatment of the subject, try *Easy Digital Photography* by Scott Slaughter (Abacus, 1999), which combines basic principles with practice. For a practical guide, try *The Digital Photography Handbook* by Simon Joinson (Metro Publishing, 2000), centred round Photosuite II, an alternative to Photoshop. Tom Ang's *Digital Photography* (Mitchell Beazley, 1999) is an admirably lucid account, with some splendid photography. A pity that many of the finest shots were taken using conventional equipment, but an excellent manual. All the amateur photographic periodicals carry information on the latest digital equipment, but the most reliable by far is the professional *British Journal of Photography,* which carries an in-depth study of a new digital camera almost every week.

Chapter 14 Halftone, electrostatic and digital printing

Continuous tones with printer's ink

The earliest photographers made all their prints by contact, but as camera formats became smaller, contact prints became less attractive, and as early as the 1860s enlargers were in regular use. An enlarger is simply a camera that produces an image at a magnification greater than unity, and an enlarger lens is basically a camera lens that is optimised for close-up work and used back to front. Indeed, the pre-Second World War Leitz enlargers were designed to use the standard 50 mm Elmar lens from the Leica camera mounted in this manner.

A problem arose when there began to be a need for photographs in newspapers and magazines. Certainly, there were methods of producing continuous-tone reproductions on plain paper without the use of silver halide, but they were slow and messy, and usually involved special preparation of the paper surface.

The difficulty is that printer's ink is black, and printing is a binary process: black ink is either present or absent on the paper. Making steel engravings from photographs is a skilled and time-consuming business; but before the halftone process was invented this was the standard method of producing pictorial matter for printing in papers and books. By making the engraved lines very fine and close together, a skilled engraver could produce a fairly convincing illusion of the presence of shades of grey in the picture. And this is the clue to the next step.

An exception was *Woodburytype*, a process that involved a gelatin relief image which produced an indentation in a soft metal plate when clamped to it under heavy pressure, the plate then being used for intaglio printing. With the advent of photogravure the process lapsed, but there are still some books with Woodburytype illustrations to be found in antiquarian bookshops.

The halftone principle

The idea that a grid of cross-hatched lines, or a pattern of black dots, would look grey when viewed from a distance such that the pattern would not be discernible, led to the principle of the halftone illustration, invented in 1881 by Frederick Ives. A halftone, or middle grey, could be achieved by a mosaic of tiny black and white squares of equal size, like a microscopic chessboard. A light grey would result from a mosaic where the black squares were smaller than the white ones, and a dark grey if they were larger. A continuous-tone image could thus be reproduced with nothing more than black ink and white paper, provided the original image could be coded into black and white dots of varying sizes, area for area. The solution, when it came, was ingenious. An opaque screen bearing a fine mosaic of tiny transparent squares, made by ruling opaque lines on a glass plate, is installed a short distance in front of a high-contrast photographic plate. Each of the apertures then acts as a tiny pinhole camera, forming a slightly blurred image of the iris diaphragm aperture of the camera lens at the film plane. As the emulsion is of very high contrast, it behaves like a binary switch. Below a certain exposure threshold nothing happens; above the threshold it gives total black. Since the pinhole images are blurred, only the centres of the darkest parts of the original will receive an over-the-threshold exposure, whereas in the brightest areas of the original subject much more of the dot image will be over the threshold. The developed dot size will thus be proportional to the brightness of the tiny area imaged (Figure 14.1).

The Science of Imaging

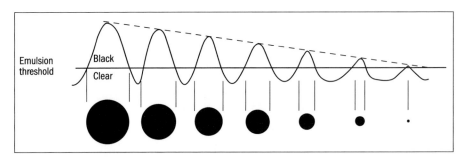

Figure 14.1 Principle of traditional halftone screen system.

This method of printing is called *letterpress*, and is still used for very long print runs such as those of a newspaper. It has been largely supplanted by lithography, a process in which the dots are made ink attractive and the spaces ink repellent. The inked image is transferred to paper via a soft rubber roller, so the printing plate is under little stress and can be made of thin aluminium sheet or even tough paper.

When this negative is printed on to a copper plate coated with a photoresist (dichromated albumen in the old days) the exposed areas become insoluble. The unexposed areas are washed clean and the plate etched with acid. The protected dots remain proud of the surface, and once the whole plate has been cleaned it can be wrapped around a printing cylinder and used to print the halftone reproduction. Figure 14.2b shows a magnified section of the halftone print in Figure 14.2a.

Figure 14.2 (a) Halftone print; (b) part of print, magnified.

Owing to their expense and fragility, separated screens are no longer in use. Contact screens with graduated dots do the same job. They are often magenta rather than black, used with a green-sensitive process emulsion. If you want to make a halftone negative and your photo lab does not boast of a process camera, you can obtain 'auto-screen' film. This has been pre-exposed to the threshold point under a contact screen, so that any further exposure will bring up the dot pattern just as though you had used a screen in the conventional way. However, in the present computer age, it is a simple matter to turn a scanned picture into a halftone with a few keystrokes, and this is now becoming the most common method.

Screen sizes The largest dot spacing in regular use is 80 dots per inch (dpi), but such a coarse screen is used only with low-quality newsprint. 125 dpi is more usual for good-quality newsprint, and, for the highest quality coffee-table books of photographs, 250 or more dpi may be used. Owing to the high sensitivity of the human eye to vertical and horizontal lines, the dot alignment on a black-and white halftone is usually tilted at 45°.

Printing in colour

If you want to print in colour (and these days even the local free newspapers carry colour pictures) you need a basic three printings, cyan, magenta and yellow, in register. The screens have to be set up at different angles; otherwise there will be blotchy coloured moiré patterns. To help avoid the risk of this, the yellow printer is often made with a somewhat coarser screen than that of the other colours. Yellow is chosen because it contributes little to the overall resolution. The printing plates are made in the usual way from separation negatives, which may be masked as described in Chapter 10; the masking is usually achieved electronically.

Because some of the coloured dots are adjacent to one another and some of them overlap, the resulting image is partly additive and partly subtractive. This places restrictions on the spectral content of the inks, and inevitably these colours together do not produce a good black, or even a neutral grey. It has therefore been the practice for many years to include a 'black' printer (actually grey) made from the colour original using a medium yellow filter. A later system uses undercolour removal (see Chapter 12), which replaces all equal combinations of the three inks with 'black'. The four-colour combination (cyan, magenta, yellow and black) is given the abbreviation CMYK, the 'K' being used for 'black' to avoid possible confusion with 'blue'.

Electrostatic copying

Electrostatic printing depends on the properties of certain types of semiconductor substance that become conductive when illuminated. The main application of the process is in xerographic document copying. The xerographic process was invented by Chester Carlson in 1938 and patented by him in 1942. A typical xerographic photocopier uses a large metal drum coated with a layer of amorphous selenium as the light-sensitive medium. A strip light illuminates the original document, and a lens focuses the strip image on the drum; the image moves in synchronism with the rotation of the drum.

The word *xerography* comes from two Greek words meaning 'dry writing'.

There are seven steps in the production of a photocopy:

1. The drum passes under a grid of fine wires charged to a potential of 5–10 kV in the dark. The resulting corona discharge puts a uniform electrostatic charge of several hundred volts on the drum surface.

2. The drum passes under the exposing slit and the projected image discharges the electrostatic charge in proportion to the illuminance. The drum now carries a latent image in the form of an electric charge.

3. Electrostatically charged toner is introduced to the drum surface, where it accumulates on the areas that have retained their charge. The toner usually consists of particles of carbon black about 5–20 µm mean diameter, in a resin binder.

4. The receiving paper is charged to a high potential, and fed into contact with the drum surface. The toner particles are attracted to the paper surface.

5. The receiving paper passes out through heated rollers, which press the toner particles into the surface, fuse the image into the paper, and eject the finished print.

The Science of Imaging

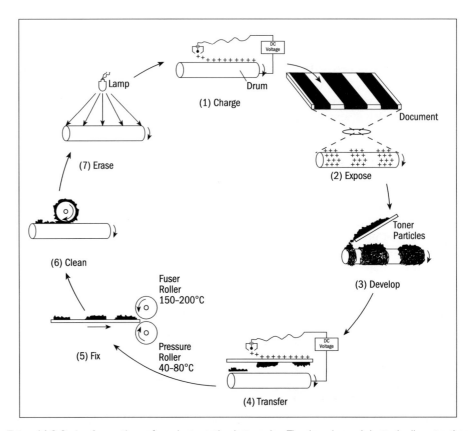

Figure 14.3 Cycle of operations of an electrostatic photocopier. The drum is much larger in diameter than shown, and all the operations take place in a single revolution.

6   Any remaining toner particles are removed from the drum by a brush.

7   The drum is exposed to light to dissipate any remaining charge.

All these steps take place in a single revolution of the drum. There are several types of optical system that may be employed to obtain the synchronised moving image, involving the movement of the illuminating lamp or the lens system or the platen, or all three. A schematic diagram of the system is shown in Figure 14.3.

For colour copying, three or four toners are needed to produce a CMY or CMYK image. The exposing light can be three or four successive strip lights with appropriate filters, or a row of RGB light emitting diodes (LEDs). In the simplest version there are three or four separate passes, one for each toner.

Copiers intended for rapid runs of several hundred copies use a charged belt system instead of a drum. As amorphous selenium is brittle, organic photoreceptors are used for the belt coating. These consist of strong electron donors (or acceptors) in a polymer substrate, usually containing a pigment to even out the spectral sensitivity. This type of photoreceptor gives better reproduction of continuous-tone material than selenium, and it is now becoming universal in both colour and black-and-white photocopiers.

Xerographic photocopiers have now become standard equipment in printing sections in universities and other institutions where runs of several hundred copies are needed in a hurry. Recent models can turn out up to 100 copies a minute (somewhat fewer for colour copies).

Halftone, electrostatic and digital printing

Printers

A printer takes the stored image from a computer, pixel by pixel, and translates it into digitally coded hue, density and saturation values, which it then turns into a printed image.

There are four basic methods of achieving this: direct mechanical contact, controlled deposition, xerographic and quasi-photographic.

Direct mechanical contact   The oldest and simplest of these is suitable only for text and simple line decorations (as for menu cards). It uses a standard typewriter format and a daisy wheel. The *dot matrix* printer grew up alongside the daisy wheel. It uses a block of up to 24 pins filling a rectangle the size of a typescript capital. Like the daisy wheel printer it uses an ink ribbon, but can operate at more than ten times the speed. It can be used for pictures, but the quality is poor, and its main advantage is its simplicity. It is generally used for calculator printouts and till receipts.

Controlled deposition   The most commonly used system, at least in non-professional printing, is the ink-jet system. The ink jets themselves are near-microscopic in size, up to 256 of them fitted into a space the size of a typescript capital. They operate by forcing out a tiny drop of ink either by heating the nozzle (which is finer than a human hair) or by squeezing its upper part with a piezo element.

A piezo element, you will remember, is a crystal or ceramic block that expands when a voltage is applied across it.

The two other systems are *dye sublimation* and *thermal wax*. Both systems use a contact ribbon, and transfer is by the application of heat. The printer head resembles the dot matrix head, except that instead of 24 pins it has several hundred microscopic needles, all of which are temperature controlled. 'Sublimation' means 'vaporisation without melting', and in the dye sublimation process the application of a hot pin to the ribbon causes the pigment to vaporise into the paper, the density of the dot depending on the temperature of the needle. (As the needles are so fine their temperature can be varied very rapidly.) The thermal wax system is similar, but at a lower temperature. The dyes are in a wax base, which under a hot pin melts into the paper. Both systems require four passes to produce a CMYK image. Some printers can be adjusted to take either dye sublimation or hot wax ribbons.

Xerographic printers   The sole representative is the laser printer. It is a logical extension of the electrostatic copier. Instead of exposing the drum to an optical image, the drum is scanned across its width by a tightly focused diode laser beam. The laser is fixed, and the scanning is accomplished by a rotating polygonal mirror. The beam intensity is varied, as appropriate to each pixel, during the scan. Figure 14.4 shows the schematic layout.

Many so-called laser printers do not use lasers at all, but a row of hundreds of near-microscopic light emitting diodes (LEDs) instead. This set-up does away with the need for a rotating mirror. The diodes are selectively switched according to instructions fed into the printer from the computer.

To produce CMYK prints from a laser printer it is necessary to use four passes, one for each colour. This can be done on a single drum with four changes of coloured toner, or, in a printer intended for fast production runs, with four drums.

Quasi-photographic printers   At the time of writing there are two printing systems that use photographic-type material, both originating from Fuji. In the

The Science of Imaging

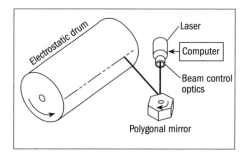

Figure 14.4 Principle of laser printer.

thermo-autochrome process the print material is somewhat similar to traditional colour print paper, with RGB-sensitive layers containing CMY colour-generating material. The scanning is by a three-colour LED row, and the print is developed by the application of heat and fixed by UV radiation. The *Pictrography* process (the name is Fuji's coining) covers printers that use red, green and blue diode lasers to make a single pass over a dye donor material, the dyes of which are transferred to a final base by contact, in a manner somewhat similar to Polaroid professional colour material.

Both of these systems yield prints of photographic quality, and it seems likely that variants of these techniques are likely to be introduced by other manufacturers, as they have the advantage over other processes of not using ribbons or printing inks. In addition, the systems operate directly from RGB files without the necessity for CMYK conversion. Of the other methods described above, dye sublimation at present provides the nearest approach to the quality of silver halide based prints, though the permanence of the colour images is still in question.

Digging deeper

There is a vast amount of information on conventional halftone printing. Perhaps the best overall review is U M Adams's *Printing Technology* (Delmer, 1996). This also contains a fairly full account of the digital technologies outlined in this chapter. If you are interested in printing *per se*, you could not do better than hunt for old copies of *The Penrose Annual*, which used to chart the progress of traditional printing year by year, and always contained breathtakingly beautiful examples of the printer's craft. And if your interest is historical, Bamber Gascoigne has written a splendid survey of printing up to 1860 in *Milestones of Colour Printing* (Cambridge University Press, 1997). Frank Romano's *Dictionary of Digital Printing* (Delmer, 1997) steers you through the forest of jargon surrounding the subject, and Ron Graham's *Digital Imaging* (Whittles Publishing) takes you up to 1998 in technology and equipment. As far as new equipment is concerned, you will need to consult the better-quality computer magazines and professional photographic journals such as the *British Journal of Photography* and the *Professional Photographer*, which regularly publish detailed surveys of equipment. Best of all, visit the Focus on Photography exhibition, which takes place every February at the National Exhibition Centre, Birmingham. Here you can get a full update, and see the results from the latest equipment for yourself.

Chapter 15 Television

Beginnings

The concept of television as a practical technology began with the work of a Scotsman, John Logie Baird, in the early 1920s. His work built on an idea of Paul Nipkov, who in 1883 had taken out a patent on a scanning disc which contained a spiral of holes positioned so that when it was spun the holes would scan an entire image, line by line, in one revolution. The brightness of the image at each point would be transmitted by wire and reconstructed by a similar disc synchronised to the first. Nipkov never exploited his patent, and it was left to Baird to realise the system, using a radio link. In Baird's transmitter the luminances tracked by the rotating Nipkov disc were recorded and coded as modulations of a radio signal (Figure 15.1). The receiver decoded the signal and fed the modulations to a light; these were transmitted either to another synchronised Nipkov disc or to a rotating polygonal mirror, which generated a flying spot that matched the luminances to the transmitter.

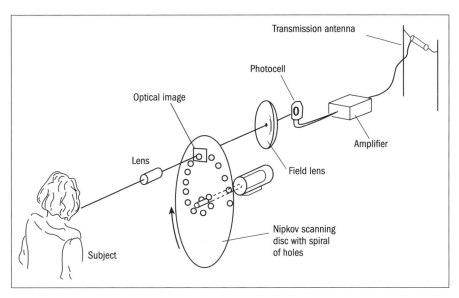

Figure 15.1 Baird's original television system. The receiver used a similar system in reverse, with a lamp replacing the photocell.

In 1887 another German, Karl Ferdinand Braun, had invented the cathode ray tube. A number of workers in both Germany and Russia played with the idea of transmitting pictures using this device, but the first practical demonstration of a working electronic television system was by another Scotsman, Archibald Campbell-Swinnerton, in 1908. This eventually led to collaboration between the two companies Marconi and EMI (Electrical and Musical Industries) to produce a commercially viable system. By 1936 the two rival systems were sufficiently developed for the BBC to begin television broadcasts, transmitting by the Baird and Marconi–EMI systems on alternate days. The following year the Baird system transmissions were abandoned.

A curious decision, in the circumstances. At that time Baird's system was visibly superior to the Marconi–EMI system, and the receivers were simpler. Unfortunately, at about the same time Baird lost much of his equipment, including his first electronic camera, in the disastrous Crystal Palace fire. But Baird did not give up. At that time he was already working on a colour camera. Early in 1940 he was seconded to the War Office, and spent the whole of the war working on radar and other advanced electronic equipment. Because of the secret nature of his work, few people knew about it at the time. He died in comparative obscurity in 1946, his contribution to the birth of television almost forgotten.

The Science of Imaging

The EMI transmissions continued until the outbreak of the Second World War, then ceased for six years. By the time they resumed the electron beam tube was king.

The electron beam tube

The electron beam tube is admittedly a cumbersome and potentially dangerous piece of equipment; but for many decades there was no substitute. Like the internal combustion engine (another living dinosaur), it has been continually developed and improved so as to outclass any rivals, and today it still dominates the field of television receivers, computer monitors, visual display units (VDUs) and, of course, cathode-ray oscilloscopes (CROs). So any chapter on television necessarily begins with an account of the way it works.

The electron beam   If you heat up a metal filament in a vacuum and give it a negative charge it will give off electrons. If you leave the filament neutral and instead surround it with a negatively charged cylinder coated with a powerful electron donor it will emit electrons much more efficiently. This arrangement is known as a *hot cathode*. If you now position a positively charged metal screen some distance away (still in the same vacuum chamber) the electrons will fly towards it at high speed. If its surface is coated with a *phosphor* (a substance which emits photons when hit by electrons) the screen will glow. The pioneer experimenter William Crookes discovered the electron current, and called the effect 'cathode rays'; the name has survived in the term *cathode ray tube* (CRT).

You can focus the stream of electrons into a narrow beam by surrounding it with a charged cylinder (the *focusing electrode*), so that the beam reaches the screen as a small spot. A second cylinder (the *control grid*), carrying a variable negative charge, repels the slower electrons and controls the beam intensity. The cathode with its complement of electrodes is collectively known familiarly as the 'electron gun'. A fuller depiction of its components appears in Figure 15.2.

The scanning principle   We have already seen something of scanning, but it is worth looking at more closely here, as it is a fundamental technique in television and video. In Chapter 8 I discussed the phenomenon of persistence of vision, and explained that when individual pictures were presented successively at a high enough rate the sensation of flicker would be absent, and movement within the frame would appear to be continuous. In television, unlike motion pictures, we don't have a whole picture presented at once: instead, a flying spot traces out the luminances of the image, line by line. At any instant only the spot is present, but it moves fast enough to appear to fill the whole frame, on account of the persistence of vision. The spot completes 625 traverses of the screen in $\frac{1}{25}$ s (525 in $\frac{1}{30}$ s in the USA). We do not see all 625, though, as a few at the top and bottom are reserved for various synchronising pulses and for teletext. Also, only alternate lines are presented at each scan, with the frames interlaced on alternate scans: in one scan, odd-numbered lines are presented, and on the next, even-numbered lines. This means that the rate of picture presentation is 50 per second instead of 25 (60 in the USA), so that the possibility of flicker is much reduced. The intensity of the spot is continuously modulated to match the picture signal at that point, i.e., it is lowered for a dark area and raised for a light area.

Beam deflectors   How is the spot steered? There are two possible methods. The first is by electrostatic deflection. If the beam passes between two plates which bear opposite charges it will be deflected towards the positively charged plate and away

Sir William Crookes (1832–1919) was an English physicist and chemist who carried out all his researches in his own laboratory. He discovered the element thallium, and invented the Crookes radiometer, a scientific toy, the working of which was eventually explained by Maxwell. In the 1870s Crookes investigated the passage of electric current in glass tubes containing gases at low pressure, and described the behaviour of 'cathode rays' in electric and magnetic fields. He was awarded the Order of Merit in 1910.

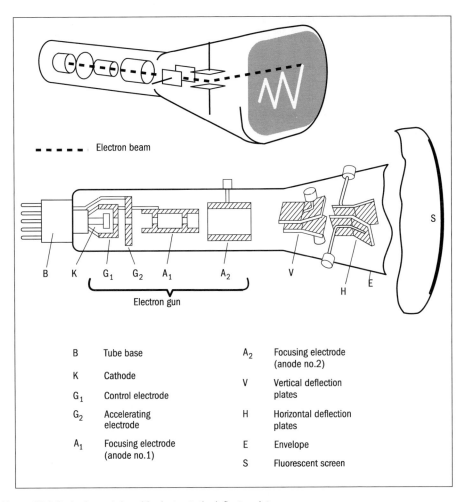

Figure 15.2 Cathode ray tube with electrostatic deflector plates.

from the negatively charged plate. Figure 15.2 shows the principle. The other method is magnetic deflection. A beam of electrons in a magnetic field is forced to travel along a circular path as long as it remains in the field. By using specially shaped coils (Figure 15.4) the beam can be deflected through a large angle. Electrostatic deflectors have less effect on the beam than magnetic deflectors, and are found only in CROs.

In a TV display the field is applied as a sawtooth waveform, that is, one that rises linearly from a negative value to a positive one, then returns abruptly to its negative state. This causes the spot to traverse the screen at a constant speed. The downward movement is provided by a second magnetic field at right angles to the first, with a much slower sawtooth, so that it returns only once per complete scan of the screen. When the spot has to fly back, at the end of each line, in order to avoid leaving a streak across the screen the beam is switched off by a *blanking pulse*. Figure 15.3 shows the way the field varies with time.

The phosphors   To create a picture, the electron beam has to generate light at the screen face, which forms the positive electrode. Monochrome TV screens are coated with a phosphor that emits white light when it is energised by a beam of electrons. (Phosphors for CROs are usually green rather than white.)

Phosphors glow for some time after stimulation ceases, and this helps to minimise flicker. For television a short decay period is appropriate, about 20–40 ms.

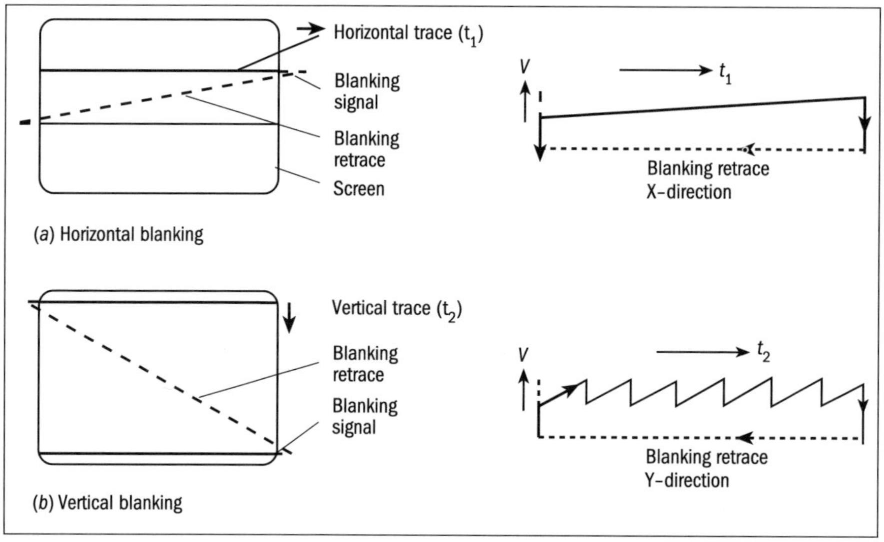

The Science of Imaging

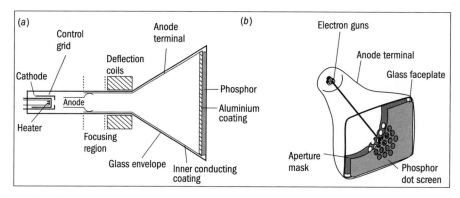

Figure 15.3 Waveforms in X and Y directions for generation of a television raster.

Figure 15.4 (a) Schematic layout of a TV tube; (b) cutaway view.

If you compare a TV tube with a CRO tube you will notice that the CRO tube is much longer and thinner in proportion. This is because an electrostatic field has much less deflecting effect on an electron beam than a magnetic field, which can be made very strong indeed. The magnetic coils on a TV tube turn the electron beams through nearly 90°, and because of this, and the pinpoint accuracy that is necessary to obtain a sharp and undistorted picture, the magnetic coils have to be designed with a cunning that is near to witchcraft.

Figure 15.4 shows a cutaway view of a TV tube.

Colour mask In order to transmit an image in colour, two signals are needed: luminance (tones) and chrominance (hue and saturation). These signals are transmitted on the same carrier. In the receiver the image hues have to be separated into red, green and blue (RGB) components, and these need to be directed separately, along with the luminance signal, to three electron guns. A perforated metal screen aligns the beams with phosphor dots that respond with light of the appropriate hue (Figure 15.5). In a monochrome TV receiver the chrominance signal is ignored.

The television camera

Image tube The optical part of a TV camera resembles that of a conventional still camera: a lens equipped with an iris diaphragm focuses an optical image on a light-sensitive surface. In the earliest TV cameras the recording device was a plate coated with photovoltaic material.

When illuminated on one surface, a photovoltaic material produces a potential difference between the light and dark surfaces. A photoconductive material is a semiconductor that becomes conductive when light falls on it. Both types of material have a largely linear response.

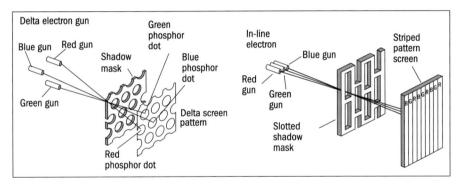

Figure 15.5 Operation of a colour mask: (a) delta format; (b) in-line format.

The optical image produced a negative charge on the rear surface in proportion to the illuminance at each point. This plate was the 'screen' for an electron beam tube. The scanning electrons were repelled in proportion to the image illuminance and were caught by an 'electron trap' which thus recorded the illuminance. This device was called an *image orthicon tube*. It was cumbersome and not very sensitive, and was prone to image artefacts, especially in scenes with bright lights. It was replaced by the *vidicon tube*, which uses a photoconductive plate. This allows the electron current from the scanning beam to be read directly from the plate (Figure 15.6). The scanning rate is the same as that of the TV receiver.

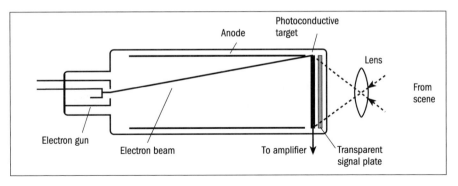

Figure 15.6 Principle of vidicon tube.

Many cameras used in professional broadcasting still have vidicon systems, but they are on the way out, replaced by the smaller and more convenient CCD system, which is faster, has a more linear response and doesn't require a vacuum tube. Amateur camcorders are now little bulkier than a traditional 35 mm camera. Indeed, TV cameras can now be miniaturised to such an extent that they can be inserted into body cavities for diagnostic purposes, or to monitor the progress of surgical procedures. They can find their way into all sorts of unlikely places, even inside a cricket stump.

CCD cameras You have already met the CCD as used in the digital still camera. These cameras mostly use a tricolour mosaic filter in front of the CCD array. TV cameras, in contrast, may use three separate CCD arrays for red, green and blue images, using dichroic mirrors (Figure 15.7).

When TV cameras use a single CCD array, the mosaic filter is sometimes the same as in the stills camera, but this arrangement is wasteful of light energy, as

The Science of Imaging

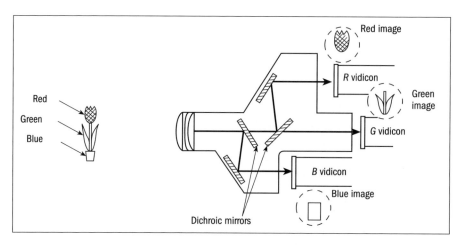

Figure 15.7 Simplified diagram of a modern professional TV camera with three CCD arrays.

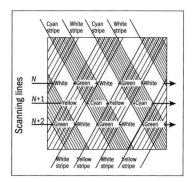

Figure 15.8 Example of a mosaic filtering system for recording RGB signals economically.

only one-third of the light energy can reach any element in the array. A more economical scheme criss-crosses cyan/white diagonal filter stripes with yellow/white stripes, giving a pattern of cyan, green, yellow and white (Figure 15.8a). The green content is recorded directly, the red by subtracting the cyan signal from white, and the blue by subtracting the yellow signal from white. This system allows at least twice as much light energy to reach the array.

An alternative filtering method that requires less processing is also used. The luminance signal is recorded (or transmitted onward) separately from the chrominance signal. The latter needs only the red and green information, as the total $(R + G + B)$ is always equal to the luminance signal, so the blue signal is obtained by subtraction: $B = (\text{luminance}) - (R + G)$. You may remember from Chapter 6 that a similar tactic was used to find blue content in the CIE chromaticity diagram.

Transmission and reception of a TV signal

The fine detail of this lies beyond the scope of this book; there are whole books on the subject (see Digging Deeper for Chapter 16), and to some extent it varies from one country to another. In addition, with the advent of high-definition television (HDTV) and digital television, new standards are being evolved, many of which are still not fully agreed upon. As far as the present analogue TV is concerned, there are three main systems, differing only in detail. The system used on the American continents is called NTSC, which stands for National Television Systems Committee. It operates on 525 lines with interlaced frames at 60 pps, two pictures to a full frame.

Colour control with this system has always been difficult, and in the early days the popular interpretation of the initials was 'Never The Same Colour'.

The PAL (phase alternating line) was adopted by most of the rest of the world. Using 625 lines at 50 pps (also interlaced) it averages out colour errors by reversing the phase of one component of the chrominance signal on alternate lines (hue depends on phase).

In France and Russia the SECAM (séquence couleur à memoire) system is used. This is an 825-line system operating at 50 pps, and uses two separate chrominance signals for successive lines.

In all these systems the luminance and chrominance signals are amplitude modulated (AM) and the sound, on a different carrier frequency, is frequency modulated (FM). This gives better quality sound, free from interference. The reason AM is used for the picture signal is that any visible 'echoes' are stationary (with FM they would fluctuate).

Dipole antennas   Radio waves are electromagnetic radiation, just as light is (see Chapter 1). At the low and medium frequencies used for sound radio (150 kHz to 1.5 MHz, or 2000 m to 200 m) diffraction enables the radiation to follow ground contours, and though higher frequencies are blocked by tall buildings and hills, they are reflected by the ionosphere at around 300 km altitude.

When the radiation encounters a conductor it sets up an electrical potential across it, and this can be used as the input to a receiver. At the frequencies used for TV broadcasts the input signal received will be very weak unless the conductor (the aerial, more properly known as the *antenna*), is facing in the right direction and is of the right length to match the transmitted wavelength. There are several ways of achieving this match, the simplest of which is the *half-wave dipole* (Figure 15.9). It consists of two conducting rods (usually aluminium alloy tube), each one-quarter of a wavelength long, orientated perpendicular to the direction of propagation and parallel to the direction of polarisation.

A strong signal produces an output of 1–2 mV from the antenna, and this is led via a cable to the input of the TV receiver. In the UK we use coaxial cable. This minimises interference from outside sources, as the outer conductor is connected to earth and shields the inner conductor. In mainland Europe they mostly use twin or *balanced* cable. The two conductors are then at opposite potentials, and the losses are lower.

As the speed of radiation is slightly lower in a conductor than in space, the half-elements should theoretically be about 5 per cent shorter than a quarter wavelength; but in practice this is not important, as the antenna needs to be able to pick up a fairly wide band of wavelengths, and its length is based on a mean value.

Multi-element antennas   There are several factors that make a single dipole antenna inefficient. The first is that it acts as a radiator itself; this means that much of the incoming energy is re-radiated into space. The second is that it can receive an equal amount of radiation from the back. It also picks up signals within an angle of around 30° to either side, so it is liable to receive unwanted signals from other transmitters. The addition of non-active *parasitic elements* reduces both problems greatly, and reinforces the signal. When an antenna array has parasitic elements, the dipole that actually feeds the signal to the receiver is called the *driven element*.

A *reflector element* is an unconnected dipole about 5 per cent longer than the driven element, positioned about one quarter-wavelength behind it. Its reflected signal is in phase with, and reinforces, the main signal, while its transmitted signal is in antiphase with the onward transmission and cancels it. At the same time it blocks signals from unwanted transmissions coming from behind the active element. Figure 15.10 shows the radiation acceptance pattern for a dipole antenna without and with a reflector element.

A folded dipole (Figure 15.10d) is a more efficient form of receiving antenna, and is almost universal in domestic TV arrays. At longer wavelengths such as those used for FM radio broadcasting, if the signal is fairly strong such a dipole may be

The ionosphere consists of a layer containing many charged particles, between about 50 and 500 km above the Earth's surface. The Appleton layer, at about 280–304 km, is particularly rich in free electrons, and acts as the main reflector for short wave radio.

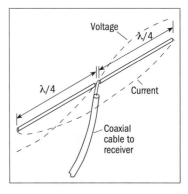

Figure 15.9 Dipole antenna. This consists of two quarter-wave lengths of conducting metal rod aligned perpendicular to the line joining the antenna to the transmitter. At the outer ends of the dipole the voltage is a maximum and the current is zero; at the centre the voltage is zero and the current is a maximum.

All TV transmitters (except for a few relay stations) transmit a horizontally polarised signal. It is the electric component that is horizontal; the magnetic component is thus vertical, which it needs to be in order to create the maximum electric potential in the antenna elements.

The Science of Imaging

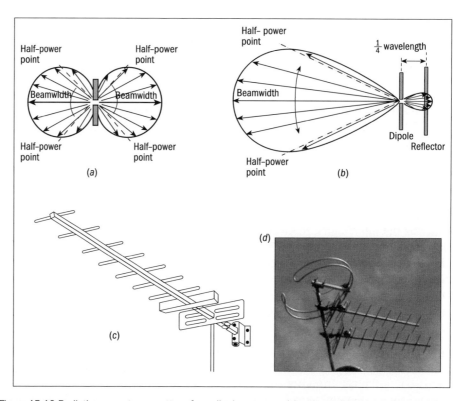

Figure 15.10 Radiation acceptance pattern for a dipole antenna: (a) without; (b) with a reflector; (c) a typical Yagi array. (d) A parallel pair of folded dipole domestic TV antennas. The uppermost antenna is a dipole matched for FM radio reception and bent into a circle for omnidirectional sensitivity.

If you spotted that the director elements were operating in an analogous manner to the quarter-wave anti-reflection coating of a camera lens, full marks. The principle is indeed the same. However, if you look carefully at a Yagi TV antenna array you will see that the parasitic elements are in fact nearer to each other than a quarter-wavelength, and often are closer together at the rear end. This is to obtain better uniformity of capture over a fairly wide band of frequencies.

constructed in a circular shape to make reception omnidirectional. You can augment the signal strength considerably by adding *director elements*. These are slightly shorter than the driven element, and are also spaced approximately one quarter-wavelength apart. Each element passes the signal onward to the next, and the overall gain in a multi-element set-up can be as much as 15 decibels (30 times). Figure 15.10c shows a typical nine-element UHF array. This is known as a Yagi array, after its inventor.

Antenna design is almost as much a craft as a science, and many variants are around. The more unorthodox designs may have been found to give maximum gain for weak sources, or maximum directionality where there is interference from other sources, or wideband reception. Some aerials can be very complicated indeed (Figure 15.11).

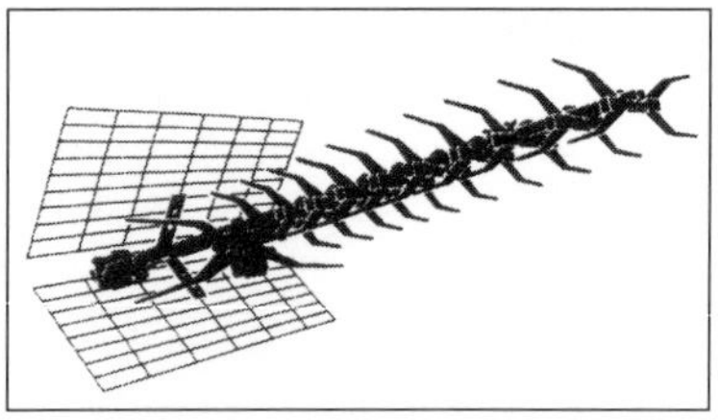

Figure 15.11 This complicated array has a gain equivalent to that of some thirty dipoles.

Transmitter antennas A transmitter also needs a matched antenna. Unlike the receiver antenna, though, it has to be omnidirectional. Low-power transmitters use a ring of dipoles, their outputs overlapping so that the power distribution is uniform in all directions. These dipoles need to be large in diameter, as they need to dissipate a good deal of heat; the greater the surface area the better. High-power transmitters sometimes have a different type of antenna that does not easily overheat. A transmitting antenna can be made omnidirectional, and dissipate more energy without overheating, if the dipole is metamorphosed into a cylinder with a narrow vertical slit. To limit diffraction the cylinder is much taller than its diameter (Figure 15.12). The slot is closed at the ends by metal strips to reduce leakage.

Television

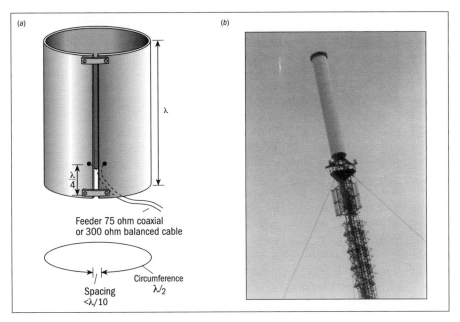

Figure 15.12 (a) Proportions of a typical cylindrical transmission antenna. (b) BBC TV mast, showing several different types of antenna. The cylindrical antennas are at the top, protected by a fibreglass shield.

Microwave relay transmission

If you want to send very high frequency (VHF) or ultrahigh frequency (UHF) signals over more than line-of sight distances you have to use some kind of relay system. In the past, relay stations were fed from enormous 'aerial farms' (such as those you can still see at Droitwich, Daventry and Rugby), and sometimes by

The VHF band, from around 97 MHz to 105 MHz, is used (at present) for analogue stereo broadcasting on FM. The UHF band, from around 300 to 600 MHz, belongs to analogue TV broadcasting. Other unused frequencies in these bands are allocated to digital radio and television.

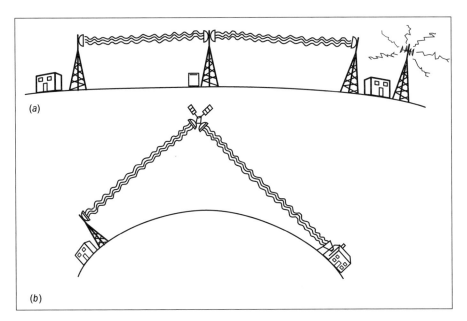

Figure 15.13 Microwave links: (a) ground relay; (b) satellite relay.

The Science of Imaging

This type of transmission is also used for mobile telephone and VHF radio relays, which is the reason so many tall buildings sport forests of antennas and dishes on their roofs.

landline. Nowadays they receive their signals via microwave links. Microwaves are electromagnetic signals with wavelengths of only a few centimetres, and behave very much like light waves. They undergo refraction, diffraction and reflection, and they travel in straight lines. If you raise the carrier frequency to several gigahertz you can focus the signal into a tight beam with a paraboloidal reflector, like a searchlight (Figure 15.13), and send the beam to a receiver (also equipped with a paraboloidal reflector) within the line of sight. This can amplify the signal as necessary, and transmit it onwards to the next relay point; and so on to the final station, where it is converted back to the original carrier frequency and passed to your TV antenna in the usual way.

In order to be able to mount receiving dishes flush to a vertical wall, they are usually designed as off-axis sections of a paraboloid. The antenna at the focus is known as a *horn antenna* from its shape, and it collects, amplifies and frequency-converts the signal before passing it to the TV receiver (Figure 15.14).

Satellite transmission

The ultimate in line-of-sight relays is satellite transmission. At a height of 36 000 km above the Equator, a satellite will remain stationary relative to the Earth, over a given point. There are a large number of satellites at present in this *geostationary orbit*, and many of them belong to the Direct Broadcast Satellite system (DBS). The Earthbound transmissions use frequencies around 14–14.5 GHz, and are retransmitted from the satellite at downlink frequencies between 11.7 and 12.2 GHz. The satellites are powered by solar energy, and radiate sufficient power to be picked up by quite small (45 cm diameter) dishes.

Owing to problems with the reception of linearly polarised signals in this mode of communication, satellite signals are circularly polarised. They are thus received in equal strength no matter what the polarisation requirement of the receiving station may be.

You may have noticed that when a newscaster on TV talks to a correspondent in some distant country, there is often a pause before the correspondent replies. This is because it takes a measurable time for the signal to travel to a distant satellite and back to the correspondent. The reply then takes the same time to return. Try working out what that time is. The answer is on the next page.

Figure 15.14 Wall-mounted off-axis paraboloidal receiving dish with horn antenna.

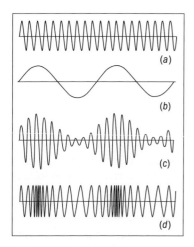

Figure 15.15 Modulation of a carrier wave: (a) carrier; (b) waveform to be encoded; (c) amplitude modulation (AM); (d) frequency modulation (FM).

The signal

The luminance and chrominance signals are encoded on to the carrier as *amplitude modulation* (AM); the sound is transmitted as *frequency modulation* (FM) on a separate carrier at a slightly different frequency.

Figure 15.15 shows the way a simple sinewave signal is coded on to a carrier by AM and FM. The decoder for AM can be a simple rectifier to cut out the negative-going part of the signal, plus a smoothing circuit. The FM decoder needs to be more complicated. In practice both decoders are quite sophisticated, as luminance, chrominance and two or more channels of stereo sound also need to be dug out of the complete signal.

The input from the antenna is fed to a transformer, which passes it to the decoding circuit. A coaxial input has one end of the transformer primary at earth potential; a balanced input has the earth in the middle.

Television

The TV receiver: types of display

The receiver is in principle just a captor of electromagnetic signals with a decoder and amplifier, and with its output connected to a picture display system and a loudspeaker. The beam scanning system is synchronised to the incoming signal by pulses added to the signal at the transmitter.

Electron beam tube display   The picture tube has a triple electron gun, one beam for each of the three primaries, red, green and blue. The perforated screen allows each gun to 'see' only its own colour, as I explained earlier. This type of display has had a long run, but we shall probably see the end of it before long, cumbersome as it is. The only recent development has been a tube in which the beam comes from the side, and is turned through a right angle before scanning begins. This does at least reduce the depth of the receiver, but does not solve other problems such as the enormously high voltages and vacuum needed for operation.

Projection systems   There are three separate small tubes, one for each primary. These give a very bright image, which is focused by a separate lens system for each tube on to a front or rear projection system.

LCD screens   These have been developed largely as the demand for laptop computers has rocketed. They are becoming common on portable TV receivers, and are universal on pocket TV sets. So-called liquid crystals are gelatinous substances that polarise light when they are subjected to an electric field. A *liquid crystal display* (LCD) is an array of pixels that can be stimulated separately by a series of contacts along the horizontal and vertical edges. For a colour picture they are covered by a mosaic of primary colour filters. LCDs can operate by either reflected or transmitted light. They are also called *spatial light modulators* (SLMs), which is a more logical name.

LED and plasma screens   Light-emitting devices (LEDs) have only recently become reliable at blue wavelengths, so LED screens have yet to make an impact; but they would appear likely to take over from both electron beam tubes and LCDs eventually, especially as they operate at low voltages, do not give off heat, are energy-efficient, can be as bright as you like, and can be installed as a flat screen. Plasma screens using gas discharge tubes are suitable for very large displays, but use masses of electrical power. A special type of LED that has recently come to the fore is the diode laser. Originally emitting only in the near infrared, in which guise it has been in use in CD players for many years, the diode laser is now available in almost any visible wavelength. Its beam is a longish ellipse with a horizontal spread of about 30° and a vertical spread of about 7°. A simple lens system corrects this beam, making possible a fully collimated, very bright image beam for projection. This would seem to be the future for projection television.

The earliest projection system used *eidophor tubes*, which (improbably) used the electrostatic distortions of a thin film of oil to produce diffraction effects that modulated the beams to produce the image.

Visual display units and computer monitors

Visual display units (VDUs) and computer monitors operate on much the same principle as television sets, but both have inputs from a computer rather than a radio signal. They need not necessarily operate on a Cartesian plan, with horizontal and vertical axes; some computer programs drive the monitor in polar coordinates like a radar screen.

Answer to question on previous page: Using the figure for the speed of light $c = 3 \times 10^8$ m s^{-1} and a distance of 72×10^6 m we get 0.24 s. This may seem a fairly short delay, but it is much more than we are used to. If the signal has to pass through several satellite relays, as it does for very long distances, the delay may reach several seconds.

Digital television

Analogue television, which is what this chapter has been chiefly dealing with until now, has a number of drawbacks. It is prone to interference from outside sources; bad weather and solar storms can cause patterns, distortion and fading; and transmission of the signal takes up a considerable fraction of the available bandwidth of frequencies. In addition, video recording by analogue techniques results in some inevitable deterioration in quality, and the videotape itself becomes degraded with repeated use. You have already had a look at the principles of digital imaging in Chapter 13. Digital television operates on similar lines to digital still photography, except that the images come faster, and are transmitted via a carrier beam rather than being downloaded into a computer or down a telephone line. The carrier beam is much as I have already described, and the only additional task of the decoding system is to transform a set of spikes into its envelope, like fitting the template to the gate in Figure 13.1. So where are the advantages of digital TV? Quite a few, as it happens.

- Narrower bandwidths allow more channels in a given bandwidth allocation.
- Higher resolution, especially with the introduction of 1250-line HDTV.
- Simpler colour coding, hence more reliable colour (this applies in particular to NTSC).
- Perfect transfer to recording systems over any number of generations of copies.
- Tape wear much less of a problem.
- Easy combination with data such as subtitles.
- Simplification of the mechanism of frame grabbers, download to computers, etc.
- Easy compression of signal without loss of information; variable levels of compression available, depending on required image quality.

Perhaps the only snag (apart from the users having to buy new equipment or ancillaries) is that when the noise becomes excessive the signal disappears suddenly and totally. But by then an analogue signal would have long been incomprehensible.

Aspect ratio and high-definition television

A perennial problem with the screening on TV of commercially produced films is that they don't fit the format. Older films made with a 4:3 aspect ration fit fairly well on the standard TV screen format of 5:4, but almost all films made for showing in cinemas are now made in wide-screen format, with an aspect ratio of from 8:5 to 5:2, depending on the process. To deal with the problem the film has to be shown with black bands at the top and bottom of the screen, or be given the Procrustean treatment of lopping off the extremities. Sometimes the presentation gets away with the ruse of squeezing the picture a little (this is common with introductory material, where the whole screen may carry important information), but on occasion whole scenes have to be re-shot with television in mind. A new generation of TV receivers is now appearing, with a screen aspect ratio of 16:9, and with these screens the problem is considerably lessened, though now the difficulty lies with those old films, where one sees either black bars at both sides of the picture, or the tops of heads cut off (you have a choice).

There is a problem of a different type concerning another innovation: the proposed introduction of high-definition television (HDTV), with approximately twice as many lines and a greater bandwidth for better overall picture quality. Different countries – and different manufacturers – have (at the time of writing) not settled on a common technology, and we may be in for a repetition of the kind of tussle that took place between rival videocassette systems when the technology was first introduced. It is to be hoped that this will not happen; if it does, it is by no means certain (as with videocassettes) that the best system will win.

Digging deeper

I have not yet discussed the way the signal is recorded on tape or disc. That belongs to the next chapter. As most of the texts on television also deal with video recording, I have left the recommendations until the end of Chapter 16. One book does deserve mentioning here, though. *Vision Warrior*, by Tom McArthur and Peter Waddell (Scottish Falcon, The Orkney Press, 1990), is a biography of John Logie Baird, and sets to right many of the myths about the man and his work. Tom McArthur is a freelance journalist, and Peter Waddell is Reader in Engineering Science at Strathclyde University, Glasgow, where there is a small but fascinating museum chronicling the history of television. If you get the opportunity, it is well worth a visit.

Another venue that shouldn't be missed by any TV enthusiast is the National Museum of Photography, Film and Television in Bradford, which has a whole floor given over to the history and workings of television, including much historic apparatus and a working example of a Baird transmission system.

The Science of Imaging

Chapter 16 Video recording and replay systems

Magnetic tape recording

Recording sound or television is necessarily a one-dimensional business. The LP record, with its single groove more than 0.6 km long, is an obvious example. Analogue audiotape, which grew up around the same time, is another. Audiotape actually began, like the first experimental $33\frac{1}{3}$ discs, in the 1930s. One of the earliest tape machines was the Blattnerphone, which used a steel ribbon passing over a magnetic head at five feet per second. The tape reels were enormous, and editing involved the use of tin snips and a soldering iron. In the 1940s steel tape was replaced by fine steel wire, and the recorders shrank to suitcase size. Soon after this, plastic tape coated with ferromagnetic iron oxide particles began to be used.

The magnetic permeability of a substance is the ratio of the magnetic flux density in the substance to the magnetic field producing it. The magnetic permeability of a paramagnetic substance is slightly higher than unity. A few substances have a permeability less than unity, and are repelled by a magnetic field. They are said to be *diamagnetic*.

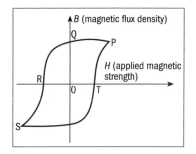

Figure 16.1 Hysteresis. The magnetic flux density (i.e., the degree of magnetisation) of a ferromagnetic substance lags behind the magnetic field strength producing it. When the applied external field falls to zero some magnetisation is retained.

Ferromagnetism, hysteresis and a.c. bias

Ferromagnetism Most elements and compounds are made up of molecules that tend to orientate themselves in a particular direction when placed in a very strong magnetic field. This is because either the electron spins or the orbital motions of the electrons are unbalanced. They are termed *paramagnetic*.

A small number of substances have a very high permeability. They are said to be *ferromagnetic*, as the main members of the group are iron and some iron alloys and compounds. In ferromagnetic substances groups of molecules, about 0.1 to 1 mm in diameter, known as *domains*, are magnetised in one direction, but the direction of magnetisation from one domain to another is random. Under the influence of a strong external magnetic field the domains that are aligned with the field grow in size at the expense of unaligned domains until eventually, if the external field is strong enough, they are all aligned, and the substance is said to be 'saturated'. After the external field is removed, some ferromagnetic substances retain their magnetism. This is called *permanent magnetism*, and is a characteristic of a number of compounds and alloys of iron, nickel, cobalt and chromium, as well as of some rare-earth elements such as samarium. The compounds most suitable for recording are those whose magnetism is easily stimulated, is retentive, and can be readily eliminated by means of an alternating magnetic field that decays to zero over a short time. Suitable materials for coating on recording tape are iron(III) oxide, Fe_2O_3, chromium(IV) oxide, CrO_2, and pure iron, Fe.

Hysteresis When stimulated by a fluctuating magnetic field at audio or higher frequencies, the magnetisation of the tape coating lags behind the stimulus. This effect is called *hysteresis* (Figure 16.1). This causes a particularly obnoxious type of distortion called *crossover distortion* (Figure 16.2a). Soon after tape recording was introduced, it was found that the distortion caused by hysteresis could be eliminated by superimposing a high-frequency signal (around 50 kHz) on to the audio signal. This is called *a.c. bias*. Figure 16.2b shows the effect on the signal of adding bias.

Video recording and replay systems

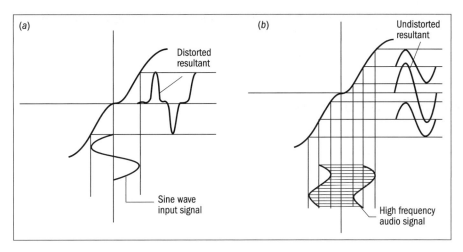

Figure 16.2 (a) A signal supplied directly to the tape suffers crossover distortion due to hysteresis. (b) The addition of a high-frequency a.c. signal (a.c. bias) allows the output to be derived only from the linear regions of the transfer characteristic, and eliminates the distortion.

Tape recording principles

Videotape was developed from audiotape; so to begin with the simplest system, we shall look at audiotape first. In a professional sound recorder there are three magnetic heads (Figure 16.3). As the tape moves from left to right, the first head it encounters is the *erase head*. This applies a high-frequency alternating magnetic field, which fades as the tape moves on, wiping out any previous magnetisation. The next head is the *recording head*, which carries the audio signal plus the a.c. bias, and impresses this on the tape in the form of a varying magnetisation. The third head is the *replay head*, which monitors what is now recorded on the tape. This head is only seen in open-reel recorders and professional and semi-professional cassette recorders. In cheaper recorders the recording head can be switched to 'replay', and monitoring is directly from the incoming signal.

There are four types of oxide coating: 'ferric', 'chrome', 'ferrochrome' (which is a hybrid of iron and chromium oxides) and 'metal', which is pure iron, treated to

The head gap is very small, 3 μm or less. This is about one-twentieth of the thickness of one of the pages of this book. The narrower the gap, the higher the resolution.

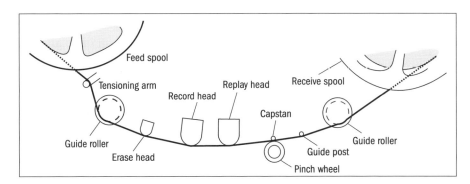

Figure 16.3 Head layout in a professional audiotape recorder.

The Science of Imaging

The recording machines notoriously used by the Nixon administration were 'crawlers' operating at $\frac{15}{16}$ in/s (2.4 cm/s), and the sound quality was correspondingly low. However, a modern analogue cassette recorder, with improved magnetic heads and better tape coatings, can equal the performance of what were once top-of-the-range open-reel recorders.

prevent atmospheric oxidation. Chrome and metal tapes require a higher erase voltage and higher bias than the other two. They also require less boosting of the higher frequency response (*equalisation*).

The standard writing speeds of audiotape show its history. They are all submultiples of 60 in/s (152.4 cm/s). Professional audio analogue recordings are made on open-reel recorders at 15 or $7\frac{1}{2}$ in/s (38 or 19.5 cm/s). Sources with less stringent requirements, such as speech, can be recorded at $3\frac{3}{4}$ in/s (9.5 cm/s). Cassette recorders record at $1\frac{7}{8}$ in/s (4.75 cm/s).

Because of the higher frequencies involved, professional open-reel digital sound recorders operate at 30 in/s (76.2 cm/s), but digital cassette recorders operate at the same speed as analogue ones, for compatibility. Rotary Head Digital Audio Tape (RDAT) uses larger cassettes with helical scanning similar to the VHS videocassette system described later in this chapter, and very low linear tape speeds of 8.15 or 12.225 mm/s.

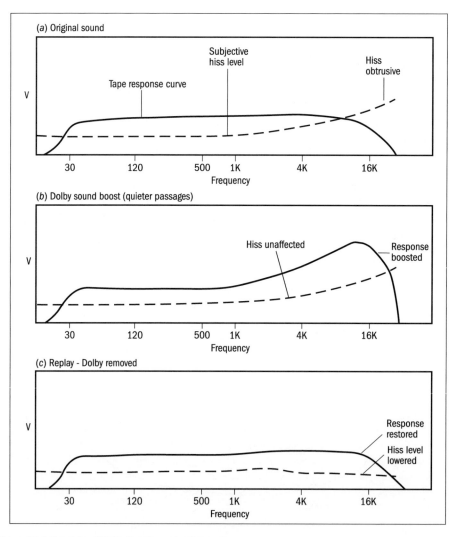

Figure 16.4 Principle of Dolby B noise reduction system.

Video recording and replay systems

Noise reduction systems

Ray Dolby, one of the pioneers of high-fidelity sound reproduction, evolved a system for reducing the effect of tape hiss, which is white noise caused by the random size and distribution of the magnetic domains in the tape coating. Dolby appreciated that this hiss became obtrusive only in the quieter passages of music, and that the upper frequencies were the most strongly perceived. He designed circuitry that would emphasise the higher frequencies in the recording of quieter passages, and de-emphasise them on replay (Figure 16.4).

This system, known as 'Dolby B noise suppression', is used in the better models of cassette recorder and player, and in all VHS videocassette sound systems. A more complicated system, Dolby A, is used in professional analogue sound recording. Most modern high-quality equipment now uses a simpler version of this, called Dolby C.

White noise is a sound containing all frequencies in equal amounts, with random phase. It sounds like escaping steam.

Videotape recording techniques

The first videotape recorders were also open-reel, but the inconvenience of their complicated lacing up system led to their being supplanted by several types of videocassette, with the lacing being carried out automatically inside the machine. These different types were mutually incompatible, and eventually the struggle for commercial survival was won by the Video Home Service (VHS) format.

The recording of an analogue video signal presented no theoretical problems that had not already been solved by audiotape technology. There was one serious practical problem, though. Whereas audiotape needs to record frequencies only up to 20 kHz (the limit of human hearing, about $2\frac{1}{2}$ octaves above the top note of a piano), video frequencies go from around 25 Hz to more than 4 MHz. This means that the writing speed on the tape has to be nearly six metres per second. The solution is a classic example of lateral thinking: move the recording head as well as the tape. All modern videotape recorders operate on this principle. There are two magnetic recording heads on opposite sides of the surface of a slightly tilted cylinder (Figure 16.5). The tape is wrapped around one half of the cylinder, which rotates at 25 rev/s in the direction of the tape movement. The information is thus recorded helically, as a succession of long oblique bands, crossing the width of the tape over a distance equal to half the circumference of the cylinder. Each traverse records one complete TV line. The tape itself moves linearly at only about 33 mm/s.

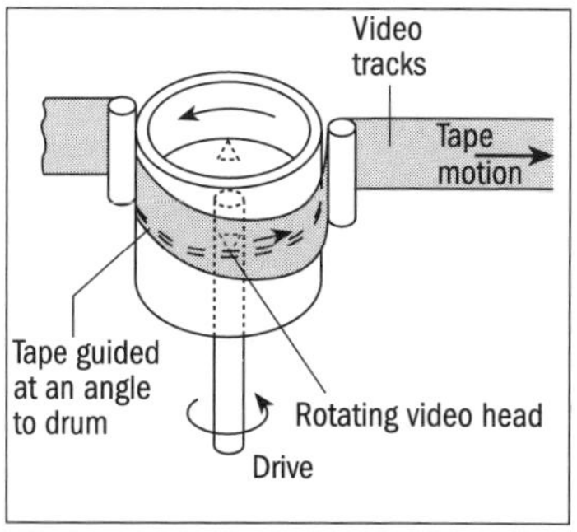

Figure 16.5 Record/replay head in a VHS recorder.

This is for 'standard play'. On the 'long play' setting the tape speed (and the image quality) is halved.

The VHS format

The standard layout of a VHS recorder transport mechanism is shown in Figure 16.6. When you feed the cassette into the recorder the internal mechanism opens the lid and pulls out a loop of tape, which it wraps around the cylinder. The rest of the guidance system then moves into place. Once threaded, the tape passes first over an erase head which removes any previous tape signal. It then passes over a so-called impedance roller, the action of which is to smooth out any possible juddering or vibration of the tape. Next, the tape passes over a guidepost, around the scanning drum in a semicircle and past a second guidepost. It next passes linear audio erase and recording heads, which can be used for subsequently adding music

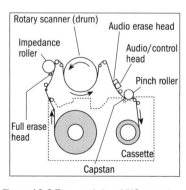

Figure 16.6 Tape path in a VHS recorder.

177

The Science of Imaging

or voice-over commentary (at one time this was the standard audio track). This head assembly also produces control pulses. The last item in the path is the capstan and pinch wheel, which maintains the correct linear speed for the tape. Finally, the tape is wound on to the receive spool.

In the modern VHS recorder with high-fidelity stereo sound, both sound and video signals are coded in FM. Two sound heads are built into the scanning cylinder. The sound is thus recorded a fraction of a second before the picture, and because the video signal is at a much higher frequency than the audio signal it is recorded much nearer the surface of the tape coating, erasing only the surface part of the audio signal. The video signal thus actually lies on top of the audio signal.

Although the VHS format has many apparent disadvantages when compared with other possible systems, it has survived (like the internal combustion engine) partly because of the very high initial sales of equipment incompatible with other systems (a self-perpetuating market), and partly because of regular marginal improvements. The most recent update has been the introduction of Super VHS (S-VHS), which uses improved circuitry and superior-quality tape to provide much better resolution and colour quality than standard VHS. The VHS principle now seems to have reached its limit, and other systems based on disc technology are beginning to take over. Nevertheless, the majority of camcorders still use a miniaturised version of VHS, with a smaller cassette and 8 mm wide tape. The signal is usually coded digitally.

Digital video recording

As I explained earlier, digital broadcasting has the advantage over analogue broadcasting that it is almost totally free from both interference and signal corruption. When digital signals are recorded, there is a further advantage: digital information can be compressed. In television pictures (and in sound) much of the information remains the same over a period of time. So you don't have to keep sending the same information over this period. You need only send a new signal when the information changes. Of course, the compression that is possible varies a good deal, from static titles (or silence in audio) to rapid action, but an average compression ratio of about 5:1 is usual. A further benefit is a simpler interface between video and computer; and precise editing is easier, too. Professional studio digital sound and video recording doesn't use compression, though field recording equipment does. The loss in quality, if any, is undetectable by eye or ear. What is important to the field worker – and the amateur – is the saving of space.

Digital videotape

With the runaway success of compact audiodiscs (CDs), which are encoded digitally (though not compressed) it quickly became clear that digital recording would eventually dominate the videotape field too. Mechanically, there is little difference between analogue and digital video recording, but the tape width can be halved, and the cassettes can be made smaller and lighter than analogue recorder cassettes. The advantages of digital video recording will also become more obvious as digital TV becomes universal; it is possible to record a digital TV programme without any loss of quality. Tape will continue to be an important vehicle for storing video images for some time, but its role is already being taken over by discs, for their ease of rapid access. This, of course, was the reason computer memories moved from tape to disc many years ago.

Video recording and replay systems

Camcorders

A camcorder, as its name suggests, is a combination of video camera and recorder. It has almost completely replaced the amateur movie camera, largely because of its low running costs and its convenience in use. Tape is cheap and reusable. In addition, camcorders operate in near silence. As the (digital) signal from the camera is recorded directly on videotape there is no need for an RF carrier, which simplifies the circuitry.

The earliest camcorders used vidicon tubes, which made them cumbersome. Today, CCD arrays are universal. Broadcast-quality camcorders use a triple array RGB system, but amateur and semi-professional equipment mostly uses a single CCD array with mosaic colour filtration. The viewfinder may be optical, but is more often an LCD display. All modern camcorders are digital, though they may have an alternative analogue output for playback on a traditional analogue TV set. Once digital TV becomes universal, analogue outputs will become redundant.

Digital videodiscs

Digital audiotape (DAT) has become the norm in sound recording studios, but attempts to launch DAT equipment for the general public have been less successful. The digital videodisc (DVD) had more of an initial success, and though there have been other formats (notably the large Laserdisc), DVD looks set to sweep the board, not least because of its enormous potential for storing information (up to well over 18 gigabytes on a single four-layer disc 120 mm in diameter). This capacity allows for up to two full-length feature films with surround sound, and dialogue available in eight languages. Another advantage of the format is that DVD players can play ordinary audio CDs.

Before the CD

Until the advent of the CD, all record players used an analogue signal in the form of a wavy spiral groove in a plastic disc (at first shellac, later PVC). The groove actually duplicated the waveform of the sound energy. This principle goes right back to Thomas Edison, the inventor of the phonograph, and Emil Berliner, who invented the flat disc. By 1960 the gramophone record had reached a high level of sophistication, as had the associated playing and amplifying equipment. Two channels could be coded into the groove, one for left and one for right, at ±45°, forming a right-angled V.

Various attempts have since been made to encode two further channels into the groove in order to provide surround sound, but without much commercial success. Another venture was binaural sound recording, using microphones in the ears of a dummy head, to be listened to through headphones. This produced startlingly realistic results, but never caught on. By the mid-1970s analogue sound recording had, it seemed, reached its limit, and the LP disc, delicate, noisy and prone to the effects of wear, dust and electrostatic charges, was becoming seen as inadequate. Something had to be done.

The stereo principle is much older than you might think. The conductor Leopold Stokowski was experimenting with stereo recording in 1934, and there are reports of demonstrations of binaural sound (using two telephones) in Paris around 1900. But the real leap forward was the 45/45 principle patented by Alan Blumlein, a genius of electronics, who lost his life tragically in an air accident in 1943 while working on airborne radar development.

The Science of Imaging

How does a DVD work? The DVD was developed using the technology of audio CDs via the CD-ROM, a random-access memory disk for computers. So the simplest way to describe the principles involved is to start with the CD.

The digital principle First, a recap on the principle involved. To describe a waveform in digital terms you measure its amplitude at a number of closely spaced sampling points. Then you draw the curve that passes through all the values. But can you really get the true waveform from those samples? What if there were some kinks in the waveform in between your measurements? The answer is that if you know the highest frequency present in the waveform (and you always do) you have to sample at more than double that frequency (the Nyquist criterion). The waveform then *has* to be smooth between the sampling points. The next thing is: how accurate does the measurement have to be? Digital measurements operate only in whole numbers. Suppose you operate on a 0–10 basis. Any of the values could be up to 0.5 too small or too large, and that might not – would not – be accurate enough. You would get steps (aliasing) in the reconstruction. But digital measurements are made in bits, not tens. You can sample to as many bits as you need. 16-bit accuracy is common: this gives 65 536 steps; 20-bit accuracy gives just over a million.

Present sampling for audio CDs is 16 bits, though the recording is 20 bits. Hi-fi enthusiasts claim that 20 are needed in the CD too. Some CD players claim to achieve comparable quality by using what is called 'over-sampling'; but whether the gain is audible is doubtful.

CDs For a CD the sampling rate is 44.1 kHz, well above twice the limit of human hearing. The record itself is in the form of a series of flat-bottomed pits, the length and spacing of which encodes the digital information. This information is read by the tightly focused beam of a semiconductor laser operating in the red or near infrared. The pattern of pits has an aluminised coating, and reflects the laser light back to a beamsplitter, where the reflected beam is monitored by a photocell (Figure 16.7).

The figure 44.1 came in the first place from the digital conversion of analogue TV signals, and was arrived at because the equipment used for mastering CDs is video based.

The depth of the pits is such that where a pit exists, the light reflected from the bottom of the pit interferes destructively with the light from its lands, but where there is no pit all the light is reflected. The difference between these two signals is picked up by the photocell and passed to the decoder and amplifier.

The discs run at a constant linear speed, tracking from the centre outwards, so that as they run they slow from about 600 rev/min to about 130 rev/min. One of the advantages of this system is that longer playing time is possible, up to 74 minutes (exceptionally, 80 minutes). The total linear length of the track is some 6 km.

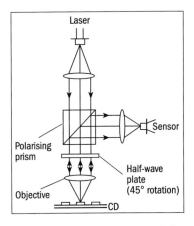

Figure 16.7 CD information retrieval. The reflected light has its polarisation turned through 45° twice, and is thus wholly reflected at the polarising prism on its return. There is destructive interference between the light reflected from the pits and their lands, and this is used to distinguish the pits from the lands alone.

CD-ROMs It was only a short step towards putting computer programs on to a CD format. A CD can contain around 650 megabytes of information, which is a good deal more than a floppy disc can hold (1.25 Mb). Music CDs are now regularly issued bearing computer-read information about their contents, and about the composers and artists.

One example is the music CD-ROM that comes with the monthly *BBC Music Magazine*.

DVDs These differ from CDs in having two layers of information, and (often) by being double sided. The inside layer is usually a normal CD that can be played with a standard CD player. The upper layer, which is ignored by the CD player because the detail is too fine and it is out of focus, contains the video and audio information. A shorter wavelength laser is used. This has a smaller focal spot, which is suitable for tracking the much smaller DVD pits. As these are only half the size of CD pits, four times as much information can go on to a DVD as on a CD. With compression, a full-length feature film with surround sound can be recorded on a single disc. Better still, a double-sided, two-layer DVD can provide more than eight hours of video of a quality far surpassing that of VHS analogue reproduction. The laser focuses automatically on the appropriate layer, and ignores the other layer. A DVD with two-layer information can't be played on a normal CD player, of course, but a CD can usually be played on a DVD player.

A DVD used as a computer memory can store 4.7 Gb of information in each layer, more than 18 Gb altogether. This is more than 25 times the capacity of a CD, which itself has more than 600 times the capacity of a floppy disc.

Digging deeper

Television and video recording is a huge subject, and I am well aware of having been somewhat superficial over both the science and the technology. I have tried to avoid cutting theoretical corners, though.

Finding really good books on the theory of television and video systems is not easy. For one thing, in this area there is no such thing as science without technology, and it is difficult (and pointless) to try to discuss scientific principles without also discussing both the technology and the hardware to which it refers. The best books are nearly all American in origin, and the technology differs in quite a few details from British and European technology.

Most books on television and video deal with either the creative side – production, editing, and so on – or with the equipment itself: servicing, fault finding, etc. Some of the best of these are to be found in the Focal Press list (there are more than 100 titles!). The standard work in this field is indeed published by Focal Press: *The Art of Digital Video*, by John Watkinson, regularly updated and splendidly comprehensive. But you need a deep pocket to buy it and a spell of weight training before you can carry it home. (There is a companion volume, *The Art of Digital Audio* also by John Watkinson, if you are interested in this aspect too.) Watkinson has also written two other companion books, *An Introduction to Digital Video* and *An Introduction to Digital Audio* for Focal Press. These eschew the tougher mathematics and stick to run-of-the-mill equipment. Make sure you buy (or borrow) the most recent edition.

Also from Focal Press is a much slimmer paperback, *Basic TV Technology: Digital and Analog*, by Robert Harting (3rd edition, 2000). This is one of Focal's *Media Manuals* series. These are laid out with one topic per double page, text on the left and diagrams on the right, a somewhat restrictive treatment that nevertheless works well here. This includes a description of the PAL system but not of SECAM. It is laudably frank about the shortcomings of NTSC.

One of the best accounts of the theory is *Basic Television and Video Systems*, by Bernard Grob and Charles Handon (6th edition, McGraw-Hill, 1999). In spite of its title it is very comprehensive, though the section on videodiscs is already becoming dated. Surprisingly for a textbook at this level, it is short on diagrams but long on (somewhat muddy) photographs. There are self-testing questions at the end of each chapter, but they are all of the true/false type and not very mind-stretching. Like the other books it is American in origin, but the authors do deal fairly thoroughly with both PAL and SECAM technologies.

In contrast to most other manuals, Charles Poynton's *A Technical Introduction to Digital Video* (John Wiley, 1996) looks at the subject from the perspective of computing and communications. It assumes a good knowledge of electronics and computing principles, and is aimed at computer system designers and television engineers seeking to broaden their horizons. Not a book for beginners, but chock full of information.

Any book that takes more than a year in the writing and publishing cannot hope to be right up with the times, so fast do things move in this field. For fully up-to-date information you have to go to the specialist magazines, where you can learn about the latest equipment, and eavesdrop on the latest behind-the-scenes horse-trading – over HDTV systems, for example.

Chapter 17 Three-dimensional imaging

How we see depth

Monocular clues   Looking out of a window, you can see that your image of the world has depth. You don't need to move around to see this. Even if your window is only a peephole, so that your viewpoint is fixed (and one-eyed), the world still doesn't look flat. Why should it? After all, it doesn't look flat in a photograph, or even in a sixteenth century painting. There are four clues to depth in a *monocular* (one-eyed), fixed viewpoint image:

- *Relative image size*   Near objects appear larger than far objects, and parallel lines appear to converge towards distant points.

- *Image overlap*   Near objects overlap far objects.

- *Aerial perspective*   Haze causes far objects to have a lower contrast and colour saturation, and to appear more bluish, than near objects.

- *Modelling and texture*   Highlights and shadows indicate the characteristic shapes and surface textures of objects.

Figure 17.1 (colour plate) gives an example of these clues that give the illusion of depth to a flat picture. The ancient Greeks and Romans were familiar with them, and according to contemporary accounts they produced *trompe l'œil* murals of startling realism. The laws of perspective were lost during the Dark Ages, but were rediscovered in the fifteenth century by Italian painters.

Renaissance painters achieved some remarkable effects with wall and ceiling paintings. In conventional paintings the rendering of light and shade on objects, and the texture of fabrics, reached a peak with the Dutch and Flemish masters of still life.

There are nevertheless important clues to depth in real life that are missing from a painted picture or a photograph. In real life, if you change your viewpoint, the relative position of near and far objects changes too. When you look at a scene your eyes are moving constantly, and so, probably, is your head. A change of viewpoint, however small, produces these positional changes in the visual image: near objects appear to move in the opposite direction to the way your head moves, while far objects remain stationary. This effect is called *parallax*.

Parallax probably gives the most powerful sense of depth in real situations. A static picture does not possess this property, of course, but a cine or TV image does. And the camera doesn't need to move, either: the subject can move itself, or simply rotate, a property often exploited in computer imaging.

A further clue to depth is the necessity to re-focus your eyes between near and distant objects. This is called *accommodation*. This is relevant to objects nearer than about two metres. It complements the other clues, but as the action of accommodation is automatic and largely unperceived, it probably contributes little to the sensation of depth. In any case, by the time they reach the age of around 45 most people have lost much of their ability to accommodate.

A *trompe l'œil* painting is one that deceives the eye into believing that the painted perspective is real depth. There are many examples, particularly in Italian churches, of false alcoves and domes painted on flat surfaces. The rediscovery of the laws of perspective is usually attributed to the painter Brunelleschi, but this is an over-simplification: many other painters were involved, particularly Uccello and Piero della Francesca.

It was said at the time that birds would peck at the grapes, and bees would visit their painted flowers. I am not altogether convinced about the latter: see Chapter 20.

The effect of parallax is most marked when you watch the countryside go by through the window of a train or car. The nearest objects flash past; middle distance objects move more slowly; and features in the far distance seem to scarcely move at all.

The Science of Imaging

Proprioception is the sense of position. It tells you, for example, whether your left arm is bent. It tells you too, even with your eyes shut, whether your left forefinger is about to scratch your nose rather than miss it and scratch your chin instead. Eye convergence is one of its more subtle perceptions.

Binocular clues   There are two further clues to the perception of depth. These depend on the possession of two forward-facing eyes, and this is important in more than one way. Whereas the first set of clues to depth was based on geometry, these two (along with accommodation) are physiological.

- *Ocular convergence*   As an object approaches you, your brain tells your eye muscles to turn the optic axes of your eyes inwards so that they meet in the plane of the object. The feedback from the eye muscles tells your proprioceptive system how much convergence has been necessary, and this information is matched to a pre-existing model in your brain, to keep tabs on the distance of the object.

- *Stereoscopic fusion*   This is a quite complex concept, as it is associated intimately with the visual cortex of the brain. Because your eyes are around 6–6.5 cm apart, the images on your two retinas are slightly different, owing to the difference in viewpoints (parallax). By a process that is still only imperfectly understood, your visual cortex compares the two images, and interprets the difference between them in terms of depth. Physiologists call this facility *stereopsis*.

Stereopsis begins to develop in human babies at the age of a few months, and is usually fully developed between one and two years, though there is some variation. What is certain is that if something prevents its development before the age of about 4–5, it will fail to develop fully. This problem was at one time common in individuals who had had an uncorrected squint as small children because of unbalanced eye muscle development.

Nowadays this condition is quickly diagnosed and rectified, but if it is ignored it leads to lazy eye syndrome, with vision in the non-dominant eye disappearing almost completely.

Stereoscopic perception varies greatly between individuals. Probably only about five per cent have the sense developed to the highest degree, this perceptual elite including all top-flight cricketers, baseball players and others whose talent demands the instantaneous and exact assessment of the position of a fast-approaching object. Of the remainder of humanity, around 20 per cent have only feeble stereoscopic perception, while an unfortunate 1 in 20 has effectively monocular vision.

Not all that unfortunate, though. These individuals may be cut off from the appreciation of stereoscopic pairs of images; but they are by no means deprived of three-dimensional vision. After all, people with only one fully functional eye can still play a decent game of tennis.

The limitations of stereo pairs of images

Let us recap on those clues to depth:

1. Relative image size.
2. Image overlap.
3. Aerial perspective.
4. Shadows, modelling and texture.
5. Static and dynamic parallax.
6. Accommodation.
7. Ocular convergence.
8. Stereoscopic fusion.

Three-dimensional imaging

Not all of these are equally important, but they do all contribute to depth perception. It is worth noting, though, that a stereoscopic pair of photographs lacks a number of these factors, in particular accommodation, dynamic parallax and, to a large extent, ocular convergence.

Early stereoscopic images

We don't know exactly when the first stereoscopic images were made. There have been theories about drawings dating back to the fourteenth century and earlier, though such drawings may just be different views of an object. It is in fact quite easy to draw stereoscopic pairs of simple geometrical objects (see Digging Deeper). You need, though, to develop some ocular skill, to be able to fuse the two images without some sort of optical aid. A pair of stereoscopic images appears on Figure 17.2 (colour plate) for you to try.

Probably the first model of a working stereoscope was that of Charles Wheatstone in 1860. This has survived virtually unchanged to the present day, and is still used for entertainment by enthusiasts, and (not for entertainment) by photo-interpreters of aerial reconnaissance and survey photography. (More about the latter usage, later.)

Wheatstone's stereoscopic principle is simple and elegant. Two photographs taken of the same scene from viewpoints the same distance apart as the distance between the eyes are mounted with their optical centres the same distance apart, and viewed through lenses that present a magnified image at infinity (Figure 17.3). Viewing the images with the eyes relaxed (i.e., not converging) produces binocular fusion and the sensation of three-dimensionality. Accommodation and dynamic parallax are not involved, and though there may be some variation in convergence when examining objects in the foreground, these do not necessarily correspond with those that would have been involved in viewing the original scene. In fact, out of the eight clues listed earlier, only the last one (stereoscopic fusion) is added to the monoscopic clues I began this chapter with, and this sense is weak in a large minority of people. So it is not surprising that an interest in stereoscopy by the general public only appears as a brief burst of enthusiasm when a new development comes along. Public interest always seems to wane, and each system in turn becomes a case of 'Whatever happened to...?'

Stereoscopic camera formats

Conventional cameras You can easily make stereo pairs with a conventional camera. All you need is a small platform fixed to a tripod, with a batten you can slide the camera along. A piece of a metre rule is a good idea, as you can then measure the distance between the two viewpoints. For general stereoscopic photography you simply slide the camera 65–70 mm horizontally between the two exposures. If your camera has the conventional left-to-right film wind, and you want the two negatives to be in the correct order, you need to take the left eye view first: remember, the images on the film are inverted (Figure 17.4). For a 35 mm negative, you will need to enlarge the image by a factor of a little less than ×2 to have the optical centres of the images about the same as the distance between your eyes. For a 4.5 × 6 cm format you will need about ×1.4 enlargement, and for a 6 × 7 cm (landscape) or 4 × 5 inch (portrait) format you can

Sir Charles Wheatstone is probably better known by generations of physics students for the Wheatstone bridge (which he did not invent, but merely improved; cf. Robert Bunsen and the Bunsen burner). Wheatstone made numerous contributions to optics, acoustics and electrical theory. He was also a trained musician, and invented the concertina.

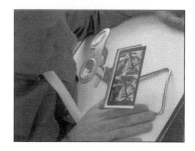

Figure 17.3 Wheatstone's stereoscopic principle in use. His original design included mirrors (see Figure 17.9). (Photograph courtesy of David Burder, 3-D Images Ltd.)

These developments include variants on Wheatstone's original stereoscope, from the 1860s onwards, the Viewmaster transparency viewer of the 1940s, the polarised films of the 1950s, the anaglyphic presentations on TV in the 1960s, the Nimslo camera of the 1970s and the interlaced computer-drawn images of the 1990s. All of these do have their merits, and I shall be explaining how they work.

The Science of Imaging

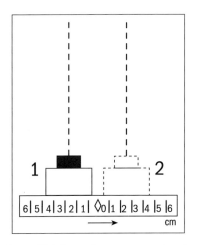

Figure 17.4 Set-up for a stereoscopic pair for a conventional camera, using a calibrated slide bar.

simply set up contact prints side by side, and view the results directly through a Wheatstone stereoscope.

Cameras and camera conversions In the early days of stereoscopic photography there were quite a few cameras designed especially for stereophotography, and many still survive. They were effectively two cameras built into one body. Some took double-width plates, and others used standard film stock. At the outbreak of the Second World War there were at least ten models in production, and several more appeared after the war. These cameras are no longer made commercially, but a number of modern versions are available, and some camera manufacturers offer stereo converter kits for conventional cameras, prismatic systems with double lenses to give portrait-format stereo pairs on 35 mm film. The principles underlying the two most common optical arrangements are shown in Figure 17.5.

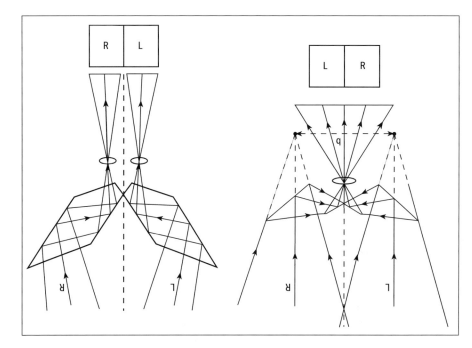

Figure 17.5 Two optical adaptors for making one-shot stereophotographs with a conventional camera.

Some cameras have more than two lenses, and these produce a different type of stereogram, which I shall come to later.

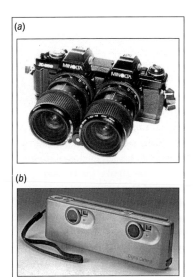

Figure 17.6 (a) Two SLR cameras Siamese-twinned for stereophotography. (b) Compact digital camera. (Photographs courtesy of David Burder, 3-D Images Ltd.)

Some specialists make stereoscopic cameras to special order. These may be near-replicas of earlier cameras, but more often they are standard 35 mm cameras that have been turned into Siamese twins. In order not to waste film the exposures are interleaved, and the right and left images are in reverse order (Figure 17.6).

Changing the base length Stereoscopic imaging has far wider applications than simply providing a quasi-three-dimensional experience. In particular, it earns its keep in applied photography. Almost all practical applications involve changing the base length from the normal interocular distance.

Aerial reconnaissance and survey photography

When an aircraft is employed to take a series of vertical photographs in order to prepare a map of a ground area, it makes a series of parallel runs, taking

Three-dimensional imaging

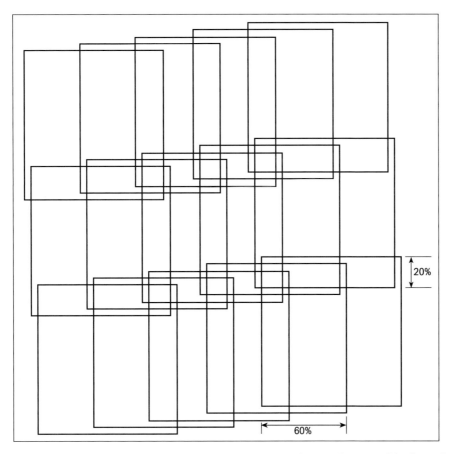

Figure 17.7 Overlap in aerial survey photography. In this diagram the forward alignment of the frames has been staggered for clarity.

photographs that usually have a side overlap of around 10–20 per cent, to avoid the possibility of missing any detail. But the longitudinal overlap is much larger, typically 60 per cent (Figure 17.7). This is to allow stereoscopic viewing. When you view two prints taken with this kind of overlap through a suitable stereoscopic viewer, you can see the image in three-dimensional detail. You can actually see more detail this way than you can from either of the two photographs seen singly.

Hyperstereoscopy in aerial photography

However, when you view the pairs with a simple viewer, the apparent size and distance of the perceived image is much reduced, so that it appears like a Lilliputian model viewed from a comparatively short distance. The reason for this is that the base is far larger than your interocular distance: anything from a couple of hundred metres to several kilometres. This effect is called *hyperstereoscopy*. It can seem a bit odd the first time, but you soon get used to it. There is a special viewer with magnifying lenses that puts the image back to infinity and eliminates the 'model' effect, but it also exaggerates the relief, so that objects appear about four times as high as they should be. The actual factor can be calculated and optimised using a rather complicated piece of geometry, but for most purposes the best stereoscopic rendering is when the ratio of base length to aircraft altitude is

I remember, many years ago, a photo-interpreter colleague spotting someone flying a kite in one such pair of images, a detail wholly invisible in either of the single prints.

The Science of Imaging

about 0.6. The vertical exaggeration of long-base stereoscopy is actually useful in the interpretation of aerial reconnaissance photography. Photo-interpreters often work directly on the negatives rather than on prints, as the resolution and tone separation are better.

Most aerial photography is for survey purposes and for map making. Stereo pairs of photographs are particularly useful here. As early as the 1930s there existed equipment for plotting contours from stereo pairs. The Zeiss Stereoplanigraph did this semi-automatically, requiring the operator only to steer a floating spot so that it coincided on both prints. Modern computer-driven equipment now does this job with much less fuss, and at far less cost.

The stereoplanigraph was said to be the most complex, and the most accurate, piece of optical equipment ever built. As far as I know, only three were ever made.

Hypostereoscopy

Some stereoscopic adaptors for cameras have lenses that are closer together than 6.5 cm, and when you examine the results in a viewer in the usual manner they show diminished depth. The familiar term for this effect is 'cardboarding', from the cut-out appearance of mid-distance objects. Even when the inter-lens separation is the full 6.5 cm, the effect can be present in mid-distance shots. Close objects may have the opposite appearance, with exaggerated depth. As a general rule, for a satisfactory three-dimensional reconstruction the subject matter should lie at a distance of between 30 and 50 times the base, i.e., 2 to 3.25 m for a 6.5 cm base. For landscapes it is advisable, if possible, to increase the base by about 50 per cent.

The reverse is true for extreme close-ups and macro work, where the parallax is too great for a normal base, and the camera lenses have to be shifted or toed in (Figure 17.8). Shifting is better than toeing in, as the latter results in keystoning, and there may be difficulty in fusing the images satisfactorily.

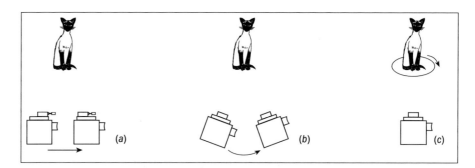

Figure 17.8 Hypostereoscopy. (a) Parallel camera movement using shift lens; (b) swinging the camera; (c) rotating the subject. The parallax angle should not exceed about 7° (the diagram is exaggerated for clarity).

In microscopy, particularly electron microscopy, it is not usually possible to take two suitably spaced views of the subject. Even in binocular microscopes with double objectives, which do permit live stereoscopic viewing, the photographic facility is invariably monocular. To obtain stereo pairs of images you have to rotate the specimen between exposures. This is exactly equivalent to using a pair of toed-in cameras, but the keystoning effect is minimal if you can keep the parallax angle down to about 7°. There will now be a Brobdingnagian effect: the image will seem very large, and some distance away. This is usually unimportant, as the viewer is well aware of observing something that is enormously magnified, and will be expecting such an impression.

Viewing methods for stereo pairs

The inventor of stereoscopic photography was, as I said earlier, Charles Wheatstone (he of the Wheatstone bridge). His stereoscopic viewer appeared in 1838, more or less coinciding with the first public appearance of photography itself. At the time there were no enlargers, and camera formats were sizeable, so the images were viewed through mirrors. The prints had to be made laterally reversed because of these mirrors (Figure 17.9).

Wheatstone's viewer was designed for large prints, but the mirrors made it cumbersome. When cameras became smaller, the viewer became less popular than the simpler one that had been introduced by David Brewster (he of the Brewster angle). This could be used with the then popular quarter-plate size prints ($3\frac{1}{4} \times 4\frac{1}{4}$ inches). There are two versions of the Brewster stereoscopic viewer (Figure 17.10). The split-lens version allows the viewing of somewhat larger prints and a small amount of eye convergence, which makes for more comfortable viewing. Both types are still extant.

The Viewmaster stereoscope of the 1940s was an enclosed Brewster stereo viewer that took a disc bearing small stereo pairs (about 8×11 mm, made on 16 mm Kodachrome film) at opposite ends of its diameters, with a lever advance mechanism. It proved immensely popular, and the catalogue of discs ran into many hundreds, not just of scenery, but often also of serious educational material, such as proton collisions in an accelerator chamber.

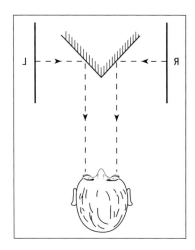

Figure 17.9 Wheatstone's original stereoscopic viewer for large prints.

As a result of the introduction of his viewer, it was claimed on behalf of Brewster that he had himself invented stereoscopy. Brewster did nothing to counter this story, and this resulted in some acrimony between him and Wheatstone, which was never resolved.

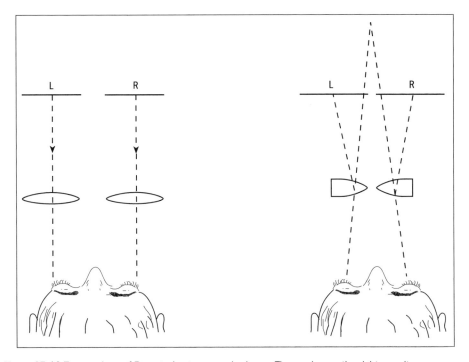

Figure 17.10 Two versions of Brewster's stereoscopic viewer. The version on the right permits some eye convergence and can be more comfortable for viewing.

The Science of Imaging

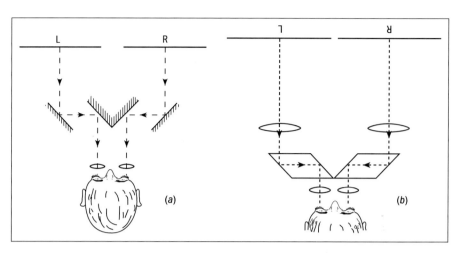

Figure 17.11 (a) Helmholtz's stereoscope for viewing large prints; (b) modern stereoscopic viewer with telescope optics and TIR rhomboids (adjustable for alignment).

A type of viewer employed for large aerial survey photographs (250 mm square) uses a double mirror. Its design is due to Hermann Helmholtz (he of the coil and the resonator), later modified by Caze (who he?). A more recent version employs 45° TIR rhomboids and telescope optics to obtain a highly magnified image (Figure 17.11). The rhomboids can be swivelled to align the images accurately.

Viewing without optical aids

Practised viewers can view a stereo pair without optical aids, by controlling their ocular convergence. This is a physical trick that has to be learned, like wiggling your ears or raising just one eyebrow. The method usually suggested for this is to look at a distant object, then to bring the stereo pair up in front of the eyes without at first re-focusing. With a little practice you can now focus sharply on the images without converging and losing binocular fusion. You will then see a three-dimensional image in the centre of your field of vision, with a vague two-dimensional image on either side. Try this on Figure 17.2 for a start.

If you can't manage this no matter how hard you try, don't feel you are a failure. At least a quarter of the population can't do it either. It *is* an unnatural muscular action, and if you practise it repeatedly for some years (as RAF photo-interpreters did during the Second World War) you may finish up with bouts of double vision.

Coincident image stereograms

If we could print a stereo pair of images one on top of the other, we could make more or less full use of ocular convergence. In fact, there are several methods of achieving this. The important thing is that by this method we can make any plane of the image coincide with the plane of the print itself. This will be the plane where the two images are in exact register on the print. Instead of the image appearing at infinity as with the optical viewers, it will lie across the plane of the photograph itself. For landscapes and distant objects the most realistic effect will be not a print but a projected image on a distant screen. This will also permit the greatest range of depth. There are two ways of making coincident image stereograms: anaglyphs and vectographs.

- *Anaglyphs* This is the better-known method. One image is printed in red and the other in blue or cyan. You view the result through colour filters – red for one eye, blue or cyan for the other. The red-filter eye sees only the blue image, and the blue-filter eye sees only the red image. The two images fuse to give a three-dimensional image in something approaching neutral grey. There may be a small amount of colour fringing for near and far objects, but your perceptual mechanism will probably ignore this. David Burder has patented a system for colour stereophotographs whereby a colour photograph is separated into red for one eye and green plus blue for the other. When you view the compound image through cyan/red spectacles you see the scene stereoscopically in full colour. There is still some colour fringing, and saturation is compromised to some extent, but the effect is convincing, and the system has been piloted on British television with some technical success. One of the merits of the system is that it permits monocular viewing, albeit with some colour fringing and loss of resolution.

- *Vectographs* A somewhat more satisfactory system is based on the optical phenomenon of polarisation. In the simplest method the two images, which may be in full colour, are projected through two polarising filters set orthogonally at ±45° to the horizontal, on to a metallic screen. The viewers wear polarising spectacles also set at ±45°, so that each eye receives only the appropriate image. This is probably the most satisfactory system to date, as there is full colour to each eye and negligible crosstalk. This system can also be used in prints and transparencies, but it involves some technical expertise in the processing. The system was evolved during the Second World War by the Polaroid Corporation.

Interlaced images An adaptation of the unaided viewing principle for viewing larger prints without eyestrain has been made possible by computer graphics. The two stereo images are interlaced in an intricate pattern (any pattern will do, as long as it is too complicated to be obtrusive). Figure 17.12 (colour plate) is an illustration. When you look at this in the same way as with an unaided stereo pair – that is, you gaze at infinity and then bring the picture up into your line of sight – the image appears in depth. The resolution is somewhat limited, and at the present stage of development the technique cannot be regarded as much more than a plaything. But it could be developed to be used as a serious form of illustration in, for example, a textbook on stereochemistry, or a thesis on Renaissance sculpture, given some properly funded development. It is an example of an *autostereogram* (see below).

Autostereoscopic systems

An *autostereoscopic* system is one in which the viewer can perceive the image with full parallax without any kind of optical aid or visual training. The supreme example of this is, of course, holography, and its technology is so important that it has a whole chapter to itself. But there are several systems that are purely photographic, and others, some of them waiting in the wings, that depend on computer-stored video. There are also hybrids of holography and photography, which, like holography itself, are dealt with in Chapter 18.

Parallax stereograms In its earliest form, the left eye image was printed through a fine screen of vertical opaque bars. The screen was then moved through the width of one bar and the right eye image printed in the interstices. The composite print was then mounted under a transparent sheet about 0.5 to 1 mm thick, with the screen mounted on top. When viewed, one eye would see only the left eye image, and the other eye would see only the right eye image. A more modern version of this uses a lenticular screen instead of the bars. The cylindrical lenses that make up

The Science of Imaging

Figure 17.13 (a) A multi-lens camera developed by David Burder for making parallax stereograms in one shot. (b) Two-view interlaced image (colour plate). (c) Full-parallax colour stereogram (colour plate).

the screen are focused on the surface of the print. This comparatively crude system provides only a single stereoscopic image within certain well defined angles, and the stereo effect reverses to a *pseudoscopic* (inside-out) image at points halfway between. It is used nowadays mainly for advertising cards that switch between two images, usually with a horizontally orientated screen for vertical switching: this avoids double images. David Burder has developed a multiple camera with twelve lenses for making full-parallax stereograms with a single exposure (Figure 17.13a).

Modern lenticular stereograms use far more images, anything from 6 to 200. These are originated by either a battery of cameras or a single camera moving along a rail. A computer-controlled printer interlaces the views so that each lenticule has a full set of right-to-left views behind it. The view is now effectively continuous, and you can perceive depth in the image even with one eye closed. Because of the large number of component images there can be a small amount of animation, if the exposures are made in succession. Parallax is limited by the focal length of the lenticular elements: the greater the focal length the better the resolution, but the less the parallax angle. The optimum is around 10–15°. There is still a brief jump into pseudoscopy as you pass the 'join', but you get several re-runs as you continue to move past (Figure 17.14, colour plate).

A lenticular stereogram does not have any vertical parallax, but as a rule this is of little importance. It would, of course, be perfectly possible to produce vertical parallax by using a vertical bank of cameras moving along a rail; you would then need to use a mosaic of microlenses instead of cylindrical lenticules. This has not (to my knowledge) been tried out commercially.

Photographic stereograms have the advantage over holograms that they can easily be made in any size, and can be viewed by a number of people at the same time, in ordinary lighting. However, the limited parallax is a disadvantage.

The Nimslo camera of the 1980s operated on the lenticular principle. The dissection and mounting of the images was carried out using dedicated machinery at special processing stations. But with only four lenses the results were jerky and the parallax limited, and as the lenses were very close together the final images showed a pronounced cardboarding effect.

Stereoscopic cinema and television

There are two well tried methods for screening stereoscopic images, namely anaglyphs and orthogonal polarisation. These methods are the same as for still photography, using two cameras set up side by side and synchronised. Anaglyphs don't need any modification to the projection system if the two images are printed on the same film in cyan and magenta and viewed through red and green spectacles. This results in a brownish monochrome picture. You can show a full-colour anaglyphic film in the same manner, using red and cyan images as described earlier for still photographs, but results are better with two projectors. The same applies to the polarising system: with two projectors you don't need special film or overlays, just two polarising filters, one over each lens; and the viewers wear polarising spectacles. You do need a special metallised screen, though, as ordinary matt screens depolarise the light.

There have been demonstrations of full-parallax cinema, using a battery of projectors and a retroreflective screen, and these appear to have worked well; but

Three-dimensional imaging

such an arrangement is enormously expensive, and would not seem to have a commercial future. In fact, stereo cinema itself has never really thrived. Every ten years or so, some film company makes a film in stereo, usually without much box-office success. The highly successful film version of 'Kiss me, Kate' was originally made in stereo, but was only distributed in mono. You can tell it was originally made in stereo from the way the actors frequently throw things towards the camera, presumably to make the viewers duck.

Polarisation is not really feasible for TV stereo presentations, though there was a somewhat impracticable design by Pye television in the 1950s, combining the images from two TV screens set at right angles, with a 45° polarising mirror as a beam combiner, and polarising filters on the screens. It was never marketed. Anaglyphs can readily be shown on an ordinary TV screen in either monochrome (red and green merging to brown, as above) or full colour (cyan and red). Some years ago the BBC carried out a week-long series of public experiments using the anaglyph method, including an open-air screening in Covent Garden Market. Technically it was a success, but the public reaction was generally apathetic, and the experiment was quietly dropped.

Further developments in stereo projection

A more recent technical development is the showing of left and right eye images alternately, the viewer wearing spectacles with synchronised LCD shutters. Helmet-mounted projection systems, one for each eye, are now being used in simulators; these may be connected to sophisticated equipment that detects head movements and changes the scene appropriately.

A much simpler approach is simulated stereopsis. This is a system that employs nothing more complicated than a pair of spectacles with one glass clear and the other heavily tinted. Perception through the darker glass is fractionally delayed compared with perception through the clear glass. This produces a strong impression of depth with horizontally moving images on the screen, when the direction of movement is from clear glass towards dark glass. Movement in the reverse direction should produce reversed stereoscopy, but it appears that most people don't notice. You can try this out for yourself, using an old pair of sunglasses with one glass removed. It works best with action movies. Of course, it is an optical illusion, not genuine stereoscopy.

Integral imaging

I mentioned in Chapter 4 that a camera lens produces an image that is three-dimensional. Not many photographers seem to be aware of this, though you can easily confirm it by setting up a camera with the back open and a small object fairly close to the front of it, with the lens at full aperture. If you focus your eyes on the film gate, you will see the image of the object floating there. It is inverted, of course, and you can only see it with one eye at a time; but if you move around a little you can see that it has parallax. And the parallax is the right way round. That is, the image is *orthoscopic*. Could we make use of this property?

There are two difficulties. The first is that to get a reasonable amount of parallax your lens needs to have a very large diameter. Merely to see the image with both eyes at the same time demands an exit pupil diameter of more than 65 mm, for fairly obvious reasons. The second is that the three-dimensional modelling of the image is correct only when the object and image are the same size, that is, the magnification is unity. This is because the longitudinal magnification (i.e., along the optic axis) is the square of the lateral magnification. What is more, for a deep object with its centre line at unit magnification, the part of the object nearest the lens will have exaggerated depth (M > 1), and the part farthest from the lens will have diminished depth (M < 1). This limits the permissible depth of the object to about plus or minus 10 per cent of the focal length of the lens. So can there be any optical system that will produce an aerial image with full depth and wide parallax that is also erect, orthoscopic and undistorted?

Gabriel Lippmann, no less, was the first to investigate this problem, almost a century ago. In 1908 he proposed a method for producing full-parallax stereoscopic images using microlens arrays in front of the photographic plate, and, after processing, reversing the light rays to re-create an image in the position of the original object. Unfortunately, when the image is viewed in the usual way, from the object side, it is pseudoscopic.

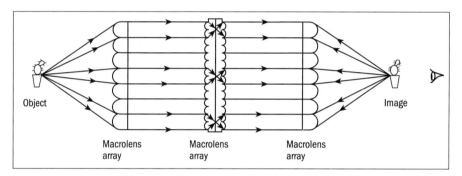

Figure 17.15 Optical principle of the integral camera.

The solution (which Lippmann didn't seek out) is to set up two such systems back to back, and do away with the photographic emulsion (Figure 17.15). A 'macrolens' array collects the light from each point on the object and collimates it so that the microlens array focuses it on what would have been the film plane. From this point the reversed system carries out the reconstruction of the image. It can be seen, geometrically exact and with full parallax and depth, and can be examined and recorded by any of the usual three-dimensional recording systems. The amount of parallax available depends on the diameter of the arrays, and (unlike a single lens) they can be made as big as you like. The system has been called 'integral photography'.

Digging deeper

For a full account of the geometry of stereoscopy, the best source is Sidney Ray's *Scientific Photography and Applied Imaging* (Focal Press, 1999), Chapter 16. Chapters 17 and 18 (Photogrammetry, Aerial Photography), also contain useful information on stereoscopy as a measuring tool. Ray gives a huge number of references to research papers and textbooks. The best of the latter are *Stereoscopy*, by N Valyus (Focal Press, 1966) and *Three-dimensional Imaging Techniques*, by T Okoshi (Academic Press, 1976). These are out of print, but are not hard to find, particularly if you have access to an academic library. Keith Henney and Beverly Dudley's *Handbook of Photography* (McGraw-Hill, 1939) has a chapter on stereoscopic photography, but the chapter on aerial photography is more interesting, because it gives a full account of the method of making aerial maps from overlapping photographs (still in use today); it also has a photograph of the Zeiss Stereoplanigraph.

The integral photography lens array system was designed, and the prototype built, by Neil Davis's team at De Montfort University, Leicester, and the research reports are N S Stevens and N Davis (1991) 'Lens arrays and photography', *J. Phot. Sci.*, **39**(5), 199–203, and N Davis and M McCormick (1992) 'Holoscopic imaging with true 3-D content in full natural colour', *J. Phot. Sci.*, **40**(2), 46–49. (Note: *Journal of Photographic Science* has been re-named *Imaging Science Journal*.) There is also a good article on the process by Charles W Smith in the *British Journal of Photography* dated 6 July 1989, under the title 'New advances in 3D imagery'. For a comprehensive review of recent research you could dip into SPIE Proceedings Vol. 3023 (1997) *Three-Dimensional Image Capture*; but give yourself plenty of time.

A useful account of stereoscopic vision appears in Richard L Gregory's *Eye and Brain* (5th edition, OUP, 1998). There are some good examples of 3-D illusions here; the whole book is well worth acquiring. There is rather more detailed information on binocular perception in N J Wade and M Swanton's *Visual Perception* (Routledge, 1991), but it is a less easy read.

Chapter 18 Holography

Holography is a unique method of producing three-dimensional images. It evolved, not from photography, but from a number of ideas associated with electron microscopy and with radar, and it was developed by physicists, not by artists or photographers. Consequently, many people still think you need a PhD and a lab full of expensive equipment before you can make a hologram. The truth is that anyone can make a hologram in a cupboard under the stairs with no more equipment than a glass plate, a laser pointer from a souvenir shop, a piece of holographic film and some processing chemicals. You don't even need a lens. Holography uses the principle of interference, which we first met in Chapter 1.

Coherence

The examples of interference we have seen previously (Lippmann photography, anti-reflection lens coating) only operate over very short distances, because after one or two wavelengths the light waves get jumbled up, like army recruits getting out of step. To make a hologram you need a beam of light that stays in step (in phase) like a well-drilled army platoon, over the whole depth of the object you are recording. This property is called *coherence*. A beam of light is coherent if it satisfies three requirements:

- *Temporal coherence*   The source must radiate at a single frequency.
- *Spatial coherence*   The light must appear to have been emitted by a point source.
- *Phase coherence*   At any point where interference takes place, the two beams must have a constant phase relationship.

The well disciplined beam from a laser can satisfy the first two conditions. The third will be satisfied only if the two beams are derived from the same source.

When two mutually coherent light beams meet head-on, they produce stationary interference planes one half-wavelength apart. You may remember from Chapter 6 that this was the principle of Lippmann's colour photography. The planes are called Bragg planes, after the Braggs, father and son, who first described them. Lippmann generated his two beams by having his emulsion up against a reflecting surface; but the coherence of his light was so low as to create at best only a few Bragg planes. When the laser was invented in 1961, Yuri Denisyuk substituted a reflective object for the mirror, and shone a beam of laser light directly on to the photographic plate, with the object behind it. The two beams formed interference planes throughout the thickness of the emulsion. This was the first reflection hologram: this type of hologram is now universally known as a Denisyuk hologram.

Denisyuk himself, with characteristic modesty, called it a Lippmann hologram.

At this point it is worth revisiting Chapter 1. Have another look at Figure 1.10a. This is what happens within the thickness of an emulsion when you have a mirror in direct contact with it. When you develop the emulsion you get a stack of planes of metallic silver locating the crests of the standing waves (the *antinodes*), each separated from its neighbour by exactly half a wavelength. Now suppose you substitute a shiny object for the mirror. The interference layers will still be formed, but they will be warped by the uneven contours of the reflecting surface. In effect, they will form a distorting mirror (Figure 18.1).

Holography

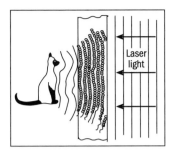

Figure 18.1 Denisyuk's hologram. The object and reference beam are on opposite sides of the emulsion and (somewhat distorted) Bragg fringe planes are formed by the uneven reflecting object.

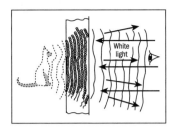

Figure 18.2 The distorted fringe planes of Figure 18.1 reconstruct the beam that was reflected into the emulsion by the object.

But this is a very special distorting mirror. When you illuminate it with the original beam, or even with white light, it reflects back a replica of the beam that came from the object! And if you look down this reflected beam you will see an image of the object, just as if it were still there (Figure 18.2).

(A few names are in order here, as you will meet them many times more in this chapter. The undisturbed beam that illuminates the emulsion directly is called the *reference beam*; the distorted beam that returns to the emulsion from the object is called the *object beam*; the beam that is used to form the image is called the *reconstruction beam*, or, more often, the *replay beam*; and the beam that carries the image of the object back to your eyes is called the *image beam*.)

It may seem a little mysterious that the 'distorting mirror' should produce an image in this way, but it is quite logical, really. The first person to appreciate it was Denis Gabor in 1947, though under somewhat different conditions. He didn't have a coherent light source, as there were no lasers at the time. The best he could manage was filtered mercury light, with a coherence length of less than a millimetre. So he had to restrict his hologram (a transparency) to pinhead size, with the reference beam travelling along the same axis. The result was not very successful, as there was a spurious real image that got in the way of the genuine image, but it was enough to prove his idea correct. His first publications on the subject used rigorous mathematical arguments that are well beyond the scope of this book. But a simple proof (see Box) needs no more than an ability to manipulate some basic trigonometrical formulae.

Denis Gabor (1900–1979) was a Hungarian-born physicist and electronics engineer. He was trying to improve the resolution of electron microscopes when he conceived the idea of a hologram, for which he was eventually awarded the Nobel Prize for Physics in 1971.

Off-axis holograms

One difficulty with Denisyuk's first holograms was that with the reference beam falling perpendicularly on the emulsion, the replay beam had to do so as well, so your head got in the way when you tried to view them. But he found that the holograms worked equally well when the reference beam was at an angle. This doesn't affect the creation of Bragg planes: it simply tilts them a little. Most reflection holograms are made today with the reference beam above the viewing line by either 45° or 56° (the Brewster angle, to help suppress reflections).

Denisyuk holograms, and other types of reflection hologram, do not have to be viewed by laser light. This is because of the Bragg condition: if you illuminate

If the reference beam is at 45°, the Bragg planes will be at 22.5° in the interference space. But within the emulsion they are at a shallower angle (about 15°) and closer together by about one-third, because of the refractive index (~1.5) of the emulsion.

How a hologram works

Although we mostly talk of sine waves, I am going to use cosines here. There is no difference in the outcome, but it simplifies the picture, as a cosine function is symmetrical about the origin, whereas a sine function is not. The only bit of trigonometrical manipulation you need to know is the relationship

$$\cos X \cos Y = \tfrac{1}{2}\cos(X + Y) + \tfrac{1}{2}\cos(X - Y)$$

Any travelling wave that is (co)sinusoidal when viewed as it passes a fixed point can be described by an equation of the form

$$a = A\cos(\omega t)$$

where a is the instantaneous amplitude at a time t. A is the maximum amplitude, and ω the angular frequency, which is related to the wavelength λ by the relationship

$$\omega = 2\pi/\lambda \quad \text{(radians)}$$

If the waveform is not centred on the origin we need to put in a 'phase' term φ that tells us how many radians it is away from a wave that *is* centred on the origin. If we have two waves, which we can call U_1 and U_2, we can write down their equations as

$$U_1 = A_1\cos(\omega t + \varphi_1)$$
$$U_2 = A_2\cos(\omega t + \varphi_2)$$

Now U_1 and U_2 are amplitudes, and the time-averaged intensity $= \tfrac{1}{2}(\text{amplitude})^2$; the intensities I_1 and I_2 are given by $\langle\tfrac{1}{2}U_1^2\rangle$ and $\langle\tfrac{1}{2}U_2^2\rangle$ respectively, where the angle brackets mean that the intensities are time-averaged. Now, if U_1 and U_2 were not mutually coherent, their combined intensity would be merely $\tfrac{1}{2}(U_1^2 + U_2^2)$. But as the beams *are* mutually coherent, their combined intensity I is $\tfrac{1}{2}(U_1 + U_2)^2$, that is

$$I = \tfrac{1}{2}(U_1^2 + U_2^2 + 2U_1U_2)$$

Writing out the equations in full gives

$$I = \tfrac{1}{2}[A_1^2\cos^2(\omega t + \varphi_1) + A_2^2\cos^2(\omega t + \varphi_2)$$
$$+ 2A_1A_2\cos(\varphi t + \varphi_1)\cos(\varphi t + \varphi_2)]$$

Using the identity $\cos X \cos Y \equiv \tfrac{1}{2}\cos(X + Y) + \tfrac{1}{2}\cos(X - Y)$ for the third term gives

$$I = \tfrac{1}{2}[A_1^2\cos^2(\omega t + \varphi_1) + A_2^2\cos^2(\omega t + \varphi_2)$$
$$+ A_1A_2\cos(2\omega t + \varphi_1 + \varphi_2) + A_1A_2\cos(\varphi_1 - \varphi_2)]$$

Notice that the final term does not contain the symbol t, and is thus independent of time.

We now have to time-average this unwieldy expression. This simplifies it quite a lot, because the time-average of $\cos^2 X$ is simply $\frac{1}{2}$, and the time-average of $\cos X$ is 0. We are left with

$$\langle I \rangle = \tfrac{1}{4}(A_1^2 + A_2^2) + \tfrac{1}{2} A_1 A_2 \cos(\varphi_1 - \varphi_2)$$

As there is no t term present, the intensity must be constant at this point in space. But how does it vary in space (i.e., with respect to the plane of the emulsion)? Well, the term φ_1 refers to the reference beam and, as this is undisturbed, φ_1 is either constant (if the beam is perpendicular to the plane of the emulsion) or varies linearly across the plane (if the beam is at some other angle). The term φ_2 refers to the object beam, and varies irregularly across the plane. So the positions of the fringes are coded in the phase relationships $\varphi_1 - \varphi_2$, and their intensities are coded in their amplitude variations A_1 and A_2. The emulsion records the positions and intensities of these fringes.

How do we replay the image? We simply direct the beam U_1 on to the hologram. We obtain the equation of the emergent wave, which we can call U_3 by multiplying U_1 at that point by the film transmittance function at that point. This is proportional to $\langle I \rangle$. So U_3 will be proportional to $U_1 \times I$. Writing out the values for these functions,

$$U_3 = \tfrac{1}{4} U_1 (A_1^2 + A_2^2) + \tfrac{1}{2} U_1 A_1 A_2 \cos(\varphi_1 - \varphi_2)$$

Replacing U_1 in the final term by its actual value $A_1 \cos(\omega t + \varphi_1)$ gives

$$U_3 = \tfrac{1}{4} U_1 (A_1^2 + A_2^2) + \tfrac{1}{2} A_1^2 A_2 \cos(\omega t + \varphi_1) \cos(\varphi_1 - \varphi_2)$$

Using the identity $\cos X \cos Y \equiv \tfrac{1}{2}\cos(X - Y) + \tfrac{1}{2}\cos(X + Y)$ for the last term, we have

$$U_3 = \tfrac{1}{4} U_1 (A_1^2 + A_2^2) + \tfrac{1}{2} A_1^2 A_2 \cos(\omega t + \varphi_2)$$
$$+ \tfrac{1}{2} A_1^2 A_2 \cos(\omega t + 2\varphi_1 - \varphi_2))$$
$$= U_1 \times \text{constant} + U_2 \times \text{constant} + \text{modified } U_2 \times \text{constant}$$

The first term is thus the reference/replay beam; the second is the object/image beam. The third term is another image beam turned through an angle $2\varphi_1$ and reversed in phase (φ_2). This last term represents the 'spurious' pseudoscopic real image in the (-1) order of diffraction.

Standing waves

The previous section used the equation $a = A\cos(\omega t)$ to describe the variation of amplitude with time at a fixed point in space. But in a full description of a travelling wave we must also consider the way the amplitude varies in space at a given time. This is, of course, also a (co)sinusoidal function. In this case we write the equation as $a = A\cos(kx)$, where x represents distance, and $k = 2\pi/T$ (radians), where T is the period of the wave. We can combine these into a single equation by writing

$$a = A\cos(\omega t - kx)$$

The Science of Imaging

> Conventionally, this describes a wave moving from left to right. If the wave moves from right to left the distance variable is reversed, and we have
>
> $$a = A\cos(\omega t + kx)$$
>
> If two such travelling waves occupy the same space, we can add their equations together to obtain the resultant disturbance:
>
> $$a = A[\cos(\omega t - kx) + \cos(\omega t + kx)]$$
>
> Using the identity $\cos(X - Y) + \cos(X + Y) = 2\cos X \cos Y$ gives
>
> $$a = 2A\cos(\omega t)\cos(kx)$$
>
> Distance x is now linked to time t, and both the instantaneous amplitude a and the peak amplitude A are linked to distance. This is a (co)sinusoidal waveform that has an amplitude varying from $+2A$ to $-2A$ with time, but does not move in space. The nodes are separated by one half-wavelength.

They are transmitted instead. If you look at the wall behind a red-image reflection hologram, you will see a cyan-coloured shadow.

This situation may have been aggravated by his refusal to join the Communist Party.

a Bragg mirror with white light it will reflect only light that is close to the appropriate wavelength. Other wavelengths will interfere destructively and will not be reflected.

Denisyuk had trouble getting his work accepted by the Soviet authorities (and by his peers). Consequently, his work went unnoticed in the West for several years. In the meantime, Emmett Leith and Juris Upatnieks, working behind closed doors at the University of Michigan on synthetic aperture radar, read Gabor's papers and immediately saw the connection between his work and the fundamental theories of

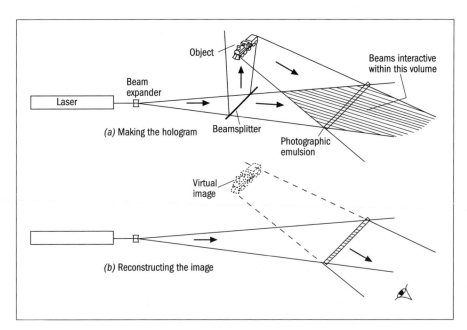

Figure 18.3 Leith and Upatniek's hologram. The object and reference beam are on the same side of the emulsion, and the fringe planes are formed roughly perpendicular to the surface.

coherent radar. They saw that Gabor's problem with the spurious real image could be overcome by offsetting the reference beam. Their holographic set-up, unlike Denisyuk's, had the reference beam and object beam incident on the emulsion from the same side. This method also produces standing waves, but instead of being more or less parallel to the emulsion surface, the Bragg planes are approximately perpendicular to it. In both cases the direction of the fringes bisects the angle between the two beams (Figure 18.3). But whereas Denisyuk's technique produces Bragg planes that resemble the pages of a book, Leith's produces a venetian blind. The result is the same, though: illumination of the final hologram with a duplicate of the original reference beam produces a virtual image in the precise position of the object.

Although you can replay a reflection hologram using white light, a transmission hologram is much less selective, and you have to use monochromatic light, preferably laser light, to display it. If you illuminate a transmission hologram with white light, all you see is a rather lumpy spectrum.

The Denisyuk and Leith–Upatnieks holograms represent examples of two broad classes of hologram, viewed respectively by reflected light and transmitted light. They are termed reflection holograms and transmission holograms. It was something of a surprise to the scientific community that neither Leith nor Denisyuk were to share the Nobel Prize with Gabor. But at least they both later received the Royal Photographic Society's Progress Medal, its highest honour.

Processing a hologram

So far, I have not said anything about methods of processing holograms. In the early days we simply developed and fixed them like photographic negatives. This worked reasonably well for transmission holograms, but the images from reflection holograms were feeble and greeny-blue, although the laser light was red. The main cause of the problem was shrinkage of the emulsion. It turned out that the way to improve the diffraction efficiency and avoid emulsion shrinkage was to bleach the developed silver fringe planes, that is, to turn them back into silver bromide, creating alternate layers of high and low refractive index (pure silver bromide has a refractive index of more than 2).

There are two possible bleaching techniques. The first, called *solvent bleaching*, resembles the bleaching stage of reversal processing of black-and-white photographic films. The developed silver fringes (the 'negative' fringes) are dissolved away, leaving the remaining silver halide (the 'positive fringes') unchanged. With this method you have to use a tanning developer such as pyrogallol with metol or phenidone, which cross-links the collagen fibres and prevents the gelatin from collapsing. The second method is subtler. It is called *rehalogenating*. The negative fringes are converted back into silver bromide, which, as it is formed, is deposited on the positive fringes. No material is lost, and the fringe structure remains undistorted. The diffraction efficiency is greatly improved: it can approach 35 per cent for a transmission hologram and 95 per cent for a reflection hologram.

Diffraction efficiency is intensity of image beam divided by intensity of (laser) replay beam, expressed as a percentage.

The real image

Viewing a primary hologram image is rather like looking at the object through a window. Indeed, Nils Abramson poetically described a hologram as 'a window with a memory'.

Because a hologram replicates the entire wavefront reflected from the object, as you change your viewpoint your eye intercepts different parts of the wavefront, and you see the image from the different points of view. Each point on the

Nils Abramson (b. 1931) is Professor of Industrial Metrology at the Royal Institute of Technology, Stockholm. He has been responsible for many advances in industrial holography, and was the first person to capture an image of a light pulse in flight.

The Science of Imaging

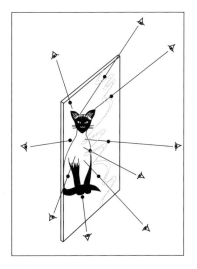

Figure 18.4 A hologram 'freezes' the entire wavefront reflected from the object. Each point on the emulsion encodes a different viewpoint.

hologram codes the view of the object from that point, so the larger the hologram and the nearer it is to the object, the greater the angle of parallax (Figure 18.4). The parallax is both horizontal and vertical: if you move up or down, you see more of the upper or lower surfaces of the object.

In addition to the virtual image, there is a spurious real image. This is what gave Gabor his trouble. With his in-line set-up it was directly in front of the virtual image he required. By having the reference beam off-axis you swing the real image out of the way: it moves to the other side of the reference beam (Figure 18.5). It is usually difficult to see, because it is at a steep angle; if the angle of incidence of the reference beam is more than 45° it will not be there at all. It is weak, too, because this image beam does not satisfy the Bragg condition. However, if you flip the hologram, so that the replay beam is effectively from the reverse direction, you will have reversed the phase of all the wavefronts coded in the hologram, and the image beam will converge to form a real image on your side of the hologram (Figure 18.6).

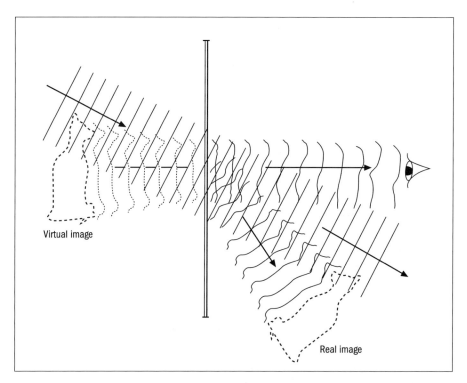

Figure 18.5 The spurious real image. It is the (−1) diffraction order. It is in fact largely suppressed, as it does not meet the requirements of the Bragg condition.

If you look at the pseudoscopic image of a holographic portrait you get the full treatment of the visual illusion I mentioned in Chapter 3. The head seems to swing violently from one side to the other as you move your point of view.

As the Bragg condition *is* satisfied for this beam, the image will be bright and clear. It is also inside out or *pseudoscopic*. If you move to the right, instead of seeing more of the right side of the image you see more of its left side. This makes the image of, say, a table tennis ball look like the inside of an eggcup, and vice versa.

If you position a ground glass screen (or tracing paper) in the image space, you will see the image projected on it, and if you move the screen in and out you will bring different planes of the image into focus.

Holography

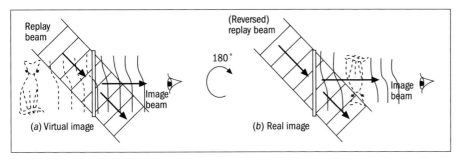

Figure 18.6 Flipping the hologram so that the direction of the replay beam is reversed, reverses the phase of the wavefront and creates a real image in the same position as the virtual image. It is pseudoscopic, i.e., has reversed parallax.

Transfer holograms

You can generate this pseudoscopic image with either a transmission or a reflection hologram. The exciting thing is that you can make a second hologram using this image as the object. You simply illuminate the primary or *master hologram* with a laser replay beam, so that it forms the real image straddling the *transfer hologram*. This, of course, needs its own reference beam derived from the same laser. Figure 18.7 shows the principle.

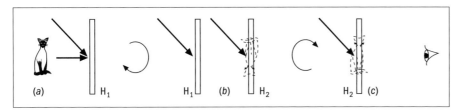

Figure 18.7 Principle of the transfer hologram. (a) Master hologram (H_1). (b) H_1 is flipped, and the real image is projected across the plane of a second film to act as object for the second, transfer hologram (H_2). (c) After processing, H_2 is flipped, and the image is now orthoscopic and across the plane of the hologram.

As the image is pseudoscopic, you have to have the transfer reference beam coming from 'below' (the set-up is usually on its side, for simplicity), because you will have to flip the finished hologram to undo the pseudoscopic effect again.

Made in this manner, the image will straddle the plane of the hologram. The transfer principle has a number of advantages.

- You can make as many copies as you like, long after the original set-up has been dismantled.
- An image-plane reflection hologram will replay as a sharp image even when the replay light is not a very good point source (i.e., has poor spatial coherence).
- You can make a good reflection transfer from a transmission master hologram (it is not easy to make a reflection copy from a reflection master hologram).
- By increasing the distance between the master and transfer hologram so that the image lies between them, you can produce a final image that is wholly in front of the hologram.

The Science of Imaging

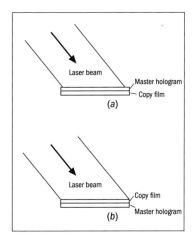

Figure 18.8 Making a copy hologram: (a) transmission; (b) reflection.

- You can view an image-plane transfer transmission hologram by white light, provided the image is shallow.
- By using converging or diverging replay beams for the master hologram, you can reduce or enlarge the image to a small extent.

Contact copies

You can also make a replica of any hologram by contact copying by a single-beam transfer process. You simply clamp the master and transfer films in close contact and aim the replay/reference beam at the sandwich. For a transmission copy the transfer film goes on the side away from the laser; for a reflection copy it goes on the same side. Figure 18.8 shows the arrangement for both types of copy.

Focused-image holograms

A reversed master hologram produces a focused image. In a way you can think of the master hologram as a kind of lens, though a somewhat peculiar one: in a way, a lens with a memory. But a conventional lens also produces a real image with depth, as we saw earlier. We can therefore use the optical image produced by a lens as the 'object' for a hologram. And as this image can straddle the plane of the film, we can produce an image-plane hologram in one step. A notional set-up for either a transmission or a reflection hologram is shown in Figure 18.9. The image is inverted but orthoscopic, so it doesn't need flipping.

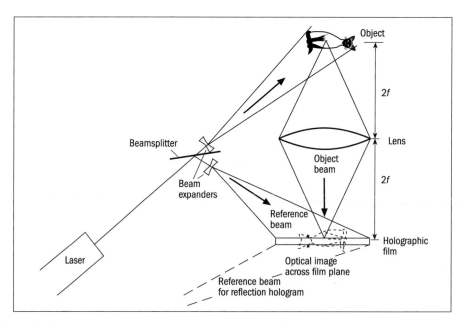

Figure 18.9 Layout for a focused-image transmission hologram. For a reflection hologram the reference beam is from below left (broken lines).

Rainbow holograms

Rainbow holograms (also called Benton holograms, after their inventor) are transmission transfer holograms that can be replayed using white light, even though the original subject matter may have been quite deep. The technique relies on the fact that when you view the image in a transfer hologram you are viewing it through the real image of the frame of the master hologram.

Benton masked his master hologram down to a narrow horizontal slit. This meant that a viewer of the transfer image would seem to be viewing it as though through a narrow letterbox. With such an arrangement you would get full horizontal parallax, but as soon as you moved your viewpoint up or down the image would disappear. That is, if you were replaying the hologram with a laser beam.

If, on the other hand, you replay such a hologram using a white spotlight, the result is quite different. Because red light is diffracted more than blue, you get a whole vertical spread of letterboxes going right through the spectrum (Figure 18.10). So as you move your viewpoint upwards you see the image in pure colours that range from violet to deep red – hence the name 'rainbow hologram'. Because you are viewing the image by only one wavelength there is no colour blur, and the image can be very deep.

Stephen Benton is Head of the Spatial Imaging Group at the Media Research Laboratories, Massachusetts Institute of Technology. His main research has chiefly been directed at realising holographic television, and his invention of the rainbow hologram was a by-product of this research.

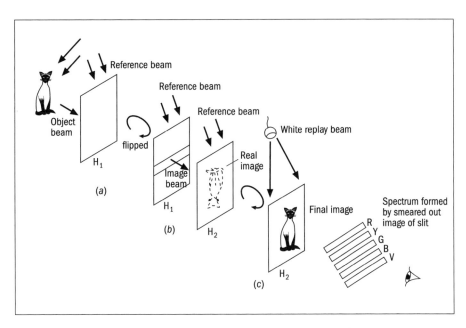

Figure 18.10 Making a rainbow hologram. (a) Conventional master hologram H_1 flipped and masked down to a horizontal slit. (b) Image-plane transfer hologram H_2 made. (c) When illuminated by a white spotlight, images of the slit form a spectrum, through which the viewer sees the final image in one or other spectral hue depending on vertical position.

Pulse laser holograms

Most lasers used for making holograms emit a continuous beam of some 3 to 35 milliwatts for amateur work, and up to 5 watts for large professional holograms. Holographic emulsions are very slow, and one of the major headaches is the need to keep the subject matter stable to within a tenth of a wavelength during an exposure that may be anything from a few seconds to several minutes. This rules out portraiture and most live subjects, even flowers. (Snails may be OK.)

The solution is to use a pulse laser. This is a laser that emits its light in a giant flash lasting only about 25 nanoseconds (ns). The total light energy in the flash is only a few joules, about the same as a small studio flash on half power, but the actual power, the rate of energy emission, runs into megawatts, so for safety's sake the subject illumination has to be well diffused. Otherwise, a holographic portrait studio is much like a photographic studio, except that all the 'floods' are reflectors, because the light has to come from a single source. Until recently, all pulse lasers used an artificial ruby rod as lasing material, and the wavelength was 694 nm, very near the infrared. Unless Hollywood style make-up was applied, complexions looked very waxy. The recent introduction of green pulse lasers has improved matters greatly.

Embossed holograms

The little shiny holograms you see on credit cards, £20 banknotes and other items as a security device are rainbow holograms backed by a mirror. They are produced mechanically by a process resembling that used for producing CD records. The master hologram is made in the usual way, but the transfer hologram (often containing more than one image) is made on positive photoresist instead of a photographic emulsion. The interference pattern thus consists of surface ridges (Figure 18.11).

Positive photoresist is a material that is insoluble when kept in the dark, but becomes soluble in certain solvents when exposed to blue or UV light. This is the reverse of the (negative) photoresist used in making halftone letterpress plates.

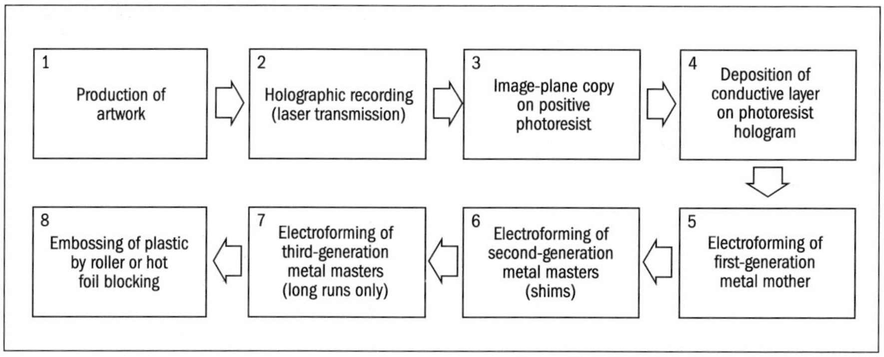

Figure 18.11 The production of an embossed hologram. All embossed holograms are basically rainbow holograms. There is an embossed hologram in the back pocket of the book. You will see an image of sorts in any light, but the image is best when it is lit by a small spotlight about 45° above it.

This intermediate master is then nickel-plated. The nickel 'mother' is stripped, and replicated in hard nickel, which can then be fixed to a hot roller and used to stamp out copies on very thin plastic material with an aluminium foil backing. This in turn is backed with heat-sensitive adhesive, and can then be flush-mounted on documents or cards.

Holographic stereograms

You will probably have seen holograms with two or more images that change suddenly from one to the other as you shift your viewpoint. There is no magic

Holography

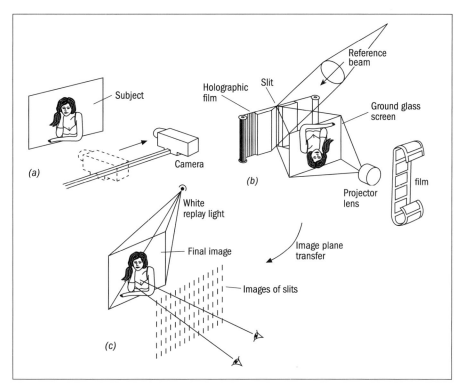

Figure 18.12 Process for making a holographic stereogram. (a) A series of photographs is taken by a camera moving along a rail. (b) The photographic images are used as objects to make adjacent narrow holograms on a film masked by a slit. The film moves one slit width between exposures. (c) The resultant intermediate hologram is transferred so that the image is across the plane of the final hologram and the image of the slits is in the plane of the viewer's eyes and is thus invisible. The initial image may also be a cine clip or a computer animation.

about this: it simply involves transferring two or more holograms side by side. When you view the final hologram you are looking at the image through the image of the master hologram's aperture. As you move across the field you see the image of the first hologram through its own space, then the image of the next hologram appears through *its* space, and so on. The change is sudden, because the plane of the 'window frame' is close to your eye.

By making a series of rather narrow vertical master holograms and setting them up butted for transfer, you can produce a jerky sort of animation effect. The logical extension of this idea is to make a series of master holograms only a millimetre or so wide, on a single film masked with a slit, moving one slit width for each exposure. The individual holograms have no horizontal parallax (they are vertical letterboxes), so the 'objects' do not need to be three dimensional. Usually they are ordinary cine frames, showing either a changing viewpoint (the camera runs along a rail), or motion (a stationary camera records movement), or a little of both. Computer animation stills are also a popular subject. Figure 18.12 summarises the whole process.

Holograms in natural colours

There are two ways of producing holograms in full natural colour: three-colour holography and holographic stereograms.

The Science of Imaging

For example, a pure yellow object that reflected neither red nor green light would appear black. Fortunately, few natural colours are pure.

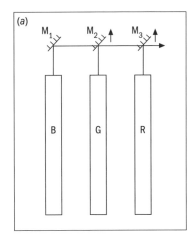

Figure 18.13 (a) Making a natural-colour hologram using three lasers. The mirrors M_2 and M_3 are removed in turn for three successive exposures. (b) Photograph of a three-laser hologram by Hans Bjelkhagen (colour plate).

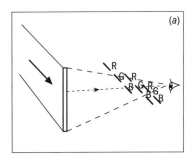

Figure 18.14 (a) Natural-colour rainbow stereogram. The original colour photographs have black-and-white separations made, and these have to be aligned so that the spectra overlap in correct register (in practice the 'windows' are close together). (b) Holographic stereogram (in the back pocket of the book) made in the manner described on the previous page. The two-colour effect is produced by making a second exposure for the background with a different reference beam angle. The image is best viewed using a spotlight at about 45° above the hologram.

The sandwich hologram was another of Nils Abramson's inspirations.

Three-colour holography   If you use three lasers, one red, one green and one blue, you can make a reflection hologram containing three independent sets of Bragg planes, replaying with white light to give an additive colour image. The triangle enclosing the three wavelengths, as plotted on the CIE chromaticity diagram, needs to enclose as large an area of it as possible, in order to produce good, saturated colours. A good combination is a krypton-ion laser at 647 nm (red), a solid-state laser at 532 nm (green) and a helium–cadmium laser at 442 nm (blue-violet). This combination produces good results for pastel or desaturated colours (Figure 18.13a, b colour plate), but some saturated colours are inevitably somewhat falsified in hue.

Colour holographic stereograms   This is one way to achieve true colour. Theoretically, you get the best colour reproduction when your taking filters cover the whole spectrum without gaps and with minimum overlap, and the printing filters are as narrow-band as possible. This method approaches that ideal. The original is a run of colour transparencies, from which you need to make a set of red-, green- and blue-separation positives in black and white. You now need to make a holographic stereogram master from each. You then make a transfer rainbow hologram from all three, simultaneously. They need to be set up so that the 'red' separation views as red, the 'green' as green and the 'blue' as blue, all in register. Not an easy task: most modern equipment uses computer control. But this process does give very fine colour, though the viewing has to be from a very specific height or the hues will be incorrect. Figure 18.14 is a schematic diagram of the set-up.

Holographic interferometry

All holograms are interferograms, of course: their images are produced by interference. It was noticed fairly early in their history that any very slight movement of the object during the exposure produced an image overlaid with stripes. This turned out to be a moiré pattern generated by the two sets of primary interference fringes from the two positions of the object. The first person to take these patterns seriously was Karl Stetson, who in 1965, with R L Powell, published a paper on the analysis of vibration from secondary fringes. Holographic interferometry has now become an important tool in industrial and scientific research. The principle is simple: if an object is slightly distorted between two holographic exposures, the distortion will be contoured by dark and light fringes each representing one half-wavelength of distortion. There are three basic kinds of technique.

Real-time   You make a hologram of your test piece, using a holographic plate and a very accurate holder, and after processing you position the plate back in the holder. The test piece now coincides with its image, and any force on it will produce a strain that will show in the form of secondary fringes (Figure 18.15).

Double-exposure   Here you make one exposure before applying the stress and another one after applying it, on the same plate or film. The strain will be contoured by fringes.

By employing separate plates for the initial exposure and for several subsequent exposures, and using the special plateholder, you can examine the effect of progressive stressing by examining the plates in pairs. This method, called 'sandwich holography', also enables you to see whether the distortions are inwards

Holography

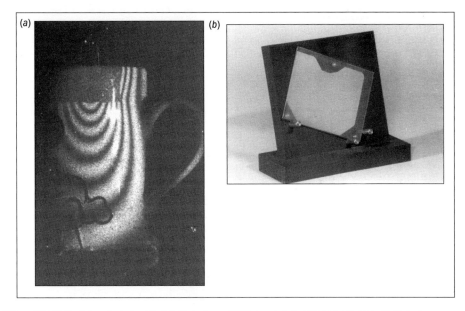

Figure 18.15 Real-time holographic interferogram. (a) The cup is positioned precisely in its own holographic image space. The fringes contour the distortion, here caused by a bulldog clip clamped to the rim. (b) The self-locating plateholder designed by Nils Abramson.

or outwards, in cases of doubt. To record dynamic stress (e.g. a blow on a crash helmet) a double-pulse laser is employed, with the two pulses a few milliseconds apart, timed just before and just after the blow. Double-pulse holograms can also be used to record non-stable vibrations (Figure 18.16).

Time-average When a component is in a state of stable vibration, it is momentarily stationary at the two extremes, and these two positions provide the secondary interference pattern. This lends itself to Denisyuk techniques, and the contour fringes are approximately one quarter-wavelength apart. The bright patches indicate nodes, where there is no movement (Figure 18.17).

Holographic optical elements

I said earlier that you could think of a master hologram set up for making a transfer hologram as a kind of lens. This is not just being whimsical. If you make a hologram of a single bright point, using a collimated reference beam, you can use a reversed replay beam to recreate that point. Put simply, the hologram *is* a lens, and the point image is at its principal focus. What is more, it obeys all the lens laws, and it is, truly, that otherwise mythical object, a 'thin' lens. If you go through a similar process with a reflection hologram set-up, you will produce a holographic mirror that will focus light just like the mirror of a telescope. The layouts for making these two types of hologram are shown in Figure 18.18.

These holograms are called *holographic optical elements* (HOEs), and they have brought about something of a revolution in optical equipment. Not only can you copy *any* shape of lens or mirror, but you can do it on a flat plate (or a curved one, if necessary); but you can also put more or less as many elements as you like into the same space. To take a simple example, suppose you make a hologram of three points at different distances and different angles. You now have a lens with

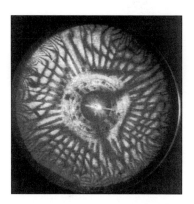

Figure 18.16 Double-exposure holographic interferogram of the Rolls-Royce RB211 fan rotating at speed, showing vibration patterns. This image was made using a modified 35 mm camera to make a focused-image hologram. The image was stabilised by an Abbe inverting prism rotating in the opposite direction to the fan. (Interferogram by Ric Parker, courtesy of Rolls Royce.)

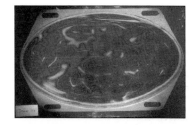

Figure 18.17 Time-average holographic interferogram of a loudspeaker fed with a pure sinewave at 5 kHz. The bright areas are the stationary modes; the antinodes are contoured by fringes.

The Science of Imaging

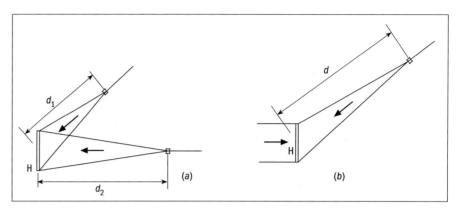

Figure 18.18 Set-ups for making (a) a holographic lens of focal length $d_1 d_2/(d_2 - d_1)$, (b) an optical mirror of focal length d.

three separate foci. No glass lens can do this. The new technology of optical computing relies heavily on HOE technology. So do head-up displays in aircraft, simulators, virtual reality sets and more worldly items such as elevated rear lights for cars and focusing screen brighteners for studio cameras.

The main disadvantage of the HOE is its high dispersion, which means that it can often only be used for applications using lasers (such as bar code scanners); but in some applications the dispersion is actually useful, as in diffraction gratings. Because the dispersion characteristics are in the opposite sense to those of optical glass, HOEs can be used in achromatic doublets.

As the interference patterns in HOEs are comparatively simple, modern microengraving techniques can copy them mechanically from computer programs. Such *diffractive optical elements* (DOEs) (the name now often subsumes HOEs) are more robust than HOEs, and can be made thinner as they do not need sealing and do not deteriorate. DOEs engraved on metal masters can be used to mass-produce copies, in the manner of embossed holograms. The original can also be produced on photoresist using microlithographic techniques.

> ## Zone plates
>
> The simplest kind of DOE is the *Fresnel zone plate*. This is a transparent plate on which are drawn concentric circular bands alternating with clear bands. The ratios of successive cycles are in proportion to the square roots of 1, 2, 3, 4 etc. (Figure 18.19).
>
> When a collimated (monochromatic) light beam is incident perpendicularly on a zone plate, the clear rings all provide constructive interference at a point on the axis a distance r^2/λ from the plate, where r is the outside diameter of the innermost disc (it doesn't matter if it is clear or opaque). The zone plate thus acts like a lens of focal length r^2/λ.
>
> It can be shown that by engraving the zones with a slope, like a Fresnel lens, the opaque portions can be dispensed with, and the diffraction efficiency raised to nearly 100 per cent.

As an example, with light of 500 nm wavelength, a zone plate with a central disc of 0.2 mm diameter would have a focal length of 80 mm. Such a device is well within modern micromachining capabilities.

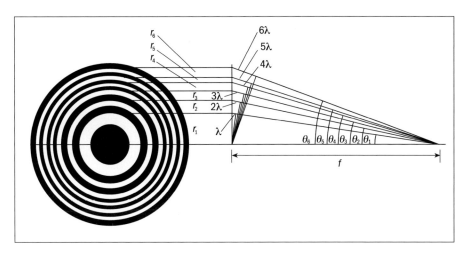

Figure 18.19 Principle of the Fresnel zone plate.

Computer-generated holograms

The interference pattern generated by a point source and a collimated reference beam is, in fact, the pattern of a zone plate. The intensity profile over the light and dark bands, though, is sinusoidal, not rectangular as in a Fresnel zone plate. It is termed a *Gabor zone plate*, after Denis Gabor, the father of holography. When you make a hologram, you are actually making a Gabor zone plate for every point on the object; so when you shine the conjugate beam back through the hologram, you are focusing the light at every point that existed on the surface of the original object.

Now, we can generate a zone plate from a single source by applying some simple calculations, so we should be able to generate other zone plates in the same way. We can indeed do this for a small number of points, but to do it for a whole continuous three-dimensional object is too complex a calculation for even a large computer. So we tackle the problem in a different way.

The Fourier-transform hologram If you position an object at the front focal plane of a lens and illuminate it with coherent light, then hold a screen at the rear focal plane, you will see a diffraction pattern – a very small one for a large object, and a much bigger one for a very small object. It represents the *optical Fourier transform* of the object wavefront. If you set up a similar lens system on the far side of the pattern (and remove the screen), you will see an image of the object, the same size, but inverted. If you add a reference beam and make a hologram of the optical Fourier transform, you can use the hologram to reconstruct the image, via the second optical system (Figure 18.20).

Now, the interference pattern of an ordinary hologram is too complicated for a computer to calculate (except for very simple objects), but a Fourier transform is a different matter. You can calculate the Fourier transform of a simple function such as a single square pulse on the back of an envelope. For a more complicated, three-dimensional object with shading you need a computer, and a fairly powerful one. But it will not only calculate the Fourier transform, it can also draw the interference pattern that characterises the hologram, and feed the result into an electron-beam microlithography set-up or into a megapixel spatial light modulator. You can now use the result as a hologram to create a three-dimensional image of an object that never existed, such as a car body design, a bridge or a proposed new

This is an alternative, and valid way, to describe the formation of a holographic image. I think the wavefront approach is easier to grasp, though, particularly with complicated types of hologram.

Fourier transforms are explained in more detail in Appendix 3.

This set-up is for purposes of explanation. In practice a different, simpler configuration is used for making Fourier-transform holograms, but it is not so easy to see what is going on.

A spatial light modulator (SLM) is what we used to call a liquid crystal display (LCD). 'Liquid crystal' is a poor name for these substances, and new developments in SLMs need not necessarily employ them.

The Science of Imaging

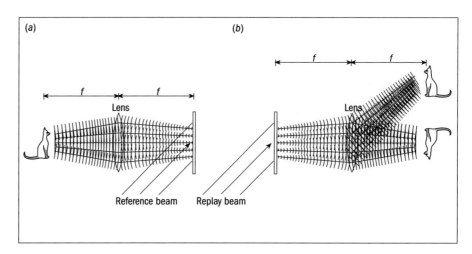

Figure 18.20

city centre. At the time of writing we are only just getting there, but the ability to achieve such a result is well within reach of the new generation of computers.

Digging deeper

Literature on holography is limited to a fairly small selection of books, which cover a wide spectrum from the near-frivolous to the more or less unreadable. If you want the full theory with no holds barred, the best book is P Hariharan's *Optical Holography* (Cambridge University Press, 1996). The maths is not too difficult provided you can cope with complex numbers and exponentials. The most approachable text on applied holography is Nils Abramson's *The Making and Evaluation of Holograms* (Academic Press, 1981, temporarily out of print). If you want to try making your own holograms, there are several instructional manuals. Fred Unterseher, a brilliantly innovative fine-art holographer, has collaborated with Bob Schlesinger and Jeannine Hansen to produce *Holography Handbook* (revised edition, Ross Books, 1996), which is friendly and amusing, though some of the set-ups are less than ideal. If you want to build your own holographic studio/lab in your garage, Don McNair shows you how, in *How to Make Holograms* (Tab Books, 1983, now out of print), which is excellent on concrete mixing, less so on making good holograms. My own *Practical Holography* (2nd edition, Prentice-Hall, 1994) explains how to make every type of hologram in detail, with a full account of the principles of holography and a discussion of its applications. It is now available from Pearson Education, on the Internet from www.amazon.co.uk to special order. A new edition is in preparation. An earlier book of mine, *Manual of Practical Holography* (Focal Press, 1991, now out of print) is aimed at the less ambitious enthusiast. *Holography for Photographers* by John Iovine (Focal Press, 1997) is a promising replacement, but is not really adequate on its own as a home tutor. It also contains some unchecked errors. (The books noted here as out of print can, of course, be borrowed via your local library.)

If you want to keep up with the latest developments, you will need to get hold of conference proceedings. The best collections come from the SPIE, who hold a conference on practical holography at least once a year. At the time of writing, the most recently published was SPIE Proceedings Vol. 3956 (2000), *Practical*

Holography XIV and Holographic Materials VI. The best way to get copies of these proceedings is via the Internet at http://www.spie.org. The mailing address is The International Society for Optical Engineering, PO Box 10, Bellingham, WA 98227, USA. The proceedings of the triennial Symposiums on Display Holography at Lake Forest College, Illinois, are a goldmine. The first four were published by the College; you can get copies from Integraf, PO Box 586, 745 North Waukegan Road, Lake Forest, IL 60045, USA. All subsequent proceedings are published by the SPIE under its own volume numbers. You can obtain copies of all SPIE publications either directly from Bellingham, or via the Internet from www.amazon.com.

A research team at QinetiQ, Malvern, headed by Chris Slinger, is currently working on computer-generated holographic real images, sponsored by the Ford Motor Corporation. Their Internct address is www.iq.com/industries/transport/automotive/case_study_holographic_imaging/index.asp.

Chapter 19 The high and the low

This chapter is concerned with extreme situations: wide and narrow angles, long and short distances, large and tiny images, images from above the atmosphere and from under the sea. All these have their special qualities and problems.

Panoramic images

The original 'panorama' was a cylindrical chamber painted with a 360° landscape (Figure 19.1). As you stood in the centre and turned around you could see the whole scene in correct perspective. The term has become re-interpreted by modern camera manufacturers to mean any angle of view wider than about 90°, or alternatively, any format with an aspect ratio greater than about 1.5:1.

Ultrawide and fisheye lenses   I discussed wide-angle lenses and their associated perspective problems in Chapter 4. One of the earliest ultrawide-angle lenses was the Goerz Hypergon, which had a field of 130° at $f/22$, with correct drawing. It consisted of a pair of uncorrected hemispherical meniscus glasses, in which the otherwise appalling $\cos^4$ fall-off was taken care of by a little star-shaped windmill in front, blown round by an air bulb (or, more often, by the photographer's breath) (Figure 19.2). There were other panoramic lenses, too, one of which was filled with water, and had baffles on the (fixed) aperture to reduce the fall-off.

Most modern wide-angle lenses for conventional photography are *rectilinear*, i.e., free from distortion. The maximum angle of view for wide-angle lenses is about 140°, but at such angles there is severe illumination fall-off, even with the compensations described in Chapter 4. If distortion is less important than covering the widest possible field, you can increase the angle of view to 180° and beyond. This is the realm of the *fisheye lens*.

Figure 19.1 A panorama was originally a room painted with a 360° scene.

Figure 19.2 The 130° Goerz Hypergon lens of 1850. (Illustration courtesy of Sidney Ray.)

Why 'fisheye'?

This takes us back to basic optics. A fish at the bottom of a stream has an effective angle of view of 180°, though its visual field may be much less. The critical angle of refraction for water is $\sin^{-1}\frac{1}{1.33}$, which is 48.75°. So if the fish has a 97.5° visual field, it can see everything that is above the surface. As a bonus, everything beyond the critical angle undergoes total internal reflection, so outside the disc of view there is a reflection of the bottom of the stream (Figure 19.3).

The first true fisheye lens was designed in 1924 by Robin Hill for the Meteorological Office for cloud studies. It was basically a standard camera lens with an enormous negative meniscus lens in front, producing an extreme retrofocus configuration. As it was to be used with a deep red filter it was not colour corrected. It was operated at $f/16$ to $f/32$. Modern fisheye lenses are designed on a similar principle. The distortion follows one of three possible characteristics: *equidistant* (radial distance $= f\theta$), *orthographic* (radial distance $= 2f\sin\frac{1}{2}\theta$), or *rectilinear* ($r = f\tan\theta$). These types of distortion are shown schematically in Figure 19.4. The rectilinear geometry has $r = \infty$ for $\theta = 90°$, and is not a true

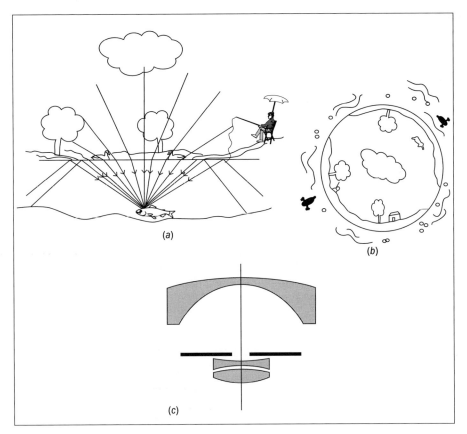

Figure 19.3 What a fish sees: (a) the geometry; (b) the view. The angle of view (about 97.5° for fresh water) is known as 'Snell's window'. (c) The Robin Hill Sky lens of 1924, the first true fisheye lens.

fisheye, though often so described. Most true fisheye lenses provide equidistant projection.

Panoramas with a conventional camera lens If you have a 35 mm camera with a

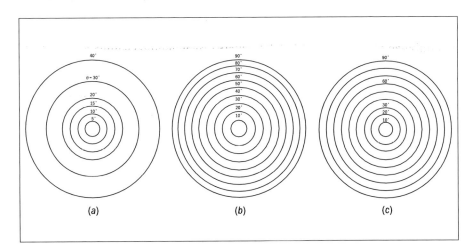

Figure 19.4 Fisheye lens geometry: (a) rectilinear ($r = f \tan \theta$); (b) orthographic ($r = 2f \sin \frac{1}{2}\theta$); (c) equidistant ($r = f\theta$).

The Science of Imaging

28 mm shift lens, you can take a pair of photographs (with a tripod) without moving the camera, but with the shift at its two sideways extremes. You will then have a pair of images covering something like 100°, and you can butt-join the prints with a perfect match. The perspective will still show the characteristic stretching of near objects at the edge of the field, but it should not be apparent in landscapes lacking near objects. Alternatively, you can use an ordinary lens and swing the camera between exposures. This time you may find some difficulty in matching the edge detail, especially in the foreground, and where a straight line (such as a house roof) appears in two adjacent images. You can minimise this by having about 50 per cent overlap, and cropping each image down the halfway overlap point. But you can't get rid of it entirely.

True panoramic cameras   There are two ways of achieving a true panoramic perspective. Both employ the slit principle. In the first the film is drawn around the inside of a cylinder centred on the rear nodal point of the lens. The lens itself is pivoted about its rear nodal point, and a narrow slit close to the focal plane is linked to the lens mounting. During the exposure the lens system rotates through its maximum swing angle, about 150°. In the second system the entire camera rotates about the rear nodal point of the lens. The slit is fixed, and the film is advanced in the same direction as the image movement. This system is shown schematically in Figure 19.5.

In a true panoramic photograph parallel horizontals are recorded as hyperbolas (Figure 19.6). This is why it is difficult to fit parts of a house roof together in a panoramic montage of a town taken from high ground. In theory you would have

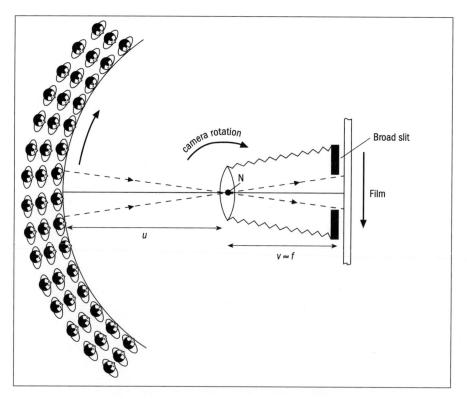

Figure 19.5 Swivelling panoramic camera. N is the rear nodal point of the lens.

The high and the low

Figure 19.6 Distortion in a swivelling panoramic camera. The pupils are in a circle centred on the camera, and the horizontal lines of the building behind have become hyperbolas. (The author – aged 12 – is in the back row, eleventh from the right.)

to take an infinite number of overlapping shots to get them to fit perfectly, and you would still finish up with hyperbolas. Incidentally, you make things worse if you don't keep the horizon centred in the frame.

Pinhole photography

Pinhole photographs are undoubtedly the cheapest form of photography, needing no more than a cardboard box, a milk bottle top, some photographic film or paper, and double-sided adhesive tape to hold the film in place. Pinhole cameras can earn their keep, though. They are standard equipment for imaging cosmic X-ray sources; they are used for photographs of aircraft cockpit layouts and images of solar eclipses, and in the sixteenth century a pinhole image was the catalyst that resulted in the adoption of the Gregorian calendar. In the field of fine art, the pinhole camera assisted Brunelleschi and his contemporaries in the rediscovery of the laws of perspective. Some modern photographers have become intrigued by the odd and interesting characteristics of the pinhole photograph, and have exploited its properties pictorially. In 1996 the Royal Photographic Society used a pinhole photograph (Figure 19.7, colour plate) on the cover of its annual print exhibition catalogue.

So what does a pinhole camera have that a conventional camera doesn't have?

- It has infinite depth of field.
- It has an extremely wide angle of view.
- The overall slight unsharpness of the pinhole image is more acceptable than the unsharpness in out-of-focus areas in a lens image.

A pinhole camera, simple as it is, has some optical details that are worth examining.

Optimum pinhole size In conventional geometry, a point object at infinity should produce a uniform disc of confusion the same size as the pinhole. This implies that the smallest pinhole produces the sharpest image. But in practice, as the pinhole becomes smaller the effect of diffraction takes over. Now, you will remember that the formula for the diameter d of the central disc of the diffraction image for a circular aperture (the Airy disc) is given by

> In the Tower of Winds in the Vatican, a pinhole in the South Roof casts a solar image that crosses a meridian mark at noon every day. By 1582 this pinhole image showed that the calendar date of the spring equinox had become more than ten days early. Pope Gregory decreed that 5 October that year would become 15 October, and that a new, properly corrected calendar would be inaugurated.

The Science of Imaging

$$d = 2.44\lambda N$$

where λ is the wavelength of the light and N is the f/no (the f/no of a pinhole is the image distance divided by the pinhole diameter).

The two functions have to be convolved, that is, the Airy disc has to be dealt out to every point on the pinhole function. This is not a very easy task, but as an approximation we can simply make the diameter of the Airy disc equal to the pinhole diameter c. then

$$c = 2.44\lambda \times f/c$$

or

$$c^2 = 2.44\lambda f$$

The formula recommended by Rayleigh in 1891 was $d = 2\sqrt{f\lambda}$, which gives a pinhole roughly 1.3 times the diameter of mine. Take your choice!

i.e.,

$$c = \sqrt{2.44 \times 0.55 f} \times 10^{-3} \text{ m}$$

$$= \sqrt{1.342 f} \text{ mm}$$

Figure 19.8 shows the variation of optimum pinhole diameter with image distance.

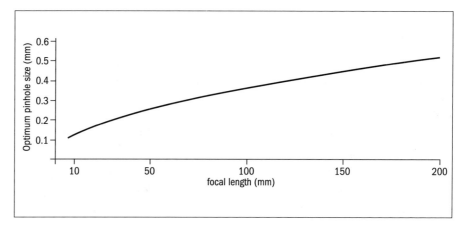

Figure 19.8 Optimum pinhole sizes.

Aberrations of a pinhole

Perhaps surprisingly, pinholes have aberrations the way lenses do:

- *Chromatic aberration* As red light is diffracted more than blue light, there will be both longitudinal and lateral chromatic aberration. Small points of light in the central area will show reddish outer edges, and in outer parts of the field they will show a red fringe on the outer side and a blue fringe on the inner side, the effect being proportionally greater for smaller pinholes and greater field angle.

- *Astigmatism* The oblique view of the pinhole from the edges of the film is an ellipse with its short axis radial from the optical centre. There is therefore more diffraction radial from the optical centre than tangential to it. As with chromatic aberration, this effect is proportional to the angle of field.

You can alleviate the fall-off by folding the shim and straightening it out before making the pinhole in the remaining kink. Mount the pinhole with the kink inwards.

The high and the low

> ## Making a pinhole
>
> The best material is shim brass, the thinnest you can find, but the aluminium sheet from milk tops or takeaway meal covers is almost as good. Having worked out your optimum pinhole size, find a needle of the right diameter (measure the shank with a micrometer), and push the eye end into a cork. Place the shim on a piece of glass or Formica, position the needle on it and give the cork a sharp tap with your fingers. Turn the shim over and polish the bump with an emery board. Now, using a magnifying glass or watchmaker's loupe, centre the needle on the bump and repeat the tap. Turn the shim over again and polish it with the emery board. Continue this procedure until you have made a tiny hole. Now enlarge the hole cautiously, twirling the needle and polishing down any burrs, until the needle goes through cleanly to its shank.
>
> To check the quality of the pinhole, shine a laser pointer through it in a darkened room on to a card held some distance away. You will see a small Airy pattern. It should be perfectly circular, with no irregularities. (If you don't have a laser, a powerful focusing torch will do, but the patch will show colour fringes.) Once the pinhole is to your satisfaction, blacken it by holding it very briefly in a candle flame.

- *$\cos^4$ fall-off* You cannot compensate for this effect, which follows directly from the obliqueness of the pinhole and the increased distance to the edges of the film. It is aggravated if the pinhole edges are thick, as you then get the equivalent of lens barrel cut-off.

- *Coma, curvature of field, spherical aberration and distortion* These are all, mercifully, absent.

To keep the exposure down you need to use fast film. With a pinhole of effective aperture $f/250$ and ISO 400/27 film, the exposure in bright daylight is around six seconds, but if you are using the full 160° angle of view that a well made pinhole is capable of, you will need to at least quadruple this exposure. The best camera for pinhole photography is a technical camera with a pinhole cemented to a spare lens board, but the true enthusiast uses whatever comes to hand, from cereal packets to soft drink cans – good for semicircular panoramas (Figure 19.9).

Figure 19.9 A pinhole camera from a drinks can. (Photograph by Justin Quinnell.)

High-level aerial and satellite photography

Aerial photography At one time, high-level aerial photography was probably the most important component of military intelligence. Even after the emergence of satellite photography in the 1960s it remained an important element for mapping purposes. Reconnaissance aircraft carried fans of five or even seven long-focus cameras, covering a swathe of some 50°–60° wide, with a lateral overlap of 20 per cent and a longitudinal overlap of 60 per cent. An additional wide-angle camera acted as 'jigsaw puzzle box lid' to match the mosaic together. In the interpretation phase the prints were examined in pairs using stereoscopic viewing systems. Owing to the tilt of the outer cameras it was necessary to rectify the keystoning if a laid-down mosaic was required.

Rectifying enlargers were scarce in those days. Mostly we just soaked the prints in hot water and stretched them until they fitted.

The Science of Imaging

A narrow-angle lens can have the near-axis aberrations (spherical and chromatic aberration and coma) very fully corrected, as the more peripheral aberrations (astigmatism, curvature of field and distortion) are less important. Emulsion graininess is also less important with a large format. The late George Brock, doyen of Second World War aerial camera design, used to assert that there was *no* substitute for focal length.

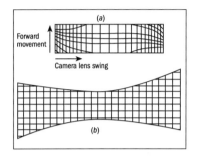

Figure 19.10 Bow-tie distortion combined with forward movement compensation in a swing-lens satellite camera. (a) Distorted grid; (b) shape of corrected area.

Aerial reconnaissance camera lenses for high-level photography typically have a focal length of a metre or so, with a format of up to 25 cm square.

The reason for the high altitude was not so much to keep out of range of anti-aircraft missiles, but to make use of what became known as the 'ground-glass effect'. If you are flying just above the haze level (about 800 m) the haze fills the whole intervening space, and the effect is similar to trying to read a newspaper through a piece of ground glass. The scattered light reduces the contrast to virtually nil. But at a height of several kilometres, nearly all the intervening space is clear, and the scattered light misses the camera lens. A deep yellow (minus-blue) filter helps to eliminate what is left of the bluish haze.

Satellite photography During the Cold War period high-level reconnaissance photography reached its ultimate stage, with aircraft flying at heights of 20 km and more; but already satellites were taking over the job. The photography itself was conventional. The satellites operated at heights of 100–200 km, with very long-focus lenses and ultrafine-grain film, which was ejected after exposure and retrieved by parachute. The authorities have been understandably cagey about the performance of these cameras, but assuming that the lenses were good enough to be aperture-limited at about $f/8$, with an emulsion resolving power of about 200 lp mm^{-1}, it would be theoretically possible to pick out objects only about 15 cm across. With these very long focus lenses (up to 3 m focal length) only a few square kilometres are imaged at each shot, and more general views are taken using a panoramic slit camera. This produces the usual hyperbolic distortion, and the image is also twisted owing to the forward motion of the satellite (Figure 19.10a). This is corrected by special printing optics matched to the camera, giving a bow-tie shape to the print (Figure 19.10b).

Much important satellite survey work is now done by means of *multispectral imaging*. Subjects can be identified in the images and analysed by their spectral signature, as we saw in Chapter 6 (false-colour infrared). The photography can be as simple as the one-shot colour camera described in Chapter 6, or as complex as a synchronised array of nine cameras equipped with narrow-band filters covering ranges between 400 and 900 nm.

Photomacrography and photomicrography

These two photographic techniques have often been called 'macrophotography' and 'microphotography' in lay journals. As a consequence, people are sometimes still confused by the nomenclature. So perhaps I should begin by giving the correct definitions.

- *Photomacrography* is the making of an image that is larger than the object, using conventional camera techniques.
- *Macrophotography* covers the techniques of producing giant prints and transparencies, sometimes called *photomurals*.
- *Photomicrography* is the technique of imaging through a microscope.
- *Microphotography*, nowadays more usually called *microimaging*, is the technique of making very tiny images from normal-sized originals.

Photomacrography Strictly, the term covers only imaging techniques where the magnification is greater than unity. Many modern zoom lenses are claimed to have a 'macro' facility, although their maximum magnification may be as low as 0.3. A true macro lens needs to be specially corrected for ultraclose-up work, as a general-purpose lens is designed to give its best performance at subject distances of several metres. When used for close-up work such lenses suffer from spherical aberration and curvature of field, both of which can be corrected only for specific image conjugates. So-called zoom macro lenses usually have a special setting for close-ups, which alters the relationships of the floating elements to provide some compensation, and this usually works best when the lens is set to a longer focal length. Lenses intended for operation at around 1:1 scale, such as process lenses, are usually symmetrical in design, giving a flat field and zero distortion.

You can use an ordinary camera with either extension tubes or a bellows extension for macro work at magnifications around ×1, but at ×2 or ×3 the usual conjugate positions are reversed, and you will get a better result by reversing the lens in its mount. The conditions concerning depth of field and depth of focus will also be reversed. There will be little depth of field and much more depth of focus, so you are better off to focus by moving the camera back rather than the lens. In order to obtain a reasonable depth of field you will need a small stop, and that brings further consequences:

- The effective f-number will be increased, as it is no longer focal length ÷ exit pupil diameter, but image distance ÷ exit pupil diameter. The image distance v is given by the formula $v = f(1+m)$, where m is the magnification, so the effective f/no is equal to the marked f/no $\times (1+m)$.

- The increase in exposure because of the change in effective f/no will be $(1+m)^2$. So a magnification of 1:1 requires four times the indicated exposure; a magnification of 2:1 requires nine times the indicated exposure.

- When the extra exposure is combined with a small aperture you may be looking at exposure durations of more than a minute; so you will experience reciprocity problems, and your subject may move during the exposure. If you are using artificial light there is a risk of overheating. You would be well advised to use an optical fibre system or electronic flash (or both).

- The easiest way to calculate depth of field is from the magnification. It is given by the formula

    $$D = 2cN(1+m)/m^2$$

    where c is the diameter of the acceptable disc of confusion and N is the f/no.

Most beginners intending to do ultraclose-up work automatically select their shortest focal length lens because it will produce the largest image with their extension tubes. This is not as logical as it may seem. For one thing, such a lens covers a much larger area than the film format, and is therefore not operating at its best. In addition it puts the lens very close to your subject, which is often awkward, and unnecessarily exaggerates the perspective. Given the choice of a 50 mm or a 200 mm macro lens, you would be well advised to choose the 200 mm. It will put you farther from the subject (important if it is, say, a live butterfly), and the perspective will be greatly improved. Don't expect an increase in depth of field, though. As you can see from the formula above, focal length doesn't come into the equation (Figure 19.11, colour plate).

> ## Resolution criterion
>
> Because of the long extension, the expression for the diameter d of the Airy disc becomes $d = 2.44\lambda N(1+m)$, where N is the marked f/no. The resolution R is half this figure, and for light of wavelength 500 nm it is given by
>
> $$R = 1.22 \times 5 \times 10^{-4} N(1+m)$$
> $$\approx 6 \times 10^{-4} N(1+M) \, (\text{mm})$$
>
> so
>
> $$N = 1667R/(1+m)$$
>
> and for a disc of confusion of diameter c this becomes
>
> $$N = 1667c/(1+m)$$
>
> For a 35 mm format with $c = 0.03$ mm and $m = 3$, this represents a (marked) aperture of around $f/4$, far too large for adequate depth of field. At a workable aperture, say $f/11$, resolution is totally governed by diffraction, and the depth of field is still only 1.5–2 mm.

Photomicrography Microscopy is a big subject, big enough to have gathered a huge literature; but photomicrography itself is nothing more than simply photography through a microscope. However, there are several different kinds of image in both light and electron microscopy, and we need to look at them separately.

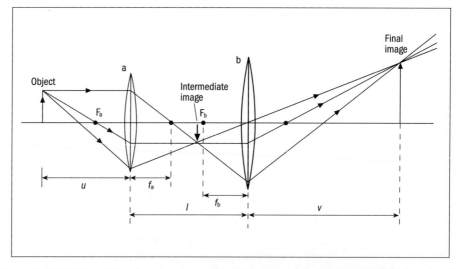

Figure 19.12 Principle of the microscope (set up for photography). For visual observation the intermediate image is focused to the right of F_b, and the eyepiece b produces a magnified virtual image.

A light microscope has the same basic layout as a refracting (Kepler) telescope, in that the magnified real image produced by the objective lens is examined by a magnifying eyepiece lens (Figure 19.12). The main difference is that the objective is set to a close (macro) focus instead of at infinity, and has a short focal length, typically 2–10 mm. For viewing, the microscope is focused so that the virtual image is formed by the eyepiece at a normal close focus for the eye of about 250 mm. For photography it is focused to form a real image in the focal plane of the camera. If you want to use an SLR camera you need to have an adaptor that replaces the camera lens and attaches the camera to the eyepiece. You then focus via the viewfinder.

Empty magnification   Ernst Abbe worked out the mathematics of microscopy in the mid-nineteenth century. Using diffraction theory, he showed that the performance of a microscope depends on its resolving power and not its magnification. In particular, the detail resolvable is proportional to the sine of the half-angle of acceptance of the objective. Abbe coined the term *numerical aperture* (NA) for this angle:

$$\mathrm{NA} = n \sin \theta$$

where n is the refractive index of the intervening medium (1 for air).

From this Abbe derived a formula for the resolution R:

$$R = \lambda/2n \sin \theta$$

where R and λ are expressed in the same units. Abbe calculated from this and the resolution of the human eye that the maximum useful magnification of a microscope objective was about $1000 \times \mathrm{NA}$, and that any further increase would be 'empty magnification', i.e., would not show any more detail. You may recall a somewhat similar situation in aerial photography (Chapter 5).

If you fill the intervening space with cedarwood oil, which has the same refractive index as the cover glass ($n = 1.5$), you will increase the NA by 50 per cent. This is why the most powerful objectives are of the oil-immersion type.

Actual magnification   Rather than doing calculations using focal lengths and conjugate distances, microscopists use formulae involving magnifications; these are much simpler, though the answers are approximate. For example, a 4 mm objective is rated at ×40; a 25 mm eyepiece is rated at ×10. The combined magnification is ×400. If you do the same calculation using focal lengths, you start with the effective focal length of the system, which is $f_o f_e / l$, where f_o and f_e are the focal lengths of the objective and eyepiece and l is the length of the tube (usually about 160 mm). Taking the near distance of clear vision as 250 mm, which is the distance of the final virtual image from the eye in a visual examination gives us $m = 250 l / f_o f_e = \times 400$ for the figures above. These figures are also approximations, as they depend on the operator's eyesight. If you need an exact figure (which you often do) you replace the slide with a *stage micrometer*, a tiny scale graduated in hundredths of a millimetre.

The depth of field is always very small, around 10 µm. The exact figure depends on a number of factors, including the refractive index of the medium the specimen is mounted in. To a first approximation

$$\text{depth of field} = \lambda/(\mathrm{NA})^2, \text{ where } \lambda \text{ is expressed in } \mu m.$$

Note that microscopists almost always refer to depth of field as 'depth of focus'.

For photography, you need to re-focus the microscope to bring the real image out beyond the eyepiece. This usually involves only a small movement of the focusing mechanism. Alternatively, you can fit a *projection eyepiece*, which produces a real image without re-focusing. The more sophisticated microscopes have a special camera port separated from the viewing eyepiece(s) by a beamsplitter or a movable

The Science of Imaging

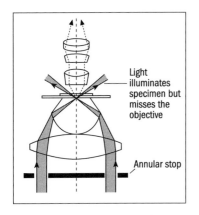

Figure 19.13 Dark-field illumination.

Frits Zernike (1888–1966) was a Dutch microscopist and mathematician who built on Abbe's work on diffraction in microscopy and was an early proponent of the Fourier model for image formation. For his invention of the phase contrast microscope he received the Nobel Prize for Physics in 1953.

mirror.

Exposure estimation is tricky: even through-the-lens metering is unreliable, and wide bracketing is in order, especially with colour transparency material. If you are working with a larger format it is a good idea to try out a test exposure on Polaroid material. Keep a full record of all exposures, including meter indications, whether correct or incorrect.

Illumination systems   The main purpose of the illumination optics is to provide uniform illumination of the specimen with no light wasted. This is more complicated than you might think, and sometimes the substage optics is more complicated than the viewing optics, including as it does condensers, aperture and field stops, filters (both colour and spatial) and polarisers.

Many specimens are hard to see under conventional lighting, because they are nearly (or even totally) transparent and are unsuitable for staining. Two important modifications to the lighting arrangement are designed to deal with this problem. The first is *dark-field illumination*. The substage optics contains an opaque annular stop that prevents direct light from reaching the objective whilst still illuminating the specimen (Figure 19.13).

A more subtle arrangement is *phase contrast*. In this system, developed by Zernike, the annular stop is positioned such that both direct light and light transmitted by the specimen reach the objective, but a quarter-wave retarding annulus is installed on (or in) the objective so that the direct light has a phase retardation. The varying refractive index of the specimen produces varying degrees of interference in the primary image, making the detail of the specimen visible (Figure 19.14).

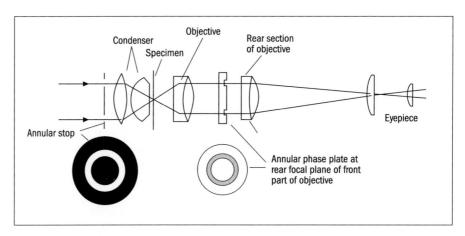

Figure 19.14 Phase contrast microscopy. The annular stop permits separate concentric beams to illuminate the specimen and the background. An annular phase plate at the rear focal point of the front section of the objective delays the phase of the 'ring' illumination so that it interferes with the object beam and turns the phase differences into amplitude differences, rendering a transparent specimen visible.

A third, rather different, way is to position the specimen between two crossed polarising filters. This is particularly useful on studies of rock specimens (petrological microscopy). Transparent sections of crystalline rock are often optically active, that is, they rotate the plane of polarisation, and you can learn a good deal by rotating the upper polarising filter (the *analyser*) and observing the colour and brightness changes.

For deep specimens where out-of-focus detail degrades the image, *confocal microscopy* is used. This is a scanning system where the image is limited by a field stop (effectively a pinhole). The out-of-focus light is greatly attenuated by the stop, the depth of field is greatly reduced, and both illumination and detection are limited to the plane of sharpness. Laser scanning microscopes usually use confocal optics.

Electron microscopy There are two main types of electron microscope, the *transmission electron microscope* (TEM) and the *scanning electron microscope* (SEM). In the TEM (Figure 19.15), the 'lens' components are similar in optical principle to those in a light microscope, but are specially shaped electromagnetic coils. The 'light beam' is a beam of electrons, and the image is formed either on a display tube or directly on a photographic emulsion. The equivalent focal length is about 1 mm and the effective aperture around $f/50$, but the wavelength of the electron beam is so short that resolution is limited by low NA and residual aberrations, not by diffraction. Resolution can be much better than 1 nm, with magnifications up to ×500 000. It sounds marvellous, but there is a snag: you can't just mount the specimen and insert it into the microscope. Not only would the actual specimen be destroyed by the beam of electrons, but in addition it is transparent to such a beam. So the specimen has first to be 'shadowed' by a very thin sputtered film of some heavy element (usually gold); and this, of course, may both hide some structural details and create unwanted artefacts.

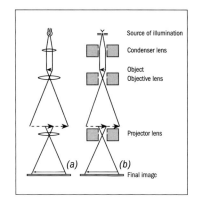

Figure 19.15 (a) Light microscope; (b) equivalent two-stage electron microscope.

The SEM uses a different principle (Figure 19.16). A very narrow-diameter (<10 nm) beam of low-energy electrons is scanned in raster form across the specimen, and the scattered electrons are collected by a system that differentiates

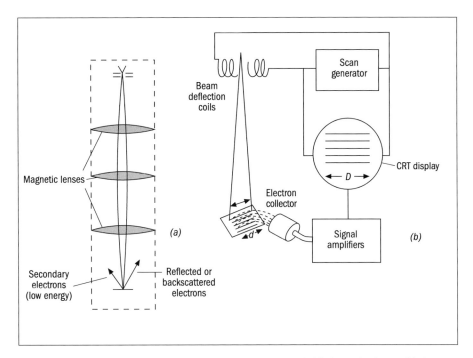

Figure 19.16 Schematic diagram of a scanning electron microscope: (a) shows the beam; (b) the scattered and secondary electrons are collected and counted separately, and the computed image is displayed on a cathode ray tube. The ratio of CRT scan width to specimen scan width gives the magnification (D/d in the diagram).

between forward and back scatter and secondary emission as well as detecting direction, so various image aspects can be presented to an image tube and display screen. There is a large depth of field, and images can have both perspective and stereoscopic depth, at magnifications from about ×10 to ×50 000. The resolution is a few nanometres. There is little or no need to prepare specimens, provided they can stand up to the high vacuum that is needed. As the image is computed, all the usual kinds of enhancement are possible.

There are other, more specialised types of electron microscope, such as the scanning tunnelling microscope, capable of imaging single atoms, but for information on these you will need to go to the references in 'Digging deeper'.

Microimaging

The techniques of microimaging evolved almost as early as photography itself. At that time sensitive materials were very slow, but they were also very fine-grained. In the 1860s jewellers set tiny images behind lenses as dress ornaments, and during the Franco-Prussian War the citizens of beleaguered Paris sent microfilmed messages by pigeon post, each bird carrying enough messages to fill a smallish book. By the end of the First World War spies were sending 'microdot' messages concealed under full stops in otherwise innocuous typewritten documents. These were made by reversed microscope optics, and were read using ordinary microscopes. The limit of resolution for fine-grain photography is usually around 1–2 µm, but because of loss of contrast owing to the shape of the MTF, words and drawings need a two-stage reduction. Although microdots may have gone out of fashion in Smiley's twilight world, microphotography is still useful for document recording, in particular for microfiches for library use. These are usually printed with either 60 or 98 frames on an A6 transparency. This is coated with a diazo emulsion, which is grainless and cheap. It is positive working, and thus produces a negative image from a negative original, white on blue.

The growing popularity and convenience of computer records on disk is contributing to the phasing out of most microfiche applications. One area that is continuously expanding, however, is that of microlithography, with special reference to the making of microchips. Lenses need to be specially designed for this purpose. As the field is small (about 10 mm diameter) the aberrations can be (and must be) very well corrected, in a near-confocal configuration. Chromatic aberration is not important, as the images are produced by exposing to single wavelengths in the blue or UV region (for the latter, special glass types are needed). The light-sensitive material is usually a positive photoresist. The component with its developed mask is then etched and doped as appropriate.

As chips become smaller and more complex, UV lasers, X-ray beams and electron beams are being increasingly used, in scanning mode. Electron beam lithography can generate patterns only a few nanometres in size.

Underwater photography

We have become so accustomed to watching superb examples of underwater photography in television documentaries that we tend to forget its technical difficulties, not to mention the hostility of the medium (and of some of its

Although much underwater photography is carried out in comparatively clean rivers and even in experimental tanks, the great majority of underwater photography goes on in the open sea.

residents). The photography itself is hampered by the turbidity of the water and the poor lighting conditions.

Light absorption and turbidity   Water has a very selective spectral absorption. At depths as little as 1 metre, roughly 50 per cent of red and orange light is lost, and from 10 m down there is little else but blue-green light. Horizontal visibility is seldom more than about 10 m, and often as little as 1 m. And although snorkelling around Greek islands with a waterproof disposable camera is great fun (I do it myself), and can sometimes lead to some striking pictorial effects, it is a very limited area of the technology. A flash on the camera is useless, as the back-scattered light degrades the contrast seriously in anything but crystal-clear water. If you need a flash, you have to have it on a long extension and to fire it well away from the camera.

Optical considerations   If you are in the habit of swimming without goggles or a facemask, you will no doubt be aware of the impossibility of seeing underwater objects sharply. Seawater has roughly the same refractive index as the cornea, so that there is no refraction at the front surface; the result is that you become impossibly long-sighted (Figure 19.17a). The flat surfaces of swimming goggles put an air gap in front of your corneas, and restore the nodal points of your eyes' optical systems to their usual position; you can now see clearly. But there is still a difference from viewing in air (Figure 19.17b). There is a fisheye effect in reverse, the rays emerging from the flat port being deviated by refraction in such a way that objects appear nearer, by a factor roughly equal to the refractive index of the water (about 1.34). They also appear correspondingly larger. One result of this is that if you are holding a flash you tend to aim it too close. In practice you need to aim it well behind where the subject appears to be.

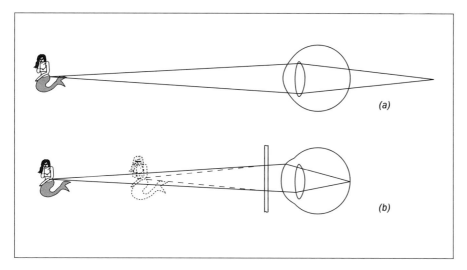

Figure 19.17 Seeing under water. (a) An underwater swimmer without goggles becomes impossibly long sighted. (b) Flat fronted goggles restore normal focusing, but the flat plate introduces distortion that makes the image appear nearer.

Some underwater housings for cameras have optically flat ports, and these not only produce this exaggeration of the image, increasing the effective focal length of the lens by one-third, but also limit the field to about 40° for a lens of normal focal length. Because of the fisheye-in-reverse configuration, there is pincushion distortion as well (Figure 19.18). Lateral chromatic aberration is also an important

The Science of Imaging

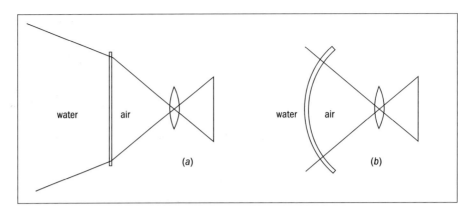

Figure 19.18 (a) Flat camera port gives the same exaggeration as swimming goggles, and the reverse fisheye effect causes pincushion distortion and limits the maximum angle of view. (b) A domed port eliminates distortion, but behaves as a negative lens, bringing the image nearer.

factor limiting peripheral resolution.

A *spherical* or *dome port* is a partial solution. Centred on the front nodal point of the lens, it eliminates the reverse fisheye distortion as well as chromatic aberration. But the curved dome also acts as a concave lens with a negative focal length about three times its radius, and forms a somewhat close virtual image of the subject, on which the camera lens has to focus. The final image has severe curvature of field. Including a positive supplementary lens can alleviate these problems. It is, of course, possible to compensate by substituting a large positive meniscus lens for the dome. A more elegant solution is to build the 'dome' into the front of the camera lens itself, and this is now the standard practice in modern specialised underwater cameras.

Digging deeper

In spite of the popularity of panoramic photography, there is little literature dealing with it, though there are plenty of coffee-table books of photographs of this type. Many articles about panoramic photography have appeared in journals, notably in the *British Journal of Photography*. Joseph Meehan's *Panoramic Photography* (Amphoto, 1990) recently went out of print. Sidney Ray has a comprehensive chapter in *Scientific Photography and Applied Imaging* (Focal Press, 1999) dealing with the subject.

Pinhole photography has had a good innings in the past, but most of the contemporary literature is unhelpful. *Pinhole Photography*, by Eric Renner (2nd edition, Focal Press, 1999), is a notable exception. Although a little short on theory, it is full of imaginative ideas.

High-level aerial and satellite photography is an awkward area, as it is easy to pick up a promising book only to find that it contains a collection of photographs from inner space with few if any clues as to the equipment or even the exposure used. An old book well worth hunting for is N Jensen's *Optical and Photographic Reconnaissance Systems* (Wiley, 1968), written in the heyday of high-level aerial reconnaissance. A more recent book along similar lines is C Etachi's *Introduction to the Physics and Technology of Reconnaissance and Survey* (Wiley, 1987). The most recent is S Drury's *Images of the Earth: A Guide to Remote Sensing* (Oxford University Press, 1998). G Thomson has written a regular report on satellite reconnaissance for the *Imaging Science Journal* (formerly the *Journal of Photographic Science*) for a number of years: 1992, **40**, 29–30, 89–90; 1993, **41**, 27–28; 1994, **42**, 40–41, 129–32; 1995, **43**, 36–37, 57–60; 1996, **44**, 66–67; 1997, **45**, 88–89; 1998, **46**, 45–48; 1999, **47**, 55–57. General aerial photography you can do yourself (with the appropriate vehicle) is catered for by Kodak's Publication O-27, *The Kodak Guide to Aerial Photography*, by Barrie Rockeach (Silver Pixel Press, 1996). You will find a more thorough treatment in W S Warner, R W Graham and R L Read's *Small Format Aerial Photography* (Whittles Publishing, 1996).

Photomacrography has always been something of a Cinderella for technical writers. The Kodak Workshop Series, which is regularly updated, contains *Close-Up Photography*, and Kodak's earlier Technical Publications Series No. N-16, *Close-Up Photography and Photomacrography*, by Henry Louis Gibson, is a gem, if you can find a copy. Two good books have appeared fairly recently: *The Art of Close-Up Photography*, by Joseph Meehan (Fountain Press, 1994), which is a guide to the photographer who wants to produce beautiful pictures of tiny things, and *Scientific PhotoMACROgraphy* (sic), by Brian Bracegirdle (Bios, 1995), which concentrates on the technology side. In photomicrography, one is spoilt for choice. In the same series as the previous book is *Scientific PhotoMICROgraphy*, also by Brian Bracegirdle with S Bradbury (Bios, 1995). The series is called The Royal Microscopical Society's Microscopy Handbooks (more than thirty of them), which between them tell you everything you ever wanted to know about microscopy. T Wilson has written the definitive *Confocal Microscopy* (Academic Press, 1990), and a newish book, from Bios again, is *Confocal Laser Scanning Microscopy*, by C J R Sheppard and D Shotton (1997). Ian Watt says the last word on *The Principles and Practice of Electron Microscopy* (2nd edition, Cambridge University Press, 1997). And if you are interested in the history of microscopy, you can sample *Microscopy from the Very Beginning*, by H Kapitze (Carl Zeiss, Jena, undated).

Although there are a fair number of books on underwater photography, not many of them can be classified as manuals on the subject. One of the best is M Edge's *The Underwater Photographer* (2nd edition, Focal Press, 1999). Another excellent book is Mark Webster's *Art and Technique of Underwater Photography* (Fountain Press, 1998). Chris Howes's *Images Below* (Wild Places Publishing, 1996) has a short chapter on photography underwater in caves, and contains many tips on how to survive in underground lakes.

The Science of Imaging

Chapter 20 Medical and scientific imaging

This final chapter brings together a group of subjects. Not long enough for a chapter on their own, they are nevertheless important areas of imaging science and technology. They are concerned with either clinical work or scientific research, or both, and they use imaging in an unusual way.

Ultraviolet and fluorescence photography

Direct UV imaging If human vision were to extend into the near ultraviolet, say to 300 nm, things would often look quite different from the way they do now. Some animals, including many insects, can see colour in the ultraviolet. The study of insect vision is important to both botanists and entomologists. To bees, for example, many flowers look quite different from the way they look to us. From UV photography we can get a good idea of what a bee sees. Figure 20.1 shows how a particular flower looks to us, and how it might look to a bee. (Incidentally, bees can see the polarisation of skylight, too.)

Figure 20.1 Flower seen by (a) white light illumination; (b) UV illumination. (Photographs by Adrian Davies.)

Another important application of UV imaging (along with holography, infrared and X-rays) is the examination of old artworks, both to establish authenticity and for conservation purposes. Among other disciplines, microscopy, astronomy and camouflage detection also make use of UV imaging.

Some films actually have an anti-UV coating.

The UV spectrum is usually partitioned into regions, in terms of its biological effects. UV-A (315–400 nm) is the radiation that gives you a suntan. It is transmitted by ordinary window glass and by Wood's glass ('black') filters. Modern optical glasses transmit UV radiation only weakly, and if you want to take your own UV photographs with a Wood's glass (Wratten 18A) filter on the camera lens, you would be well advised to choose a lens made before 1950. This

Medical and scientific imaging

region of the UV is the one that is most often used for the purposes I have outlined above; but even with old lenses you should downrate your film by a factor of about 10. UV-B extends from 280 to 315 nm. This is the region that can cause serious sunburn and can also damage your eyes. UV-C goes from 100 to 280 nm. Radiation in this region is lethal to most life forms, and is used for sterilisation and bactericidal purposes. It is radiated by the Sun, but, fortunately for us, it doesn't penetrate the atmosphere. Also, general-purpose films are almost insensitive to these regions: to get photographs you need special equipment and materials. Gelatin is totally opaque to this radiation, and you need to use a *Schumann emulsion*, which has very little gelatin, and has the silver halide crystals projecting through the surface. You also need special lenses made from fluorite and/or quartz. UV-D is ionising radiation, overlapping with soft X-rays. These very short wavelengths, from 10 to 100 nm, are sometimes called *Grenz rays*, and can be used like X-rays to show inner details of, for example, botanical specimens.

Light sources for UV imaging   Many gas-discharge sources have a UV content, and can be used for UV imaging with a filter. The best sources are high-pressure mercury–xenon lamps, which have lines down to 265 nm, plus a continuum.

UV fluorescence photography   Again, there are plenty of applications. Many substances glow in the visible spectrum when irradiated with UV radiation of wavelength 280–350 nm. In biological research, fluorescent tags are attached to organisms and their food, so that they can be tracked by UV illumination. Fluorescent labelling of goods and documents is widespread. Many chemical substances and body fluids can be identified by their fluorescence, so the uses in forensic investigation are legion. Another important use is in non-destructive testing, where the application of a fluorescent dye dissolved in penetrating oil can reveal microcracks that would otherwise be invisible. To make fluorescence photographs using UV in this region (which is the one most used for medical and scientific purposes) you don't need any special camera equipment, but you do need total darkness and a Wood's filter over the light source; also a UV-absorbing filter on the camera lens, otherwise your image may be compromised by unwanted direct UV imagery. A xenon discharge lamp is a useful portable source. To isolate specific substances you may need a spectral footprint, and this means making a number of exposures using a series of interference (narrow-band) filters on the camera in addition to the UV-absorbing filter.

Endoscopy

An endoscope is an optical device for looking inside the body by way of its orifices or through a surgical incision. Endoscopes are also used for examining the interiors of, for example, gun barrels, pipework and, on a larger scale, sewers. Clinical endoscopes are usually specialised, and have names such as *laryngoscope* (for the throat) and *otoscope* (for the ear canal). The rigid kind may use periscope optics, which consist essentially of two telescopes back to back (Figure 20.2a). A thinner tube employs relay lenses (Figure 20.2b). A third type has an extended GRIN rod, which acts as a set of relay lenses (Figure 20.2c). If the endoscope needs to be flexible, the light guides can be a coherent bundle of optical fibres (Figure 20.2d).

In these endoscopes the camera is outside the body, and you can consider the optical system as an extreme form of retrofocus lens. Illumination is usually straight down the endoscope through its optical system, the light being fed in

A coherent fibre bundle is one in which the fibres remain in the same spatial relationship throughout the length of the device. The image is thus transferred from one end of the device to the other without becoming scrambled. Note that the use of the term 'coherent' in this context has nothing to do with coherent light.

The Science of Imaging

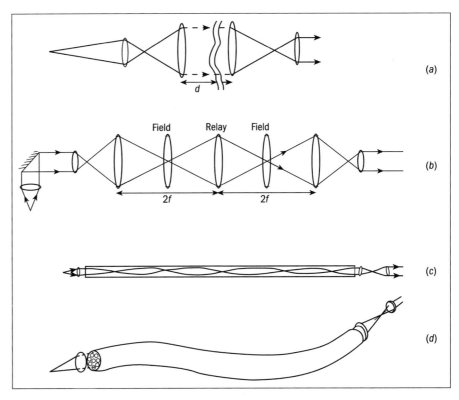

Figure 20.2 Optical systems for endoscopes: (a) periscope; (b) relay lenses; (c) GRIN relay system; (d) coherent fibre bundle.

through a beamsplitter. Recently, the advent of very small CCD arrays has made it possible to fit a tiny television camera on the end of the probe, with the image information conveyed out by cable. The resolution is better than that given by an optical fibre relay system. The cable is carried in a flexible tube, which also contains a camera steering cable and an optical fibre for illumination.

There are a number of other cameras specialised for clinical examinations. One such is the *fundus camera*, used for photographing the retina of the eye. This has a lens system that compensates for the focusing power of the eye, enabling the viewer to see the retina directly. As in modern microscopes, the camera port is separate from the viewing port. There are time-lapse facilities for studying blood flow.

> **Wilhelm Konrad Röntgen (1845–1923)** became Director of the Physical Institute of Würzburg in 1888. He made important contributions to other branches of physics, but his discovery of X-rays (which are still called 'Röntgen rays' in Germany), and of their possible applications, earned him the very first Nobel Prize for Physics in 1901. Like both du Hauron and Maddox (see earlier notes) he refused to make any financial gain from his discovery, and died in poverty.

Radiography

Radiography and its clinical applications represent a huge area of technology, though the principles of radiographic imaging are in general fairly simple.

Origins Near the end of the nineteenth century Wilhelm Röntgen was investigating fluorescence in Crookes tubes. He discovered that a mysterious radiation from the tube was causing barium platinocyanide crystals to fluoresce, although there was a cardboard barrier between the crystals and the tube; they even fluoresced when they were in an adjacent room. He called this radiation 'X-rays', and published his findings in 1895. X-rays were first used in clinical practice the following year.

Generation of X-rays X-rays are generated when a beam of high-energy electrons in a vacuum tube strike a material of high atomic number (in practice usually tungsten). The kinetic energy lost by the electrons reappears in the form of very high frequency photons (wavelength around 0.1 nm). The present-day form of X-ray tube has been developed from a design by Coolidge in 1913. The electrons are emitted by a hot cathode and focused by an electromagnetic coil to a fine beam, striking a tungsten anode specially shaped and angled to produce what is as nearly as possible a point source of X-radiation. This has a continuous spectrum with a few characteristic spectral lines due to the anode material. High (kilo)voltages produce shorter wavelengths; higher anode currents produce increased beam intensity. Figure 20.3 is a schematic diagram of a modern Coolidge tube. The entire tube is lead shielded except for a window with a filter of lead or copper compounds to absorb the longer wavelengths (soft radiation) and reduce scatter.

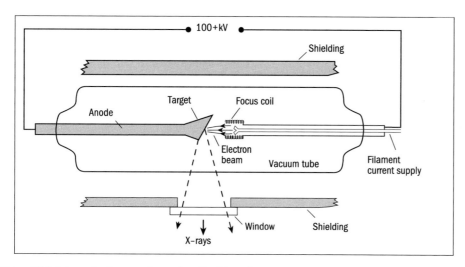

Figure 20.3 Schematic diagram of a modern Coolidge tube.

A radiograph is a shadowgraph, as X-radiation is not refracted or reflected (at least, not in the usual sense), but different materials selectively absorb it. Its clinical importance stems from the fact that bone has a transmittance to X-rays that is only one-seventh of that of soft tissue. As the beam of X-rays is divergent, it can give an enlarged image if you position the subject fairly close to the X-ray source, and some distance away from the sensitive material. By using a specially designed tube with a very small source size (~100 nm diameter) you can even obtain magnifications of over ×1000, with resolution at least as good as that of a light microscope.

Some of the radiation is scattered within the subject matter, so the sensitive material is usually wrapped in thin lead foil to reduce the effect (which resembles flare in a conventional photograph). When the subject matter is both dense and deep, as for example when imaging the cranium, a device called a *Potter–Bucky screen* is positioned between the subject matter and the detector. This is a venetian blind device constructed of lead strips interspersed with polymer spacers, and it is moved laterally during the exposure to avoid bar lines on the image.

Detection of X-rays Although Röntgen discovered X-rays through their action on a photographic plate, silver halide emulsions are not very sensitive to X-radiation

The Science of Imaging

– a fact for which peripatetic photojournalists ought to be grateful. The sensitivity falls, too, as the wavelength decreases, since shorter wavelengths, being more energetic, are also more penetrating, so most of the X-ray photons pass straight through the emulsion. As you might expect, absorption increases with increasing silver halide crystal size. Photographic films for X-ray detection are coated with a thick coarse-grain emulsion on both sides. They have a high contrast to X-radiation (maximum density ≈6) but a low contrast to visible radiation.

With plain film the necessary exposure duration would be so long that a patient would receive a possibly dangerous dose of radiation, so fluorescent *intensifying screens* are used in contact with both sides of the film. Traditional phosphors have been lead sulphide, lead barium sulphate or calcium tungstate, but compounds of rare-earth elements such as gadolinium are more efficient and are now standard.

By using the phosphor screens without a film, it is possible to view an image in real time. Where observation of movement is necessary (as in assessing blood flow or lung action), the fluorescence image is picked up by an image intensifier system. Where movement is irrelevant (as in mass screening), the exposure is as brief as possible, and the image is held by a storage tube (a kind of TV tube that holds an image for as long as required). A similar system operates at airport X-ray machines.

By a fortunate chance, the element selenium is also sensitive to X-radiation. A selenium-coated metal plate forms the sensitive surface for *xeroradiography*. The image is transferred to a transparent sheet for viewing. The edge enhancement that can be a nuisance in photocopying is a positive advantage here.

Gamma radiography The year after Röntgen had published his discoveries, Henri Becquerel, inspired by Röntgen's work, discovered natural radioactivity in uranium compounds. He was able to show that much of the emission consisted of streams of electrons, and concluded, correctly, that uranium was decaying into other elements.

The emissions are now known to be of three types: alpha (α) particles, which are helium nuclei, beta (β) particles, which are fast electrons, and gamma (γ) radiation, which is X-radiation with a range of wavelengths from 10^{-1} to 10^{-3} nm. This radiation will penetrate up to 100 mm of steel, and this makes radioactive substances suitable sources for radiography of metal structures as a part of non-destructive testing procedures. The source, which is usually an artificial radioisotope, is contained in a lead casket with a small aperture; this is positioned a short distance from the structure under investigation, the film being fixed to the other side of the structure. As the process is very inefficient, the exposure duration is often several hours.

In order to calibrate the source, a step tablet is used. This operates like the Kodak step tablet discussed in Chapter 10, but is a good deal more robust, being a set of steps machined from a steel block. Gamma radiography is a valuable adjunct to UV and ultrasonic tests in the detection of cracks and other faults in structures such as aircraft wings. Radioisotopes are also used as γ-ray sources inside metal pipes to check the quality of seam welds.

Units of measurement Like photometry, radiography was in the past lumbered with a plethora of units of measurement, but again the adoption of SI units has whittled the number down to a manageable minimum. Although not entirely abandoned, non-SI units such as the curie, rad, rem and röntgen are no longer

---

Antoine Henri Becquerel (1852–1908) was a professor at the Ecole Polytechnique in Paris. He discovered, more or less by accident, that compounds of uranium fogged photographic plates even when they were wrapped up. He investigated the nature of the radiation, and in 1903 was awarded the Nobel Prize for Physics jointly with the Curies, who had succeeded in isolating polonium and radium from uranium ores.

A *radioisotope* is an element that has a nucleus containing too many neutrons to be stable. Atoms of the isotope spontaneously decay into more stable forms, usually emitting neutrons and γ-radiation in the process. Radioisotopes occur naturally in small amounts (radium itself is one); usually they are created as by-products of nuclear reactions and they can also be produced in particle accelerators. Each radioisotope has a characteristic *half-life*, that is, the time it takes for half of its atoms to decay. This time may be anything from less than a microsecond to more than a million years.

Medical and scientific imaging

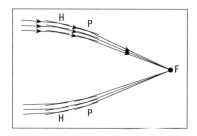

Figure 20.4 Grazing-incidence hyperbolic (H) and parabolic (P) mirrors for bringing a beam of X-rays to an optical focus (F).

Figure 20.5 X-ray diffraction image of a crystal. By comparing patterns at varying angles of incidence with known diffraction patterns, a three-dimensional picture of the crystal structure can be deduced.

persona grata with the ISO. The Becquerel (Bq) replaces the curie as the unit of radioactivity (1 Bq = 1 s^{-1}); the gray (Gy) replaces the rad as the unit of absorbed radiation (1 Gy = 1 m^2 s^{-2}); the sievert (Sv) replaces the rem and röntgen as the unit dose equivalent (1 Sv = 1 m^2 s^{-2}). The energy associated with wavelength is often quoted in electron volts (1 eV = 1.6021 × 10^{-19} J). All the other terms are now obsolete.

X-ray optics   I mentioned earlier that X-rays aren't reflected or refracted in the way light waves are. Nevertheless, there *are* ingenious ways of obtaining a focused X-ray image. The simplest way is to use a pinhole camera lined with lead sheet. This was the earliest type of camera used for identifying X-ray sources in outer space. X-rays can in fact be reflected, but only from metals at grazing incidence. One form of X-ray telescope uses a nest of narrow hyperboloids and paraboloids of revolution (Figure 20.4). This can also be set up for X-ray microscopy. X-rays are subject to the laws of diffraction, and diffracted beams of X-rays are used to deduce the molecular structure of crystals (Figure 20.5). Using microlithographic or holographic methods it is possible to construct a zone plate that will focus a beam of X-rays to produce an image. It is also possible to construct Bragg mirrors to make a reflecting telescope to investigate celestial X-ray sources, by coating metal mirrors with alternate half-wave layers of tungsten and carbon (only 2–3 atoms thick!), utilising the very small refractive indices (for X-rays) of these elements (Figure 20.6).

Energy is inversely proportional to wavelength. 1 nm corresponds to 1.234 MeV.

It was Rosalind Franklin's brilliance at X-ray crystallography that enabled Crick and Watson to work out the structure of DNA, for which they received the Nobel Prize for Physiology or Medicine along with Maurice Wilkins in 1962. Franklin had died four years previously, aged only 38.

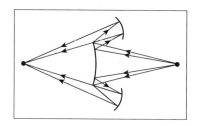

Figure 20.6 X-ray telescope employing Bragg reflection.

Tomography and scanning systems

Tomography evolved as a method of obtaining clear images of internal organs that lay within bony structures such as the skull and ribcage. The X-ray source was moved across the subject during the exposure, and the film, with a Potter–Bucky screen, was moved in the opposite direction, both movements centred on the organ to be imaged. The specific organ to be examined would have been perfused with a radio-opaque dye (Figure 20.7).

A modification of this method appears in one piece of modern equipment. If you have recently had a dental X-ray, you may have been treated to a panoramic view of your teeth. For this type of radiograph the film is mounted in an elliptical holder with its closer focus at the back of your head. The other focus is at the

The Science of Imaging

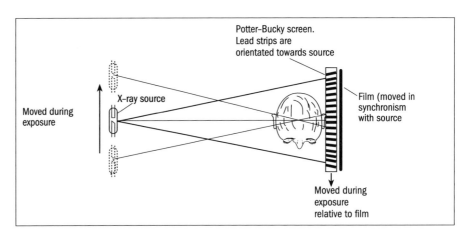

Figure 20.7 An early form of tomography.

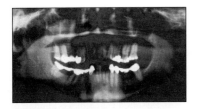

Figure 20.8 Panoramic X-ray scan of a complete dentition. The metal bridges are particularly noticeable.

front of your head, and the X-ray source follows the outline of the other half of the ellipse, while a slit traverses the film in synchronism (Figure 20.8).

X-ray body scanning   Once it had become possible to scan around a patient's body with an X-ray machine and a detector diametrically opposite one another, and to feed the acquired data into a computer, 360° *computerised axial tomography* (CAT) became a reality. Each 'slice' produces a cross-section of the patient's body, stored for display and examination. The earliest systems were similar to that of the dental panoramic X-ray, with a single source and detector swinging through the full angle. Several generations later, the most recent design has no moving parts. The X-ray emitters are in parallel rows in a semicircle below the patient, and the detectors are in similar rows above. An electron beam directed by magnetic coils scans the emitters, and scanning is complete in under a second, compared with several minutes for the original process (Figure 20.9).

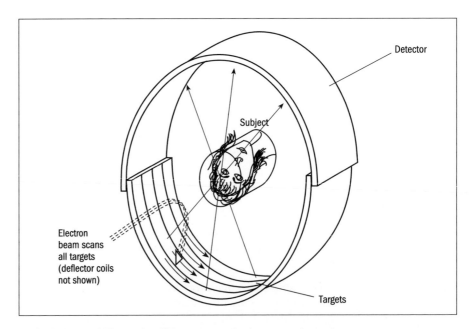

Figure 20.9 Modern CAT scanning. This newest version has no moving parts.

Radioisotope scanning   The above type of tomography also lends itself to autoradiographic scans. The organ for examination is perfused with an appropriate radioactive tracer (e.g. radioiodine for the thyroid gland). A *gamma camera*, which is basically just a pinhole camera equipped with a scintillation plate instead of a film, maps the emissions. The scintillations are recorded as the camera moves around, and the computer builds up a three-dimensional picture of the organ. The monitor can display a picture very nearly in real time, and can show movement too.

Positron emission tomography (PET) scanning   This is a similar device, but records γ-rays emitted when positrons emitted by tracers interact with electrons present in body chemicals.

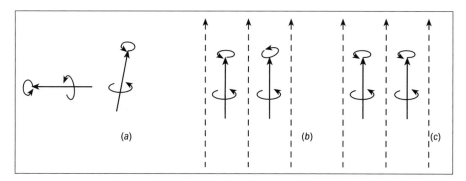

Figure 20.10 (a) Proton spin gives a random magnetic polarity. (b) A powerful magnetic field aligns the spins but not the phases of their precessions. (c) Applied radio frequency (RF) pulses of appropriate frequency put all precessions into phase. When a pulse ceases the precessions return to their original phases, producing energy that identifies the tissues concerned. A computer analyses the data and constructs the image.

Nuclear magnetic resonance imaging (NMRI)   The protons in the nucleus of an atom possess *spin*, giving them a magnetic polarity that is usually random (Figure 20.10a). In the presence of a powerful magnetic field the polarities line up. However, the spins have a wobble, a random precession like a spinning top (Figure 20.10b). An applied radio frequency (RF) pulse of the same frequency as the precessions aligns them so that the spinning protons all precess together. The whole magnetic field thus rotates in phase (Figure 20.10c). When the RF pulse ceases the spins relax into their former precessions, emitting energy. The strength, frequency, bandwidth, decay patterns and directions of the signals locate and identify the tissues concerned, each having its own signature. Watery and fatty tissues are imaged but bone is not, nor is blood. Contrast media can be injected to show up specific structures.

The magnetic field required is of the order of 1–1.5 teslas. This is an enormously strong field, over 100 times that of an ordinary bar magnet. As you can imagine, the sorting out of the information is pretty complicated, and needs a good deal of computer power. The slices are pixellated into three-dimensional elements (*voxels*), and the final image is comparatively easy to interpret, especially as it can be compared with, and superimposed on, PET slices, which show chiefly biochemical function, and CAT slices, which show bone structure.

Analysis of scanning outputs

The actual outputs of these scanning systems are complicated, especially in the case of NMRI scans, and well beyond the remit of this book (and most others). There

The Science of Imaging

are two main methods of processing the data as they accumulate. The first is an iterative method, whereby a crude 'guesswork' map is successively reprofiled until it fits all the data. The second is the Fourier approach. Each set of data is analysed into its sinusoidal components. When these have all been specified in terms of amplitude, direction, (spatial) frequency and phase, the final map is obtained by a reverse Fourier transform. (The Fourier approach to imaging is explained in Appendix 4.)

Ultrasonic imaging

In its own way, this is also a form of scanning, but it is fundamentally different from the methods above in several ways. *Ultrasound* is longitudinal wave propagation with a frequency higher than the human ear can detect. Strictly, that means any frequency higher than about 20 kHz, some two and a half octaves above the highest note of a piano. But for imaging purposes this is much too low a frequency. The corresponding wavelength is about 18 mm, and any object smaller than this will not reflect the beam. Also, diffraction means that the beam will spread out uncontrollably from the source. At frequencies of around 1 MHz the wavelength is of the order of 100 μm, the effects of diffraction are much reduced, and a broad source gives a tight, well-collimated beam. Higher frequencies are possible, but these are greatly attenuated in fluids and air. Ultrasonic imaging operates by the analysis of echoes. The three main applications are sonar, clinical scanning procedures, and non-destructive testing for hidden flaws in structures.

PZT, a ceramic material, is often used as the driver in small speakers (tweeters) and for the generation of 'beeps' in electronic equipment. It is also used in optical positioning equipment for making very small positional adjustments.

The ultrasound generator is a *piezoelectric transducer*, usually made of lead zirconate titanate (PZT). A piezoelectric transducer is a substance that changes its dimensions in response to an electric field, and, conversely, produces an electric field when it is strained.

To generate the ultrasound, a radio frequency (RF) voltage of the required frequency stimulates a specially shaped PZT transducer. As the device also operates in reverse, it can be used to detect echoes from its own emissions. So by alternating short pulses of RF stimulation with pauses to 'listen' for echoes, the incoming signal can be amplified and compared with the outgoing signal, and the result analysed.

The Doppler effect refers to the raising of the pitch of an echo when an object is approaching, and the lowering of pitch when it is retreating. Knowing the speed of sound and the change in frequency makes it possible to calculate how fast the object is moving towards you or away from you.

Sonar A transducer, directed at the sea floor, has its echo pulses displayed in real time in terms of return time (and hence distance) on an oscilloscope screen. This gives an immediate reading of the depth of the water, but also gives some indication of the nature of the sea floor (rocks, mud, coral, etc.). Shoals of fish show up as intermediate blips. Oscillating the transducer beam sideways, when combined with the vessel's forward movement, creates a map of the sea floor. Forward and side looking sonar sweeps can also detect submerged rocks and (with luck) mines. The movement of any submarines that are within range can also be analysed by making use of the change in frequency of the returning beam (the Doppler effect). Sonar is illustrated in Figure 20.11.

Non-destructive testing Ultrasound systems for non-destructive testing operate on similar principles, though here the aim is to detect air gaps in objects that should be solid. The transducer is coupled to the structure's surface by a film of grease. Any hidden cracks or inclusions give away their presence by their echoes. It is necessary to use higher frequencies, as sound travels very fast in metals, and has a correspondingly longer wavelength.

Medical and scientific imaging

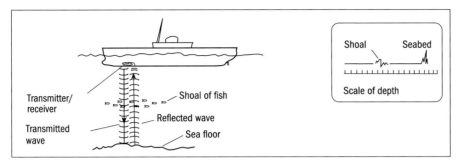

Figure 20.11 Principle of sonar.

Clinical ultrasonic imaging For clinical purposes such as brain, kidney and foetus scans, the transducer is coupled to the skin surface by a film of jelly, and moved over the whole area while its movement is tracked and, as in sonar mapping, a picture is slowly built up. Moving organs such as the heart, and blood flow, can be monitored by means of the Doppler effect (see above). In this mode the signal is continuous and the frequency changes can be compared, as in Doppler sonar. Under some circumstances the frequency differences produce audible beats.

Thermal imaging

Thermal imaging, or *thermography*, is the making of an image using the far infrared region of the electromagnetic spectrum, between 1 and 20 μm. There are two main types of camera, direct imaging and scanning. The direct imaging or *focal plane array* camera is a digital camera equipped with a CCD array of platinum silicide (PtSi), which operates in the 3–5 μm band, with a camera lens of amorphous silicon (refractive index 3.45 to IR radiation; opaque to visible light) or fluorite (RI = 1.38). The system needs to be cooled with liquid nitrogen to eliminate noise from stray radiation. As there are no moving parts the framing rate can be very high, up to 1000 pps, allowing the photography of high-speed events. Germanium (RI = 4.1 to IR radiation) optics allows an extended response, up to 12 μm.

In the *scanning imager* there is a single detector. Two orthogonal pivoted mirrors or prisms provide a scan of the scene. Figure 20.12 shows one possible combination of mirrors. Such systems can operate at television speeds.

The *pyroelectric vidicon camera* is a portable maid-of-all-work camera used by firemen and rescue workers to find trapped people, and for general surveillance at night. It is a conventional vidicon camera except that the sensitive surface is a thin sheet of pyroelectric material such as lithium tantalate ($LiTaO_3$) or PZT, which have flat spectral responses throughout the far infrared. A surface charge proportional to change in temperature builds up (Figure 20.13). The system is slow, with a refresh rate of several seconds; but it is cheap and light, and does not require special cooling.

Several types of *photoelectronic* or *quantum detector* are employed in specialised cameras such as those used in satellites for astronomical observations at wavelengths that are absorbed by the Earth's atmosphere. These all require cooling to liquid nitrogen or even liquid helium temperatures.

The Science of Imaging

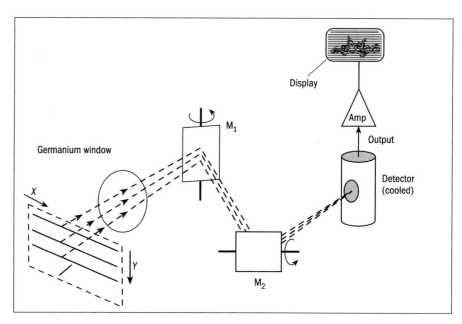

Figure 20.12 Principle of scanning thermal imager. M_1 and M_2 are oscillating mirrors.

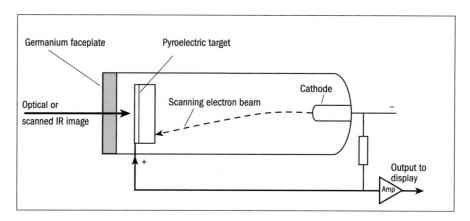

Figure 20.13 Pyroelectric vidicon camera.

Image converters An image converter is a device that changes an X-ray, UV or IR image into a visible image, which you can capture on ordinary silver halide film or a CCD array. It is coupled to an *image intensifier* (or *image amplifier*) *tube*. This is essentially a device for electronically amplifying optical images that are too weak to record directly on to film. The schematic layout is shown in Figure 20.14.

The photons that form the initial image produce electrons at the photocathode, and these are accelerated and focused to form a much brighter image at the phosphor screen, which is at 15–20 kV. This image can be read by a vidicon tube or a CCD array, or transferred directly to film.

For photographic or visual observations in very dim light (hazy starlight) up to three tubes can be cascaded, though the resultant image quality is poor and noisy.

Image intensifying low-light cameras are generally fitted with wide-aperture (f/1.5) catadioptric lenses. Conversion from X-rays, UV or IR to a visible image is by a layer of phosphor on the faceplate, which has a fibre optics disc conveying the image direct to the photocathode.

Medical and scientific imaging

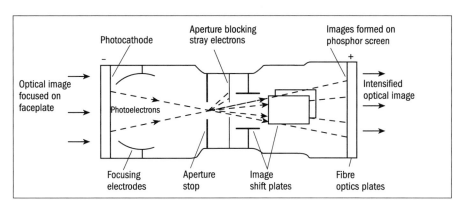

Figure 20.14 Image converter/intensifier tube (schematic).

An alternative (or additional) amplifier stage can be provided by a *micro-channel plate intensifier*. This is a flat plate of lead glass about 1 mm thick, bored with up to 10 million tiny channels a few tens of micrometres in diameter (Figure 20.15). A potential of around 1 kV accelerates the photoelectrons through the channels, and during their travel they gain further photoelectrons in an avalanche process before striking a phosphor screen. This type of device is used in most hand-held low light and IR conversion equipment.

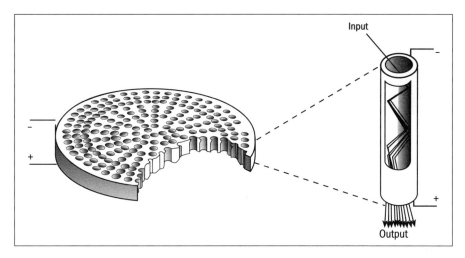

Figure 20.15 Microchannel plate intensifier.

Schlieren photography

Schlieren is a somewhat uncommon German word meaning, roughly, 'striations'. The term 'schlieren photography' was coined by Albert Töpler, who first described the technique in 1906. It is based on an idea by Foucault, who originally devised a knife-edge technique for examining astronomical telescope mirrors for defects in figuring.

The general principle of schlieren optical systems is that if you focus a collimated beam of light to a point, then position a knife edge precisely at that point, the slightest disturbance in the medium the beam has passed through results in either a cut-off of part of the beam or its enhancement.

Jean Bernard Léon Foucault (1819–1868) was a physicist at the Paris Observatory. He took the first detailed photographs of the Sun's surface in 1845, and demonstrated the rotation of the Earth in 1852 with 'Foucault's pendulum'. He followed up this line of reasoning by inventing the gyroscope. He determined the speed of light in air and water to an accuracy of better than 1 per cent, and made many contributions to telescope engineering. His observations led to the invention of spectroscopy.

The Science of Imaging

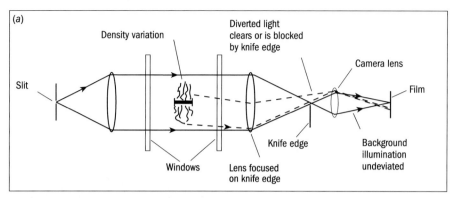

Figure 20.16 (a) Principle of schlieren photography. (b) Schlieren photograph of discharge of a .22 calibre starting pistol, made using a 0.2 μs flash. (Photograph by Ron Graham.)

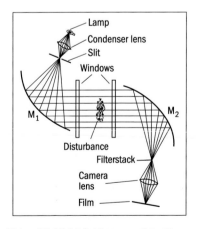

Figure 20.17 (a) Schlieren configuration using off-axis paraboloidal mirrors and colour filters. (b) A schlieren image using colour filters (colour plate). (Photographs by Jon Tarrant.)

Now suppose you pass your collimated beam of light through the test chamber of a wind tunnel containing a test piece, say an aerofoil section, and that after the knife edge you position a focusing lens and a screen, or a camera (Figure 20.16). The shadow of the test piece will be there, in focus; but the rarefied air above the aerofoil will have diverted the beam, and so will the over-pressure on its underside. In both cases the density gradient will show up as a darkening or lightening of the image. The most dramatic results are with shockwaves generated at transonic and supersonic speeds, though in ballistics research simple spark photography also gives a bright image. However, the schlieren method is also sensitive enough to show the patterns of rising air from a candle flame or even from a human hand.

In practice it is difficult to make good lenses large enough for anything but the smallest test volumes. Optical mirrors are the answer. Unfortunately, spherical mirrors introduce a double dose of spherical aberration and an unacceptable amount of coma. So, in spite of the extra expense, it is more usual to use parabolic mirrors designed for off-axis beams (Figure 20.17(a), (b) colour plate). The light source needs to be focused into a slit, for maximum resolution.

A variant of the method uses interferometry. The beam is split into two by a beamsplitter, and the two beams, only one of which has been through the test

Medical and scientific imaging

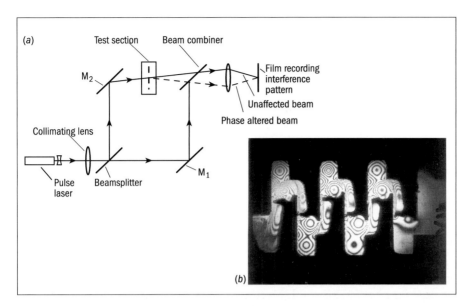

Figure 20.18 Schlieren interferometry. (a) Basic Mach–Zehnder set-up. (b) Interferometric visualisation of the gas flow pattern in a labyrinth seal. Interferogram by Ric Parker. (Photograph courtesy of Rolls Royce.)

space, are recombined on the far side by a beam combiner (Figure 20.18). You need to use a single wavelength in order to be able to see interference patterns, so laser light is mandatory. The principle is that of the *Mach–Zehnder interferometer*. If the air in the test volume is undisturbed you will see a set of straight fringes, which you can adjust to be vertical or horizontal as necessary by making small adjustments to the mirrors. Density variations are contoured by the disturbances in the fringe pattern. This system is more sensitive than the schlieren configuration, and has the advantage of direct reading, as each displacement of a whole fringe width depicts a change in the optical path of one wavelength. The interference method gives a direct reading of density variations, whereas the schlieren method shows the first derivative of the density, i.e., the density gradient, and is more difficult to analyse quantitatively. The spark method shows the second derivative and, being still more recalcitrant to analysis, is usually employed only for showing the general form of a shockwave.

We can obtain more informative (and attractive) results by substituting colour filter strips for the knife edge, for example red above, green in the middle and blue below. Then the image will be basically green, with a high-down-to-low density gradient giving a yellow band (red + green) and the reverse gradient giving a cyan band (blue + green). Figure 20.17b (colour plate) illustrates this.

Digging deeper

There is a great deal of literature on most of the subjects in this chapter; but the disciplines are still developing rapidly, and many of these recommendations are already out of print. Rather than producing new editions, the tendency seems to be to update the older ones by publishing papers in the specialist journals, until the whole business falls so far behind that some enterprising author writes a completely new book; and then the whole process begins again. Perhaps by the time you read this book some of the works recommended here will have been superseded, so I have tried to keep to the ones whose information is likely to stand the test of time.

There is a fair amount of literature on UV photography. If you can find a copy, Eastman Kodak's *Ultraviolet and Fluorescence Photography* (Technical Publication M-27, 1972) is full of sound practical advice. *Photography for the Scientist* (ed. R Morton, Academic Press, 1984) has a chapter that deals with the theoretical background in some detail. B Duffey's *Ultraviolet Radiation in Medicine* (Institute of Physics, 1982) gives a comprehensive review of the applications of UV in general. Sidney Ray's *Scientific Photography and Applied Imaging* (Focal Press, 1999) has a chapter that gives comprehensive data on light sources, filters and lenses, with a very full bibliography.

With the explosive growth of keyhole surgery techniques, endoscopy is constantly evolving, and for up-to-date information you need to consult the appropriate medical journals such as *Biological Photography*. There is an excellent book, *Biomedical Photography* (ed. I Walter, Butterworth-Heinemann, 1992), which contains a chapter on photoendoscopy by L Morris. It also contains a full account of ophthalmic photography by L Merin. For an in-depth treatment of TV endoscopy, try *Electronic Videoendoscopy* (ed. K Harubini, Harwood Academic Publishers, 1993).

Radiography, as you might expect, has spawned a vast literature; I can only point to some excellent reference works I have come across. The standard work, covering the whole field, is *Radiological Imaging* (2 vols) by H Barrett and W Swindell (Academic Press, 1981). It does need some updating, however. A more recent book, concentrating on the theoretical side, is the multi-author *X-Ray Science and Technology* (ed. A Michelle and C Buckley, Institute of Physics, 1993). Two good books on X-ray astronomy come from Cambridge University Press: G Frasier's *X-Ray Detectors in Astronomy* (1989) and P Charles and F Seward's *Exploring the X-Ray Universe* (1993). The principles of tomography are covered in Barrett and Swindell (see above); and various authors cope with the analysis of scanning data in *Computer Methods: The Fundamentals of Digital Nuclear Medicine* (ed. D Liebermann, Mosby, 1977), which needs less updating then you might expect, but could certainly do with a new edition. Probably the best book on ultrasonic imaging is *Modern Acoustical Imaging* (ed. H Lee and G Wade, IEEE, New York, 1995).

The most complete account of thermal imaging is *Applications of Thermal Imaging* (ed. B Burney, T Williams and C Jones, Institute of Physics, 1988), but there is a collection of recent papers on the subject published by SPIE Publications (1999) called *Practical Applications of Infrared Thermal Sensing and Imaging Equipment*, which is well up to date. If you are interested in the birth pangs of this fascinating area of technology you should try to get hold of a copy of the *Journal of Photographic Science*, 1961, **9**, 375–379, where there is an article by A Bouwers *et al.* 'Low brightness photography with electro-optical intensifiers'. Dr Bouwers of De Oude Delft was a pioneer of image intensification, and almost all modern developments in the technology stem from his insights into both wide-aperture optics and electronic light amplification.

Schlieren photography is an old technique, and apart from the use of lasers for schlieren and holographic interferometry, little new has developed recently. The best summaries of technique are both to be found in two Focal Press books with Sidney Ray as author or editor: Chapter 17 'Flow Visualisation' by Peter Fuller, in *High Speed Photography and Photonics* (ed. S F Ray, Focal Press, 1997), and Chapter 24 in Ray's *Scientific Photography and Applied Imaging* (see above). For a full account of all the techniques, try *Schlieren and Shadowgraph Techniques* by G S Settles (Springer, 2001). If you want to go back to beginnings and can read German, call up a copy of Töpler's original paper 'Beobachtung nach einer Neuen Optischen Methode', *Ostwalds Klassiker der Exacten Wissenschaften* (Leipzig, 1906, No. 157).

Appendix 1 Logarithms: what they are, what they do

John Napier (1550–1617) was a Scottish aristocrat and a brilliant scientific amateur. He was fiercely Protestant, and published a long treatise denouncing Catholicism. He invented a water pump for coalmines, and researched the use of fertilisers in agriculture. He produced the first calculating machine, using numbered rods ('Napier's bones'). He was also the first person to use decimal point notation.

Logarithms were the brainchild of John Napier, who invented them as a way of simplifying the lengthy calculations involved in astronomy, his chief interest. He coined the name 'logarithm' from two Greek words meaning 'ratio' and 'number'.

The ubiquity of the pocket calculator has made the use of logarithm tables redundant for calculations (the slide-rule, too, into the bargain); but the concept of the logarithm (usually abbreviated to 'log') is still important, particularly where perceptual processes are concerned. Its basic thesis is that any positive number can be expressed as a power (usually of 10), and that this power (the *logarithm* of the number) has a unique relationship with the number itself.

Let's look a little closer at this relationship. Consider two series of numbers:

	1	2	3	4	5...
and	10	100	1000	10 000	100 000...

Strictly, logarithm base 10.

Do you see the connection? The most obvious one is that the upper row is the number of zeros in the lower row. A closer connection is that 10 is 10^1, 100 is 10^2, 1000 is 10^3, and so on. The upper row is said to be the *logarithm* of the lower one. The series can continue as long as you like: you can see that the next number in the upper row is going to be 6, and in the lower row 1 000 000 (i.e., 10^6).

But what *kinds* of series are these? Well, the upper row is plainly a 'counting' or *arithmetical* series: to obtain the next number you simply add 1 to its predecessor. The lower row is a multiplying or *power* series: to obtain the next number you multiply its predecessor by 10.

You've seen how we can continue both series upwards. But what happens if we try to count downwards? The answer is that you continue both series in the same way: you subtract 1 from the upper row, and divide the number in the lower row by 10, giving the continuation downwards:

−3	−2	−1	0	1
1/1000	1/100	1/10	1	10

By analogy, the lower numbers can be written 10^1, 10^0, 10^{-1}, 10^{-2}, 10^{-3}. This is perfectly logical, since for all values of x and y, $a^x \times a^y = a^{(x+y)}$ and $a^x \div a^y = a^{(x-y)}$. Thus $10 \div 10 = 10^{(1-1)} = 10^0$, and $1 \div 10 = 10^{(0-1)} = 10^{-1}$.

There is no log of 0. Logs of negative numbers exist only within the set of complex numbers. The log (base e) of −1 is $i\pi$, where i represents $\sqrt{-1}$. This follows from the curious and interesting identity $e^{i\pi} \equiv -1$.

The next question is: Do the numbers *in between* those in the lower row have log equivalents too? The answer is yes, they do. For example, consider the number $10^{0.5}$. Now, from the same rule of adding of powers, $10^{0.5} \times 10^{0.5} = 10^1 = 10$. This means that $10^{0.5}$ must be $\sqrt{10}$, which is approximately 3.2. Using similar arguments, you can work out the log of *any* positive number.

You can also show that $10^{1.5}$ is 32, $10^{2.5}$ is 320 and so on. This indicates that the number behind the decimal point of the log (called the *mantissa*) tells is what the figures are in the original, and the number in front of the decimal point (called the *characteristic*) tells us its order of magnitude. A log value with a characteristic of, say, 2, tells us that the number lies between 100 ($= 10^2$) and 1000 (10^3), and a log value with a characteristic of 0 corresponds to a number between 1 ($= 10^0$) and

Logarithms: what they are, what they do

10 ($= 10^1$). When we count downwards from 1 (log 1 = 0.0) we write the characteristic as −1, −2 etc, but keep the mantissa the same. So the log of 0.32 is written $\bar{1}.5$, the log of 0.032 is written $\bar{2}.5$, and so on.

Pocket calculators don't do this. They actually indicate the full negative value. This makes no difference to calculations involving logs.

To the image-maker, the most important logs are the log of 10 and the log of 2. The log of 10 we already know: it is 1. The log of 2 we can find by looking at the log of 32, which you will remember is 1.5, that is, 32 is $10^{1.5}$. Now, $32 = 2^5$, so 2 is $\sqrt[5]{32}$, which is $10^{(1.5 \div 5)}$, i.e., $10^{0.3}$. So the log of 2 is 0.3. We can now write down the series

No	1	2	4	8	(10)	16	32	64	(100)	128	256	512	1000	2000
Log	0.0	0.3	0.6	0.9	(1.0)	1.2	1.5	1.8	(2.0)	2.1	2.4	2.7	3.0	3.3 etc.

Leaving out the figures in parentheses, which are put in only as an indication of where they fit in, this is a doubling-up series, and it matches the series of log values we use in plotting H & D curves. It is also the series (with the decimal point omitted) of ISO logarithmic film speed indices.

The small discrepancy at 1000, which you would expect to be 1024, is because the logarithm of 2 is actually 0.301; but an error of one-third of one per cent is not very important.

Logarithmic scales Some graphs use scales in which equal intervals along the scale represent, not equal arithmetic increments, but equal multiples. These are called *logarithmic* scales. They are often convenient for plotting purposes where there is a very large range of values, and the small values are also important in themselves. When there is a power-law relationship between the two variables, for example $y^3 = x^2$, the use of log paper gives a straight line instead of a curve. When there is an exponential relationship such as $y = a^x$, you can obtain a straight line by using log–linear paper, that is, paper in which one axis is logarithmic and the other is arithmetic.

Logs to base 2 These underlie the light-value scales found on most exposure meters. Each increment on the scale represents a doubling of the stimulus:

No	1	2	4	8	16	32	64
Log	0	1	2	3	4	5	6

We use logs base 2 when we speak colloquially of (say) giving three stops (= 8 times) more exposure.

To convert logs base 10 to logs base 2, all you have to do is divide the log by 0.3 (strictly, 0.301); and to convert logs base 2 to logs base 10 you multiply by the same amount.

Logs to other bases You can in fact have logs to any base you like; but the only base other than 10 and 2 that you are likely to come across is e. This is one of those curious constants like π that seem to crop up in all sorts of odd places in mathematics. Its value is 2.71828..., and like π, it goes on for ever. It is derived from the slope of a curve that is equal to its own y-value, an *exponential* curve, and it has a number of properties that make it particularly useful in advanced mathematics. To convert logs base e to base 10 you divide the figure by 2.303, and to convert the other way you multiply by the same figure. Logs base e are properly called *natural logarithms*; they are also sometimes called Napierian logarithms, although Napier himself had nothing to do with this constant. (Logarithms base 10 are called *common logarithms*.) In textbooks Napierian logarithms are abbreviated *ln* instead of *log*, and the expression e^x is often printed as exp(x).

In fact, Napier's logarithms were to the base $1/e$, though Napier himself had no concept of e. He had originally arrived at the concept of a logarithm by a geometrical process. It was Napier's friend Henry Briggs who suggested the base 10.

Appendix 2 The meaning of pH

The *mole* (mol) is the fundamental SI unit of quantity of substance (usually atoms, ions or molecules). It is defined as the amount of substance that contains as many entities (e.g. atoms) as there are atoms in 12 grams of carbon-12. (The actual number of atoms is 6.023 × 10^{23}, Avogadro's number.) A *molar solution* is a solution that contains 1 mole per litre of dissolved substance.

The pH of a solution is a measure of its acidity or alkalinity. A neutral solution has a pH of 7; higher figures are associated with alkalinity, lower figures with acidity. Every aqueous solution contains hydrogen ions (H^+) and hydroxyl ions (OH^-). In neutral solutions and pure water there are exactly 10^{-7} moles per litre (mol/ℓ) of each.

The product of hydrogen and hydroxyl ion concentrations in an aqueous solution is always 10^{-14}. If a solution is made alkaline by adding a substance that increases the OH^- concentration by a factor of, say, 100, increasing it to 10^{-5} mol/ℓ, the H^+ concentration falls to 10^{-9} mol/ℓ. We say that the pH has changed from 7 to 9. Conversely, if the H^+ concentration is increased by a factor of 100, bringing it up to 10^{-5} mol/ℓ, we say that the pH has changed to 5. (The OH^- concentration will have fallen to 10^{-9} mol/ℓ.)

Each step on the pH scale means a change in the hydrogen-ion concentration by a factor of 10, so a solution with a pH of 10.5 is ten times as alkaline as one with a pH of 9.5. It is a logarithmic scale (see Appendix 1), so that each pH increment of 0.3 represents a doubling of the OH^- concentration and a halving of the H^+ concentration. The pH values found in photographic processing solutions range from about 10.5 for the fiercest developers, to 5.5 for acid fixing baths and 4.5 for acid bleach baths.

Laboratory pH meters are expensive and need frequent recalibration. Soil pH meters from garden shops are not usually accurate enough outside pH values between 6 and 8. You are better off with pH indicator papers, which are strips of absorbent paper impregnated with a dye that changes colour at specific pH values. These can be 'universal', changing colour through a whole spectrum as pH values vary between 2 and 13 (you match the colour against a printed colour chart) or 'narrow band', changing colour over a given small pH range (say pH 8–9).

Appendix 3 The Fourier model for image formation

Of all the models there are for the behaviour of light, the one that gives the best and most satisfying description of the formation of an image is the Fourier model. In particular, it provides a simple and elegant explanation of two aspects of image formation: the limits of resolution of an optical system, and the optical transfer function. In this necessarily brief account I shall try to give you an intuitive feel for the Fourier approach. You have already seen something of it in Chapter 5, where I suggested that every scene could be considered as a large number of sinusoidal patterns of luminance with different orientations, amplitudes and spatial frequencies. The converse, that you can make up any sort of pattern, repeated or not, from a set of sinusoidal components, is also part of the Fourier model, and perhaps easier to grasp.

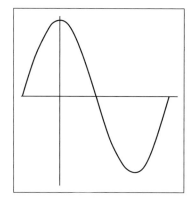

Figure A3.1

To see how this works, we can take an example that may be familiar to you if you have had some training in electronics engineering: building up a square wave from sinusoidal components. For our purpose it is preferable to use cosines instead of sines, as a cosine function is symmetrical about the origin, i.e. $\cos(-x) \equiv \cos(+x)$. Let us consider first a simple cosine function with spatial frequency q cycles per millimetre and amplitude A (Figure A3.1).

Now add a second cosine function with spatial frequency $3q$ and amplitude $A/3$, with a phase at the origin that is opposite to the fundamental cosine function (Figure A3.2).

The pattern, as you see, is no longer a simple cosine pattern, but it is still a periodic waveform, and still symmetrical about the origin. And it is more flat-topped than the fundamental cosine. Now let us add a further cosine function, this time of frequency $5q$ and amplitudes $A/5$, in antiphase to the second function (Figure A3.3).

You can see that the resultant function is still periodic and symmetrical, and is becoming progressively more flat-topped. Now if we add all the odd multiples of the fundamental spatial frequency as far as the fifteenth, the result is as in Figure A3.4.

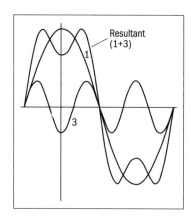

Figure A3.2

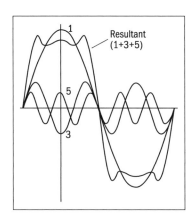

Figure A3.3

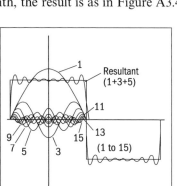

Figure A3.4 Generation of a square wave, showing synthesis up to the 15th harmonic.

The Science of Imaging

A music synthesiser does exactly this sort of thing, putting together new tone flavours by adding harmonics to the fundamental frequency in varying proportions.

The combined function is now looking much more like a square wave. If you continue to add in all the odd multiples of frequency (in electronics terminology, the odd harmonics) right to infinity, you will get a true square wave, or, to be mathematically correct, a rectangular function. This is *Fourier synthesis*.

Fourier methods also prove the converse: that a rectangular function does actually consist of those cosine components, and with those phase relationships. The proof is mathematical, not intuitive, of course; but I hope you can see that it makes sense. It applies not only to rectangular functions, but also to all kinds of function: sawtooth, triangular, trapezoidal and so on. This converse is called *Fourier analysis*.

There is a graphical method of depicting the spatial frequencies and amplitudes of the components of our rectangular function. It is called a *Fourier spectrum*, and is a graph of amplitude against spatial frequency for all the components of the function. Because of the symmetry of the cosine function, I have split the amplitudes on either side of the origin, but have kept the phases right (Figure A3.5).

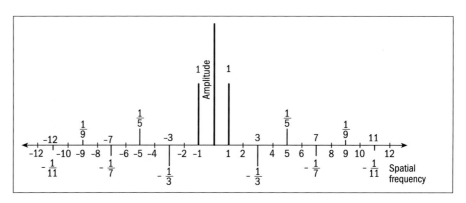

Figure A3.5 An alternative way of showing the components: the Fourier spectrum of a square wave.

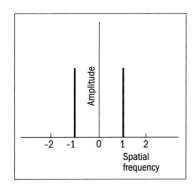

Figure A3.6 The Fourier spectrum of a cosine wave.

(Note: the reason for including a zero-frequency component is that when dealing with light energy, the total always has to be greater than zero: there is no such thing as negative energy. By analogy with electronics, this is often called 'the d.c. component'.)

In contrast to this complicated graph, a pure cosine function has a Fourier spectrum that consists of only a single spatial frequency, so we can represent its frequency spectrum by just two lines (Figure A3.6).

Now, if you pass a narrow parallel beam of light through a cosine grating, that is, one in which the transmittance varies (co)sinusoidally with distance, you get just two diffracted spots, one each side of the zero-order spot (the d.c. component) (Figure A3.7)).

But if you pass it through a rectangular grating you get a whole row of spots (Figure A3.8).

You can see what is happening here: the position of the spots is giving us the spatial frequency plot of the grating function.

The Fourier series for a square-wave function is $(4/\pi) \sum (1/n) \cos(n\pi x)$ where $n = 1, 3, 5\ldots$ There are no sine terms as the function is symmetrical.

The mathematical statement of the amplitude and frequency spectrum of a function of this type is in the form of a series of sine and cosine terms, called the Fourier series for that function. The diffraction pattern of a grating is the optical equivalent of the Fourier series that describes the grating function.

The Fourier model for image formation

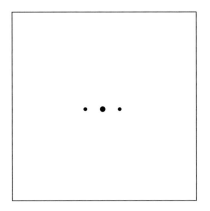

Figure A3.7 The diffraction pattern produced by a cosine grating.

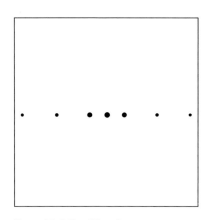

Figure A3.8 The diffraction pattern produced by a square grating.

In the pattern for the square grating you can see clearly that it is made up of discrete, regularly spaced spatial frequencies. You can work back to the values of those frequencies by applying the formula $\lambda = d \sin \theta$ from Chapter 1, and solving it for d using the measurements of the spot distances. If the fundamental pitch of the grating is w, then $d = w, 3w, 5w \ldots$, and if you measure the light intensities, you will find them to be $E, E/9, E/25 \ldots$, which are the squares of the amplitudes.

Figure A3.5 is the frequency plot for a true square function, that is, one where the width of the slits is the same as the width of the spaces between them. If we double the spacing between the slits, the frequencies in the spectrum will become twice as close together, because if you double d in the diffraction formula, you have to halve $\sin \theta$ (Figure A3.9).

If you double the spacing again, the frequencies again become twice as close together (Figures A3.10 and A3.11).

Notice that in squaring the amplitudes we lose the phase information, as $(-x)^2$ is the same as $(+x)^2$. Of course, the positive and negative phases do still exist.

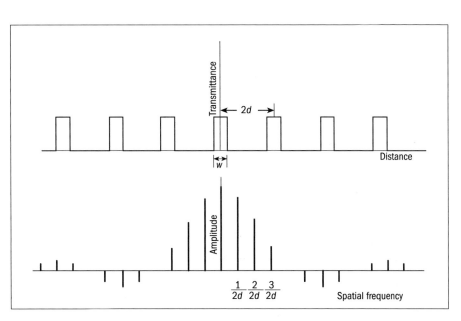

Figure A3.9

The Science of Imaging

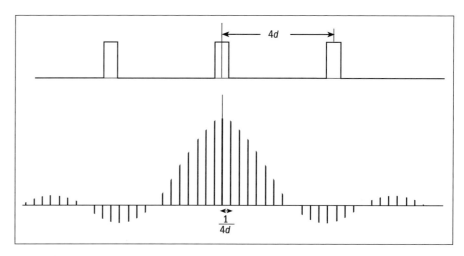

Figure A3.10

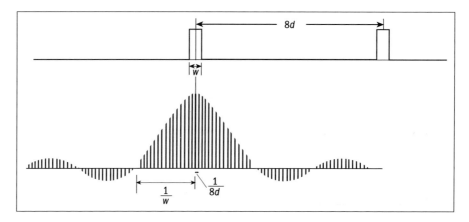

Figure A3.11

As we keep increasing the spacing between the slits the frequencies will get closer and closer together, until in the limit, when the adjacent slits have moved away to infinity, we are left with a single slit, and the frequency spectrum has become a continuum (Figure A3.12).

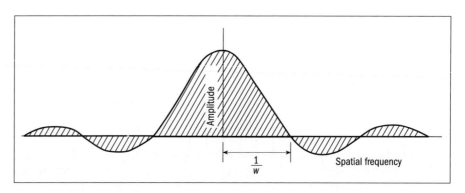

Figure A3.12 As the adjacent pulses move away to infinity, the envelope of the Fourier series becomes a continuous curve – the Fourier transform.

The Fourier model for image formation

The equation of the envelope of this spectrum is called the *Fourier transform* of the single rectangular function. This represents the amplitude distribution in the diffraction pattern of a single slit. The distance from the origin to the first zero is inversely proportional to the pulse width, and this is exactly what you see in the diffraction pattern as you change the slit width.

This last point has a further important implication. If we make the pulse narrower, the first zero moves farther away from the origin, until in the limit the pulse becomes infinitely narrow and the first zero moves away to infinity. The Fourier transform of an infinitely narrow pulse (called a Dirac delta function) contains all frequencies in equal amount (Figure A3.13).

In the mathematical description, the sigma ($\sum$) symbol has turned into an integral ($\int$) symbol. Notice that the shape of the envelope has remained unchanged throughout. This is because we have kept the width of the pulse (or slit) constant.

The Dirac delta function is obtained by taking a rectangular pulse of unit area and making it progressively narrower, keeping the area the same. In the limit the pulse becomes infinitely narrow and of infinite amplitude, still with unit area. Paul Dirac (1902–1984) was one of the giants of modern physics. He played a crucial part in the development of quantum electrodynamics. He predicted the existence of antiparticles and integrated the work of Louis de Broglie and Erwin Schrödinger on wave mechanics with Werner Heisenberg's model of the electron and with special relativity. He shared the 1933 Nobel Prize for Physics with Schrödinger.

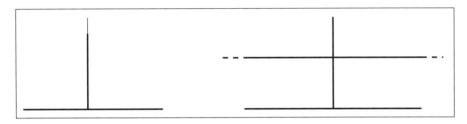

Figure A3.13 The Fourier transform of a delta function is a constant term.

Now look at this the other way around. If our slit (or pulse) is infinitely wide, its (spatial) frequency is zero. There is no diffraction with an infinitely wide slit! So this works in reverse (Figure A3.14).

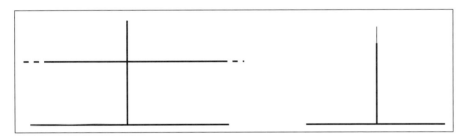

Figure A3.14 The Fourier transform of a constant term is a delta function.

This converse principle proves to be true for all Fourier transforms. If you take a Fourier transform of a Fourier transform, you get back to the original function, give or take the odd constant.

The second rule of Fourier transforms is also implicit. If you add several functions (in the same way as we added components to make a square wave) and take the combined Fourier transform, the result is the same as if you took the Fourier transforms of the functions separately and then added them together. They do not all have to be in one dimension, either; they still add in the same way. A two-dimensional function has a two-dimensional Fourier transform.

No calculation of this type ever seems to be complete without some constant like $2/\pi$ turning up to make a nuisance of itself.

There is one practical point to make. In order to produce an exact optical Fourier transform, you need a lens. The object must be situated at its front principal focal point; the optical Fourier transform is formed at the rear principal focal point. The diffracted waves from the object (which are, of course, plane waves) are focused on the screen at the rear principal focus, as in Figure A3.15.

Without a lens you still get a diffraction pattern, of course, but the electromagnetic field is not an exact Fourier transform.

The Science of Imaging

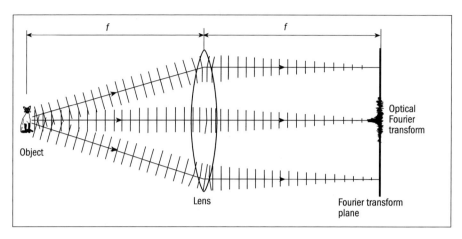

Figure A3.15 An optical Fourier transform is formed at the rear principal focal plane of a lens when the object is in the front principal focal plane.

The outermost waves are those that have been diffracted through the largest angle, and therefore represent the highest spatial frequencies present in the object, as we have already seen. But this simple fact is of fundamental importance in imaging optics, because *the relative size of the lens aperture determines the highest spatial frequency the lens can transmit*. This represents the diffraction-limited resolving power of the lens. You can work out this figure from the f/no of the lens, or (as Abbe did) from the NA of the microscope objective.

If you go through the maths of this you can prove that this is true, and even show that the image is orthoscopic and inverted.

But what about the optical image? You cannot get an image when the object is at the front principal focus – at least, the image, such as it is, is at infinity and infinitely large. But remember the rule about the Fourier transform of a Fourier transform: it goes back to the original function. So all we need to do is to put a similar lens system with its front principal focus at the Fourier transform plane, and we obtain a reconstruction of the object field in its rear principal focal plane (Figure A3.16).

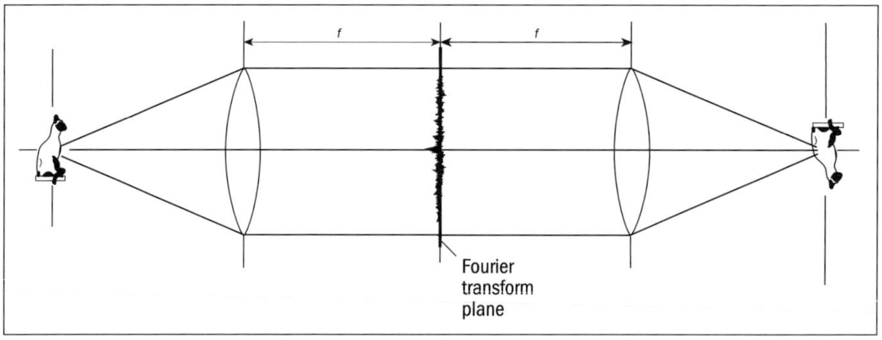

Figure A3.16 The double Fourier transform returns the image to its original form.

By blocking off (or attenuating) parts of the diffraction pattern, you can modify the final image. It is not difficult, for example, to eliminate the raster lines of the image of a TV screen or the halftone dots of a newspaper photograph, as the diffraction patterns of these are characteristic patterns of spots that you can easily see and block off. You can even remove the blur from an image showing camera shake, though this is less easy.

The Fourier model for image formation

The Fourier model evolved gradually, from work by Fresnel and, later, Abbe, on diffraction theory. But it was only in the 1960s, with the arrival of the laser, that it became possible actually to see the diffraction pattern at the rear focal plane. (With white light it *is* there, but is smeared out by dispersion.)

One final, very important point. It is not difficult to show that the optical transfer function of a lens system is the Fourier transform of its line spread function, and vice versa. The earliest derivations of OTFs (in the early 1950s) were in fact derived from LSFs by (in those days highly protracted) computer calculations.

In this appendix I have only been able to give a flavour of the Fourier model. It is a fascinating approach that goes a good deal deeper than I have been able to indicate, and gives many insights into the whole subject of imaging. You can find a more comprehensive non-mathematical exposition in an appendix to my book *Practical Holography* (2nd edition, Prentice-Hall, 1994). The *Manual of Photography* (9th edition, ed. R E Jacobson, Focal Press, 2000) has a whole chapter relating the Fourier model to digital imaging and image modification, an aspect I haven't had room for here. For a full treatment of the subject, including all the maths, the standard text is E G Steward's *Fourier Optics: An Introduction* (2nd edition, Ellis Horwood, 1989).

Index

A
Abbe, Ernst 48, 223
Abbe number 48
Aberrations
　of a pinhole 217–8
　of a single lens 46–9
Abramson, Nils 201, 209
Accommodation, ocular 183
Achromatic doublet 46
Acutance 57, 62
　developers for enhancement of 135
Adjacency effects 134–6
　in colour emulsions 136
Adobe Photoshop 139, 143
Advanced Photographic System (APS)
　format 41
Aerial photography 186–8, 219–10
After-treatment of negatives 120, 132, 133
Airy diffraction pattern 9–10, 57
Airy disc 9, 10, 57
Airy, Sir George Biddell 9
Aliasing 142, 145
Alkali 109
Amplitude modulation (AM) 170
Anaglyph 190–1, 192
Analogue–digital relationship 144
Analogue signal 144
Anastigmat 48
Angle of field 45–6
Angle of incidence, 10–11
Angle of reflection, 10
Angle of refraction, 11
Angle of view 45
Antennas 167–9
　dipole 167–9
　horn 170
　multi-element 167–8
　transmission 168–9
　　cylindrical 168–9
　Yagi 168
Argon bomb 100
Aspect ratio, TV picture 172
Aspheric surfaces 49
Astigmatism
　in a pinhole 218
　in a simple lens 48
　ocular 28–9
Atom, structure of 103–4
Audiotape recorders
　cassette 175, 176
　digital (DAT) 176
　Dolby noise reduction 175–6
　head layout 175
　tape coating types for 175–6
　writing speeds 176
Automatic exposure control 89–90
Automatic focus control 87–9
Autoscreen film 156
Autostereoscopic systems 191–2
Axial (longitudinal) magnification 42, 194

B
Babington-Smith, Constance 58, 64
Baird
　John Logie 161
　television system 161
Beam deflection
　electrostatic 162–3
　magnetic 163
Becquerel, Antoine Henri 234
Benton, Stephen 205
Best curve, drawing 116, 144
Bias, a.c. 174–5, 176
Binary arithmetic 146–7
Binocular vision 32–3, 184
Bjelkhagen, Hans 78
Black body radiation 16, 20
Blanking pulse 163–4
Blumlein, Alan 179
Bonds, interatomic 103–5, 107, 108–9
Bragg
　condition 197, 200
　planes 196, 197
　reflection (diffraction) 7, 66, 235
　Sir Lawrence and Sir William Henry 66, 196
Brewster
　angle 25, 197
　Sir David 25, 189
Bromide, effect on development 134–5
Bunsen, Robert Wilhelm 115
Burning in 133
Byte, definition 146–7

C
Callier effect 125, 127
Camcorders 179
Camera obscura 13
Camera shake 91–2
Cameras 79–101
　for still photography 79–93
Camera types 81–4
　aerial 83
　cine 95–6
　field and studio 83
　pinhole, see Pinhole camera
　rangefinder 82
　rostrum 84
　self-processing 83
　simple 81
　single-lens reflex (SLR) 82
　single-use 82
　twin-lens reflex (TLR) 82
　underwater 83
　wide-angle and panoramic 84, 214–17
Candela 15–16
Cardinal (Gauss) points 42–3
Card memories 152
Catadioptric systems 50, 240
Catalyst 107
Cathode ray oscilloscope (CRO) 162–3
Cathode ray tube (CRT) 162
Charge-coupled device (CCD) 149–50
　arrays 150–2, 153
　buried-channel 151
Chromatic aberration 46, 218
Chromaticity diagram 69
Chrominance signal 164, 166, 170
Cine projector, layout of 95
Coherence 66–7, 196
　length 66–7
　phase 196
　spatial 196
　temporal 66, 196
Collodion 103
Colour
　constancy 35
　errors in colour materials 136–7
　halftone prints 136–7
　in a digital camera 151–2
　masking 76–7, 136–7
　measurement systems 69–70
　negative–positive systems 76–7
　perception 33–5
　prints 71–2, 73, 76
　　from colour negatives 76
　　from separation negatives 71–2
　　from transparencies 73
　perceptible range 34–5
　separation negatives 70–71
　synthesis
　　additive 68, 70
　　subtractive 70
　temperature 20–21
Coma 47–8, 49
Compact disc (CD and CD-ROM) 179–81
Comparing two films 118
Complementary hues 23, 70
Compression techniques 152
Conjugate foci 39
Constancy, perceptual 35
Contact screen 156
Contrast
　arithmetic 60
　control in negatives 132
　logarithmic 60
Converging verticals 35, 53, 140
Convolution 48, 63, 218

256

Index

Critical angle 11, 214
Crookes, Sir William 162
Cross over distortion 174–5
Cross-processing 77
Curvature of field 48

D
Daylight, colour temperature of 21
Denisyuk
    hologram 196
    Yuri 196, 200
Densitometer 114, 116, 121
Density, photographic 114
    transmission 114
    range of print paper 124–5
    reduction of 120, 121, 132, 133
    reflection 123
    step tablet, Kodak 116
Depth of field 40–42
    in photomacrography 221–2
    in photomicrography 223
Depth, perception of 183–5
Developers 107
    constituents of 107–9
    effect of varying composition 117–8
Developing agents 107–8
    and molecular structure 108–9
Development 102
    principles of 106–7
    time and temperature, effect of variations in 116–7
Dielectric 149
Diffraction 8–10
    efficiency 201
    limitation of resolution 57, 222
    limited lens 62
    patterns 9–10
Diffractive optical elements (DOEs) 51, 210
Diffusion transfer 75
Digital videodisc (DVD) 179, 180, 181
Direct Broadcast Satellite (DBS) system 170
Direct positive prints 73, 129
Discharge lamps 21
Disc of confusion 41
Dither 142
Digital signal 144
Dioptre 46
Dirac
    delta function 253
    Paul 253
Direct positive materials 73, 129
Distortion 48–9, 214–15, 227
Doppler effect 238
Duffieux, P M 58
Duplicating images 129, 130
Dye bleach process 111
Dye transfer prints 71–2

E
Eberhard effect 135
Edge spread function (ESF) 62
Einstein, Albert 1, 2
Ektachrome process 72–3
    and push processing 122
Electron beam tube 162–4
    for TV receiver 171
Electron microscope 225–6
Electronic flash 19, 22
Electrostatic copying 157–8
Empty magnification 56, 223
Emulsion, photographic 102
    black-and-white, processing of 106–10
    colour, processing of 111
    sensitivity of 106
    structure of 110–11
    effect of impurities 106
    inherent contrast 114–15
Endoscopic photography 231–2
Energy 1
    electromagnetic 2
Exit pupil 41
Exposure
    meaning of 113, 115
    meters, calibration of 119
Eye 27–30
    and evolution 27
    as an optical device 27–8
    blind spot 29
    resolution of 41
    sensitivity range 30
    structure of 28–30

F
Fall-off, $\cos^4$ 49, 218–19
Faraday, Michael 2
Farmer's reducer 120, 121
Ferromagnetism 174
Field-effect transistor (FET) 148–9
Filter factors 23
Filters, light 22–5
    colour print 24
    colour separation 23–4
    contrast 23
    emulsion balancing 22–3
    infrared (IR) 24, 75
    lighting balancing 24, 138
    neutral density (ND) 24
    polarising, *see* Polarising filters
    special effects (SFX) 25
    ultraviolet (UV) 24
First excellent print 118
Fixation 102, 110
Fizeau, Armand Hippolyte Louis 7
Flash 90–91
    guide numbers 91
    synchronisation 90–91
    through-the-lens (TTL) metering 91
Flashbulbs 21–2, 99
Flicker 94, 162
Fluorescent lamps
    spectral qualities 21, 138
f-number 41, 80, 221

Focal length 13, 39, 43, 220
Focal-plane shutter distortion 80–81
Focusing screens 84–5
Forward motion compensation (FMC) 92
Foucault, Jean Bernard Léon 7, 241
Fourier
    analysis and synthesis 59, 249–50
    model for image formation 66, 248–54
    spectrum 249
    transform 62, 252–3
        optical 253
Frame grabber 154
Franklin, Rosalind 235
Frequency
    a.c. mains 2
    bandwidth 66–7
    definition of 2
    modulation (FM) 170
Fresnel, Auguste 13
    lens 13, 82, 85
Fresnel's Law 50
Fundus camera 232

G
$\bar{G}$ 115–17
Gamma 113, 115
    radiography 234
Gabor, Denis 197, 211
Gauss, Karl Friedrich 42
Gaussian optics 42–3
Gelatin 103, 106
Godowsky, Leopold and Mannes, Leopold 72
Granularity 63
Grating condition 9
Grenz rays 231
Grey card, 18% 119
Grey scale, Kodak 123
Ground glass effect 220

H
Halftone
    principles 155–7
    screen, operation of 155
    three- and four-colour 157
Halogens 102
H & D curves 113, 114–6, 117, 118, 119, 120–122
    cascading of 126–9
    for colour negatives 121
    for print materials 123–4
    for transparency materials 121
    skewed 138–9
H & D speed 113
du Hauron, Louis DuCos 70
Helmholtz, Hermann von 68, 190
Herschel, Sir John 103
Hertz, Heinrich 1
Hill Sky lens 214–15
Hologram
    computer-generated 211–12
    copying 204

257

Denisyuk 196–7
embossed 206
focused-image 204
formation of real image 201–3
Fourier-transform 211–2
how it works 198–200
in natural colours 207–8
processing 201
pseudoscopic image 202
pulse laser 205–6
rainbow (Benton) 205
transfer 203–4
Holographic
   interferometry 208–9
   optical elements (HOEs) 209–10
   stereogram 206–7
Hot cathode 162, 163
Hubel, David and Wiesel, Torsten 32, 38
Hunt, Robert W G 69, 78
Hurter, Ferdinand and Driffield, Vero Charles 113
Huyghens
   Christiaan 1
   wave model for light propagation 8–9
Hyperfocal distance 42
Hypergon lens 214
Hyperstereoscopy 187–8
Hypostereoscopy 188–9
Hysteresis 174, 175

I
Illuminance 16–17
Image
   intensifier tubes 240–41
   manipulation, digital 139
   motion compensation 92–3, 98
   orthicon tube 165
   real and virtual 40
Inertia point 113
Information storage, electronic 147–50
Infrared (IR) emulsions 73–5
   black-and-white 73–4
   colour 75
Integral photography 193–4
Intensification 120, 132, 133
Interference 5–8, 50, 65–7
   constructive and destructive 5, 196–7, 201
   fringes 6, 8, 9
   pattern 66
   thin film 7, 50
Interferometry 208–9, 242–3
Intermittency effects 141
International Standards Organisation (ISO) 114
Internegative 73
Intervalometer 83
Inverse Square Law 17, 49
Ionosphere 167
Iris diaphragm 41, 49
Ishihara colour vision test cards 33–4
ISO speed 114
   criteria for colour materials 122

J
Junction transistor 148–9

K
Kelvin, William Thomson 21
Kodachrome process 72, 111
Kostinsky effect 135

L
Lambert's Law 17, 18, 49
Latent image 102, 103–5
Lateral geniculate nucleus (LGN) 31, 32
Latitude, exposure 103, 119
   and subject contrast 119
   as applied to print emulsions 124
Lazy eye syndrome 184
Leith, Emmett and Upatnieks, Juris 200
Lens
   aberrations 46–9
   aperture limited 57
   coating 50
   covering power 45–6
   equations, sign convention 40
   fisheye 49, 214–15
   GRIN 51, 231, 232
   laws, Newtonian 39
   long-focus 45, 219–20
   meniscus 47–8
   normal-angle 45
   origins of 13
   perspective control ('shift') 53, 215–16
   prime 43
   retrofocus 44
   surface profile 14, 49
   telecentric 50–1
   telephoto 43–4
   varifocal 44–5
   wide-angle 45, 214
   zoom 44–45, 49
Lenticular stereogram 191–2
Leyden jar 149
Light
   chopper 101
   emitting diode(LED) 159, 171
   sources, types of 21–22
Line spread function (LSF) 62
Lippmann
   Gabriel 65, 194
   photography 65–7
Liquid crystal displays (LCDs) 171, 211
Logarithms 30, 31, 114, 246–7
Log-exposure range 124–5
Luminance 18
   signal 164, 166, 170
Lumen 16
Lumière, Louis and Auguste 67, 68
Luminous efficacy 18–19
Luminous energy 18
Luminous flux 16
Luminous intensity, 15–16
Lux 16

M
Mach–Zehnder interferometer 243
Mackie lines 135
Maddox, Richard Leach 103
Magnetic domains 174
Magnification 39, 43, 194, 221, 223
Marconi–EMI television system 161–2
Maxwell, James Clerk 1, 2, 68
Maxwell's equations 2
Measurement systems, international 15
Microimaging 226
Microwave relay transmission 169–70
Miller principle 97
Mirror and drum photography 96–7
Mixed lighting 138
Mirek scale 20–21
Modulation 60
   transfer function (MTF) 61–64
Mole, molar solution 248
MOS capacitor 150
MOSFET 149
Multispectral imaging 220
Muybridge, Eadweard 95

N
Napier, John 245, 246
Newtonian optics 39
Newton, Sir Isaac 1, 7, 12
Newton's rings 7–8
Nicol prism 25
   William 25
Nipkov disc 161
Nodal points 43
Noise reduction 177
NTSC TV system 166
Numerical aperture (NA) 223
Nyquist criterion 144–5, 180

O
Octal notation 146
Ocular accommodation 183
One-shot colour camera 71
Opacitance 113
Opponent-colour theory 35
Optic
   chiasma 31, 33
   nerve 29, 32
Optical fibres 231–2
   cascading of 62–3
   transfer function (OTF) 61–4, 132, 255
Orthochromatic emulsion 106
Oxidation and reduction 107

P
PAL TV system 166
Panchromatic emulsion 106
Panoramic
   cameras 216–7
   format 214
   images by overlapping 215–16
Parallax 85–6, 183, 188, 202

Index

Parasitic antenna elements 167–8
Permanent magnetism 174
Permeability, magnetic 174
Persistence of vision 94, 162
Perspective 51–4
    converging verticals 53–4, 140
    effects of long and short focus lenses 52–3
Petzval, Josef Max 48
    sum 48
pH, meaning of 248
Phase 5, 198
    contrast microscopy 224
    transfer function (PTF) 61, 63
Phi phenomenon 94
Phosphor screen 162, 163–4
Photocell types 89
Photo finish camera 92–8
Photomacrography 220–22
    depth of field 221
    resolution in 222
Photometry 15–22
Photomicrography 222–6
    confocal optics in 225
    depth of field 223
    exposure estimation 224
    illumination systems 224
    magnification assessment 223
Photon 2, 105, 106, 140–1
Photoresist 156, 206
Piezoelectric transducer (PZT) 238
Pinhole camera 13, 217–9
    aberrations of pinhole 218–9
    making a pinhole 219
    optimum pinhole size 217–8
Pixel
    arrays 145, 151–2
    size and resolution 64
Planck, Max Karl Ernst Ludwig 2
Planck's equation 16
Point spread function (PSF) 57, 58
Polar diagram 15–16
Polarisation 4–5
    circular 5
    linear 5
    p- 25
    s- 25
Polarising filters 25–6
    applications 25–6, 191, 192–3, 224
    circular polarising 25
Polaroid colour systems 75–6
Positive hole 147
Potter–Bucky screen 233, 236
Pre- and post-flashing 141
Preservative 109
Priestley, Joseph 103
Primary hues 70
Principal focus (focal point) 43
Principal points 43
Printers, types of 159–60
Printing, photographic, principles 110
Print papers 110, 125–6, 130

colour printing 110, 129, 130
    fibre based 110
    resin coated (RC) 110
    variable contrast (VC) 110, 125–6
    washing requirements 110
Prisms 12–13
    and TIR 12
    dispersion of light 12–13
Proprioception 184
Pseudoscopic image 192, 199, 203
Push processing 122
Pyrogallol 113, 114, 201

Q

Quadrant diagram 126–9
Quarter-wave plate 26

R

Radian 15
Radiation, electromagnetic 3
Radiography 232–5
    principles 233–4
    obtaining enlarged images 233
    units of measurement 234–5
Radioisotopes 234
    for autoradiographic scans 237
Radiometry 15
Rangefinders 86–7
Ray model for optics 39
Rayleigh
    criterion for resolution 57
    Lord 57
    scatter 74
Real image, *see* Image, real and virtual
Reciprocity Law 115
    failure of 121, 140–41
Rectification 140
Rectilinear propagation of light 19
Reflectance 18
Reflection
    laws of 10
    specular and diffuse 18
Reflector dish 170
Refraction, laws of 11
Refractive index 11
Relative aperture 41
Resolution 56–64
    diffraction limitation 57
    spurious 59, 92
Resolving power 56
    of a CCD array 145, 152
Restrainer 109
Retina,
    sensitivity of 16, 30
    structure and function 29–30
Reversal development 121
Röntgen, Wilhelm Konrad 232

S

Sabattier effect 142
Safelights 106, 110, 111
Satellite

    photography 220
    transmission of signals 170
Sayce resolution test object 56, 58
Scanning systems 153–4
    flat bed 153
    hand held 153
    35 mm 154
Scheele, Karl Wilhelm 103
Scheimpflug's rule 53–4
Schlieren photography 241–3
Schumann emulsion 231
Shutter types 79–81
Schwarzschild
    exponent 141
    Karl 141
SECAM TV system 166
Secondary hues 70
Selwyn, E W H 58
Seidel aberrations 46
Semiconductors 147
Sensitometry, definition 113
Shading, dodging and burning in 132–3
    with VC and colour print materials 133
Sharpness of image 40–41, 133–4
Shutter
    efficiency 79–80
    timing mechanisms 80
Siemens star test object 58
Sign conventions 40
Silver halides
    chemical properties 102–3
    crystal structure 104–6
Simple lens, definition 39
Sinusoidal
    grating 60
    test object 59
SI units 15, 16
Slow motion and time lapse photography 96
Slussarev effect 49
Smear and streak photography 98–9, 100
    lighting for 99–100
Snell, Willebrord 11
Snell's Law 11
Solarisation 141–2
Sound recording 174, 175–7
    a.c. bias 174–5
    the Blattnerphone 174
Spatial
    frequency 58, 60
    light modulator (SLM) 211
    period 9
    phase 60
Spectral power distribution 19–20
Spectral sensitivities of emulsions 106
Spectrum
    colours of 12–13
    electromagnetic 3–4
    visible 4
Speed point
    colour materials 122
    monochrome materials 115, 116, 117, 118

Spencer, D A 77–8
Spherical aberration 46–7
Square wave 59
  analysis of 59
  synthesis of 249–50
Standing waves, 66, 199–200
Steradian 15, 16
Stereograms 190–91
  lenticular 191–2
Stereoplanigraph 188, 195
Stereopsis 33, 184
  simulated 193
Stereoscope
  early versions of 185
Stereoscopic
  cameras 185–6
  cinema and television 192–3
  images
    in aerial photography 186–8
    limitations of 184–5
    modern viewing methods 189–92
    public interest in 185
    viewing without optical aids 190
Stokowski, Leopold 179
Streamers in developed film 135
Stroboscopic lighting 100–1
Surround sound 179, 180

T
Talbot, William Henry Fox 79, 103
Technicolor process 72
Television 161–73
  camera 164–6
    CCD 165–6
  digital 166, 172
  displays, aspect ratio of 172–3
  high-definition (HDTV) 166
  receivers, types of display 171
  signal encoding 170

Thermal imaging 239–41
Tomography 235–7
  analysing results 237–8
  nuclear magnetic resonance imaging (NMR) 237
  positron emission (PET) scanning 237
  X-ray body scanning 236
Tonal reproduction
  distortion in 128
  of a film 118
Total internal reflection (TIR) 11–12
Transistor 148–9
Tricolour carbro prints 71
Trompe l'oeuil paintings 183
Tungsten filament lamps, colour temperature of 21

U
Ultrasonic imaging 238–9
Ultraviolet (UV) and fluorescence photography 230–31
Unbalanced lighting 137–8
Undercolour removal 137
Underwater photography 226–8
Unsharp masking 133–5
US Air Force resolution test object 56

V
Vectograph 191
VHS format 177–8
Video recording 174–182
Videotape recording techniques 177–9
  digital systems 178–9
Vidicon tube 165
Viewfinders 84–6
Viewmaster 189
Virtual image, *see* Image, real and virtual
Visual
  constancy 35
  cortex 31, 32
  display unit (VDU) 162, 171
  field 32–3
  illusions 35–7
  pathways and processing 30–32
  perception 27–38

W
Washing and drying films and paper 110
Wavelength 3
Weber–Fechner Law 31, 113, 114
Wedge spectrogram 120
Wedgwood, Thomas 103
Wheatstone, Sir Charles 185, 189
Windows Paintbox program 139
Woodburytype 71, 155

X
Xerographic copiers 157–8
Xeroradiography 234
X-rays 232–6
  crystallography 235
  generation of 233
  intensifying screens for 234
  optics 235
  scanning systems 235–6
  units of measurement 234–5

Y
Young, Thomas 6, 67, 68
Young–Helmholtz theory 67–8
Young's slits 6–7, 8–9

Z
Zernike, Frits 224
Zoetrope 94–5
Zone
  plate 210–11
  system 130